Fundamentals of Carbanion Chemistry

ORGANIC CHEMISTRY

A SERIES OF MONOGRAPHS

Edited by

ALFRED T. BLOMQUIST

Department of Chemistry, Cornell University, Ithaca, New York

1. Wolfgang Kirmse. CARBENE CHEMISTRY, 1964
2. Brandes H. Smith. BRIDGED AROMATIC COMPOUNDS, 1964
3. Michael Hanack. CONFORMATION THEORY, 1965
4. Donald J. Cram. FUNDAMENTAL OF CARBANION CHEMISTRY, 1965
5. Kenneth B. Wiberg (Editor). OXIDATION IN ORGANIC CHEMISTRY, PART A, 1965; PART B, *In preparation*
6. R. F. Hudson. STRUCTURE AND MECHANISM IN ORGANO-PHOSPHORUS CHEMISTRY, 1965
7. A. William Johnson. YLID CHEMISTRY, 1966
8. Jan Hamer (Editor). 1,4-CYCLOADDITION REACTIONS, 1967
9. Henri Ulrich. CYCLOADDITION REACTIONS OF HETEROCUMULENES, 1967
10. M. P. Cava and M. J. Mitchell. CYCLOBUTADIENE AND RELATED COMPOUNDS, 1967
11. Reinhard W. Hoffman. DEHYDROBENZENE AND CYCLOALKYNES, 1967

Fundamentals of Carbanion Chemistry

by DONALD J. CRAM

DEPARTMENT OF CHEMISTRY
UNIVERSITY OF CALIFORNIA
LOS ANGELES, CALIFORNIA

1965

ACADEMIC PRESS □ New York and London

ACADEMIC PRESS INC.
111 Fifth Avenue, New York, New York 10003

United Kingdom Edition published by
ACADEMIC PRESS INC. (LONDON) LTD.
Berkeley Square House, London W. 1

LIBRARY OF CONGRESS CATALOG CARD NUMBER: 65-22772

Second Printing, 1968

PRINTED IN THE UNITED STATES OF AMERICA.

Preface

A natural outgrowth of the study of organic reaction mechanisms was the discovery that many superficially unrelated reactions involved intermediates of the same general structure. These intermediates stand at the crossroads of structure and reactions, and their generation, structure, and capabilities embrace much of organic chemistry. Of the four major reaction intermediates (carbonium ions, carbanions, carbon radicals, and carbenes), the primary research literature of carbonium ions and carbon radicals has been best gathered and collated in the review literature. Carbene chemistry has developed largely since mid-century, and is more specialized than that of the other intermediates. Although carbanions were recognized earlier than most of the other reaction intermediates, reviews of the literature on these negative ions have been rare. The sections on carbanions in advanced textbooks are microscopic. For obscure reasons, only recently have physical organic chemists in numbers turned their attention to carbanions.

The purpose of this monograph is to stimulate interest in carbanion chemistry. The research literature since about 1959 has been rich in this subject, and although a much more comprehensive work could be written in the future, the fundamental problems are now visible. Here we will examine these problems, evaluate the advances, and point to research of the future.

The monograph has the following organization. In Chapter I, the variation of the thermodynamic and kinetic acidities of carbon acids with substituents and environments serve to introduce the subject. Chapter II treats the modes of carbanion stabilization by substituents. The stereochemistry of hydrogen–deuterium exchange and of organometallics is discussed in Chapter III, and is followed in Chapter IV by a more general description of carbanion stereochemistry. Chapter V is concerned with unsaturated anionic rearrangements, and Chapter VI with other rearrangements.

A number of topics have been treated only implicitly. The general chemistry of organometallic compounds and ylids is much too extensive to be encompassed by this work. Only material judged (by the author) to be fundamental to carbanion chemistry has been covered.

A limitation of this book arises from the dependence of the theory involved on qualitative and, in many cases, fragmentary evidence. The main justification for developing carbanion theory at this point is that many facts require correlation and the basic problems need definition. Finally, the development of a field feeds on the challenge presented by speculation. Carbanion chemistry is no exception.

I am greatly indebted to my colleagues who are active in the carbanion field. Doctors A. Streitwieser, Jr., J. D. Roberts, A. Nickon, F. A. L. Anet, F. G. Bordwell, H. M. Walborsky, E. C. Steiner, C. D. Ritchie, R. E. Dessy, S. W. Ela, M. C. Whiting, and D. E. Applequist contributed helpful suggestions, and Drs. A. Streitwieser, Jr., R. E. Dessy, J. D. Roberts, E. C. Steiner, C. D. Ritchie, and F. A. L. Anet were kind enough to provide me with results in advance of publication.

DONALD J. CRAM

Los Angeles, California
May, 1965

Contents

PREFACE v

Chapter I

Carbon Acids

Thermodynamic Acidity Scales 1
Correlations between Thermodynamic and Kinetic Acidity 8
Kinetic Acidities of Weak Hydrocarbon Acids 20
Effect of Medium on Kinetic and Equilibrium Acidity of Carbon Acids 32

Chapter II

Carbanion Structure and Mechanism of Stabilization

s-Character Effects 48
Conjugative Effects 52
Inductive Effects 55
Homoconjugative Effects 63
Aromatization Effects 65
Negative Hyperconjugative Effects 68
d-Orbital Effects 71

Chapter III

Stereochemistry of Substitution of Carbon Acids and Organometallic Compounds

Carbon Acids That Form Symmetrical or Near Symmetrical Carbanions 86
Systems That Form Unsymmetrical Carbanions 105
Special Case of Carbanions Confined in Three-Membered Rings 113
Homoconjugated Carbanions 114
Stereochemistry of Substitution of Organometallic Compounds 116

Chapter IV

Stereochemistry of Carbanions Generated with Different Leaving Groups

Carbon as Leaving Group in Electrophilic Substitution at Saturated Carbon 138
Other Leaving Groups in Electrophilic Substitution at Saturated Carbon 159
Comparison of the Stereochemical Capabilities of Carbanions and Carbonium Ions 170

Chapter V

Isomerization by Proton Transfer in Unsaturated Systems

Intramolecularity 176
Geometric Stability of Allylic Anions 193
Hydrocarbon Substituent Effects on the Rates of Allylic Rearrangements of Alkenes 196
Olefin Equilibria 200
Collapse Ratios for Allylic Anions 204

Chapter VI

Molecular Rearrangements

Ring-Chain Anionic Rearrangements 211
1,2-Rearrangements 223
Rearrangements with 1,3-Elimination Reaction Stages 243

AUTHOR INDEX 257

SUBJECT INDEX 266

CHAPTER I

Carbon Acids

Perhaps the simplest view of carbanions is as conjugate bases of what will be called here *carbon acids*. A carbon acid is an organic substance that, when treated with a suitable base, donates a proton to that base by fission of a carbon–hydrogen bond. This definition is made in terms of one of the most important of all chemical reactions, that of proton abstraction from carbon. Since reactions are now known in which protons are abstracted from saturated hydrocarbons, and since most organic compounds contain carbon–hydrogen bonds, most organic compounds are potential carbon acids. This definition is profitable because it allows vast numbers of organic compounds to be classified as to their acid strength, which, in turn, is related to the base strength of their conjugate bases and, therefore, to the stability of carbanions.

The acid strength of the carbon acids has both a thermodynamic and kinetic aspect. Thermodynamic acidity deals with the *positions of equilibria* between acids and their conjugate bases, whereas kinetic acidity pertains to the *rates* at which acids donate protons to bases. Acid–base theory was largely developed in terms of the oxygen acids and bases. The rates of proton transfer were so fast that they have been measured only in the 1950's and 1960's with the use of modern instruments. Thus, the concept of kinetic acidity has been applied most frequently to carbon acids, since the rates at which they donate protons to bases frequently can be easily measured.

The first sections of this chapter treat the thermodynamic and kinetic acidities of the carbon acids and the relationship between the two. Later sections deal with the general effects of structure and medium on these two aspects of acidity.

THERMODYNAMIC ACIDITY SCALES

In a classic paper, Conant and Wheland[1] ranked a number of weak carbon acids in an acidity scale, which was later expanded and made

[1] J. B. Conant and G. W. Wheland, *J. Am. Chem. Soc.*, **54**, 1212 (1932).

more quantitative by McEwen.[2] The McEwen scale, in spite of shortcomings, has survived and has served as a point of departure for other investigators.

In these studies, sodium or potassium salts of the carbon acids in ether were prepared and treated with other carbon acids. The equilibrium constants between the two salts and the two carbon acids were estimated colorimetrically through use of the differences in color of the two salts. Most of the color changes were fast, but a few, particularly the sodium salts of the weaker carbon acids, took weeks to come to equilibrium. When no excess carbon acid was added to the salt, the authors presumed that the two carbon acids differed by 2 pK_a units when the visually detected metalation reaction proceeded to 90% completion. When a fivefold excess of carbon acid was added, an 0.4 pK_a difference was attributed to the two carbon acids when the proton transfer was estimated by color change to be 90% complete at equilibrium.

Equation (1) formulates the equilibrium involved in the metalation reaction, and Equations (2) and (3) the dissociation constants of the two carbon salts. The authors assumed that the equilibrium constants for dissociation of the two carbon salts, K and K′, were approximately equal. They pointed to the fact that others[3a] had observed that the conductivities in pyridine were approximately the same for salts of several substituted triarylmethanes of differing acidity. These assumptions, coupled with the definition of pK_a, allow one to write Equations (4) and (5), the latter of which relates the pK_a differences of the carbon acids to the visual color changes associated with equilibrium in the metalation reaction. More recent work[3b] has demonstrated that at room temperature in tetrahydrofuran, sodium and cesium fluorenide existed as contact ion pairs, which exhibited complex spectra in the ultraviolet and visible region. The lithium salt at 25°, or the sodium salt at −80°, was largely in the form of solvent-separated ion pairs, whose ultraviolet spectra were distinctly different, but whose visible spectra were more similar to that of the contact ion pairs. Conductivity measurements demonstrated that only trivial amounts of dissociated material were present. These facts indicate that ion pairs were responsible for the color changes used by McEwen to develop his acidity scale, and not dissociated ions. Thus Equation (5) more nearly represents the basis for the scale.

$$\text{RH} + \text{R'M} \underset{}{\overset{\text{Ether}, N_2}{\rightleftarrows}} \text{RM} + \text{R'H} \tag{1}$$

[2] (a) W. K. McEwen, *J. Am. Chem. Soc.*, **58**, 1124 (1936); (b) see also A. A. Morton, *Chem. Rev.*, **35**, 1 (1944).

[3] (a) K. Ziegler and H. Wollschitt, *Ann.*, **479**, 123 (1930); (b) T. Hogen-Esch and J. Smid, *J. Am. Chem. Soc.*, **87**, 669 (1965).

(2) $$R'M \underset{}{\overset{K}{\rightleftarrows}} R'^{-} + M^{+}$$

(3) $$RM \underset{}{\overset{K'}{\rightleftarrows}} R^{-} + M^{+}$$

(4) $$pK_a - pK_{a'} = -\log\frac{[R^-]}{[RH]} + \log\frac{[R'^-]}{[R'H]} \quad \text{(equilibrium condition)}$$

(5) $$pK_a - pK_{a'} = -\log\frac{[RM]}{[RH]} + \log\frac{[R'M]}{[R'H]} \quad \text{(equilibrium condition)}$$

Many pairs of carbon acids were compared with one another, and cross checks were provided through intercomparisons of the pK_a differences of sets of carbon acids which differed from one another by one or two pK_a units. An additional type of crude confirmation of the general ranking of the carbon acids was provided by carbonation experiments. The position of equilibrium between two carbon acids and their respective metal salts was estimated by carbonation of the mixture, isolation of the resulting carboxylic acids, and determination of the amount of each acid present. The ratio of the two acids isolated was assumed to reflect the ratio of the two metal salts in the equilibrium. In other words, the authors assumed that k_2 and $k_3 \gg k_1$ and k_{-1} in Equation (6). Use of these techniques provided an internally consistent pattern of pK_a differences between carbon acids.

(6) $$\begin{array}{ccc} RH + R'M & \underset{k_{-1}}{\overset{k_1}{\rightleftarrows}} & RM + R'H \\ CO_2 \big\downarrow k_2 & & CO_2 \big\downarrow k_3 \\ R'CO_2M & & RCO_2M \end{array}$$

40–70% Yields

These techniques related the pK_a's of the carbon acids to one another, and a third technique related these pK_a's to those of the oxygen acids. The optical rotations of menthol and sodium menthoxide in benzene differ considerably and are easily determined. The positions of the equilibria (Equation (7)) between sodium menthoxide and various carbon and oxygen acids on the one hand and menthol and sodium oxygen and carbon salts on the other were determined polarimetrically. This technique allowed the acidities of the oxygen and carbon acids (and a few nitrogen acids) to be interrelated, and a continuous acidity scale ranging from methanol to cumene was constructed. The pK_a of methanol was taken to be 16.[4] McEwen[2] was careful to point out that the pK_a's of the acids between 18 and 37 are minimal because of the assumptions

[4] A. Unmack, *Z. Physik. Chem.*, **133**, 45 (1928).

TABLE I

McEwen's Acidity Scale

Compound	pK_a	Compound	pK_a
CH_3OH	16	$C_6H_5C{\equiv}CH$	21
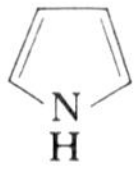 N, H	16.5	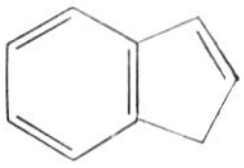	21
$C_6H_5CH_2OH$	18	$(C_6H_5)_2NH$	23
C_2H_5OH	18	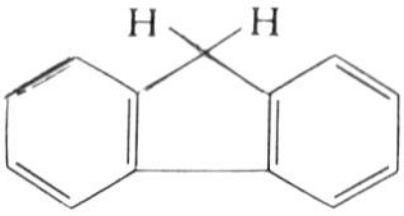 H, H	25
$(C_6H_5)_2CHOH$	18	$C_6H_5NH_2$	27
$(CH_3)_2CHOH$	18	p-$CH_3C_6H_4NH_2$	27
$(C_6H_5)_3COH$	19	p-$CH_3OC_6H_4NH_2$	27
$(CH_3)_3COH$	19	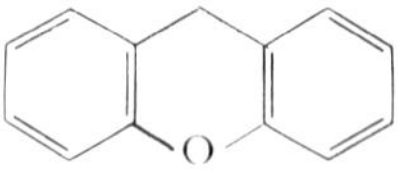 O	29
$C_2H_5(CH_3)_2COH$	19	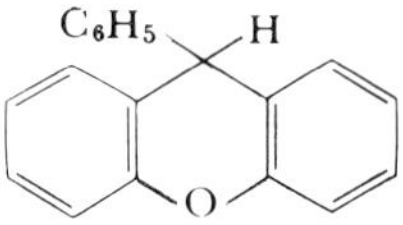 C_6H_5, H, O	29
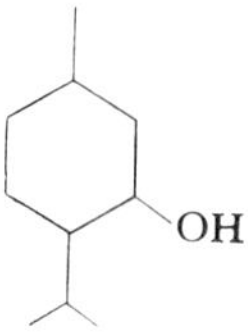 OH	19	p-C_6H_5—$C_6H_4(C_6H_5)_2CH$	31
$C_6H_5COCH_3$	19	$(C_6H_5)_3CH$	32.5
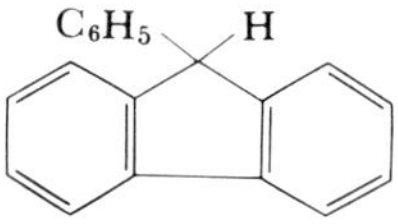 C_6H_5, H	21	α-$C_{10}H_7(C_6H_5)_2CH$	34
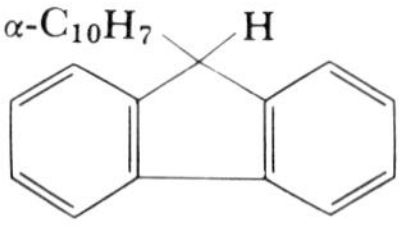 α-$C_{10}H_7$, H	21	$(C_6H_5)_2CH_2$	35
		$(C_6H_5)_2C{=}CHCH_3$	36
		$C_6H_5(CH_3)_2CH$	37

made in the colorimetric measurements. Table I records the structures and estimated pK_a's of McEwen's acids.

(7) 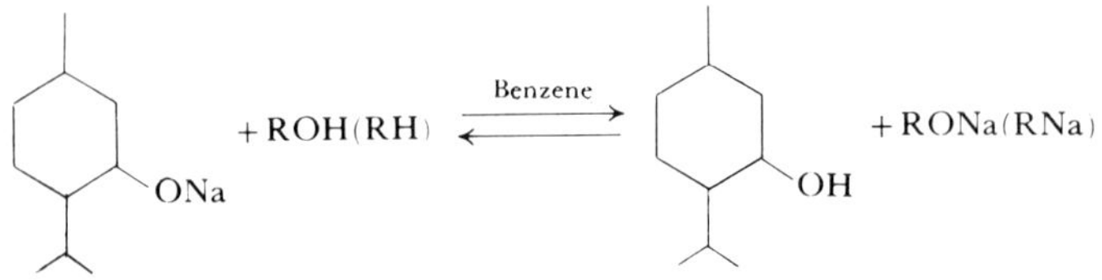 ONa + ROH(RH) $\underset{}{\overset{\text{Benzene}}{\rightleftarrows}}$ OH + RONa(RNa)

A second thermodynamic acidity scale has been developed by Streitwieser and co-workers.[5] These authors measured the equilibrium constants between lithium or cesium cyclohexylamide and the carbon acid on the one hand, and cyclohexylamine and the lithium or cesium carbon salt on the other in cyclohexylamine as solvent (Equation (8)). Through-

$$\text{RH} + c\text{-C}_6\text{H}_{11}\overset{-}{\text{N}}\text{H}\overset{+}{\text{M}} \underset{}{\overset{K}{\rightleftarrows}} \overset{-}{\text{R}}\overset{+}{\text{M}} + c\text{-C}_6\text{H}_{11}\text{NH}_2 \tag{8}$$

out this book, "*c*-" is used as a symbol for "cyclo." The concentrations of the colored carbon salts were determined spectroscopically. The pK_a values of the carbon acids examined were based on that of 9-phenylfluorene, which Langford and Burwell[6] found to be 18.5 in aqueous tetramethylene sulfone (sulfolane). The spectra of the carbon salts shifted only from 0 to 10 mμ to shorter wavelengths when the cation was changed from lithium to cesium with all compounds examined except di- and triphenylmethane. This spectral insensitivity to the cation is strong evidence that little covalent interaction occurs between metal and carbon in these salts. Table II lists the pK_a values.

Four compounds, fluorene, *p*-biphenyldiphenylmethane, triphenylmethane, and diphenylmethane are common to the two scales. The difference in pK_a for fluorene on the two scales is 2 units, and the difference for each of the other compounds is less than one pK_a unit. The similarity between the two scales is striking, particularly since different bases, solvents, and anchor compounds were involved.

The drawback of both of these scales is that they do not extend all the way to the saturated hydrocarbons. Severe experimental obstacles attend solution of this problem. Two approaches to this problem have been made. In one, Applequist and O'Brien[7a] measured the equilibrium constants between various alkyl-, alkenyl-, and aryllithiums and the appropriate iodides in ether, and in mixtures of ether–pentane (Equation (9)). The values of the equilibrium constants varied remarkably

$$\text{RLi} + \text{R'I} \overset{K_{\text{obs.}}}{\rightleftarrows} \text{RI} + \text{R'Li} \qquad K_{\text{obs.}} = \frac{[\text{RI}][\text{R'Li}]}{[\text{RLi}][\text{R'I}]} \tag{9}$$

little with changes in solvent (by a factor of about 3). Although some effect on $K_{\text{obs.}}$ might be due to the states and kinds of aggregation, a

[5] A. Streitwieser, Jr., J. I. Brauman, J. H. Hammons, and A. H. Pudjaatmaka, *J. Am. Chem. Soc.*, **87**, 384 (1965).

[6] C. H. Langford and R. L. Burwell, Jr., *J. Am. Chem. Soc.*, **82**, 1503 (1960).

[7] (a) D. E. Applequist and D. F. O'Brien, *J. Am. Chem. Soc.*, **85**, 743 (1963); (b) R. M. Salinger and R. E. Dessy, *Tetrahedron Letters*, **11**, 729 (1963); (c) R. E. Dessy, private communication.

TABLE II

Streitwieser's Acidity Scale

Compound	pK_a	Compound	pK_a
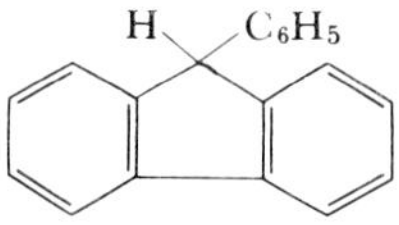 H C_6H_5	18.5	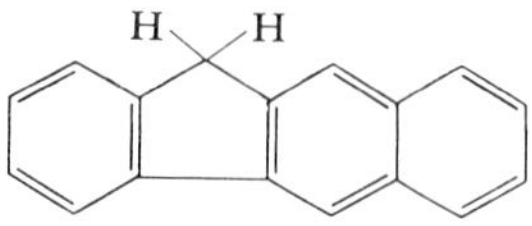 H H	23.2
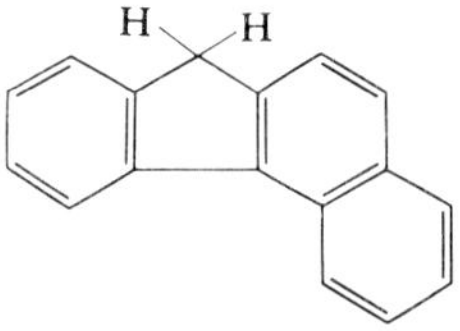 H H	19.4	H $(C_6H_5)_2C—CH{=}CHC_6H_5$	26.5
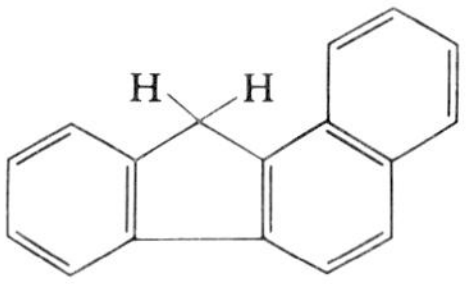 H H	20.0	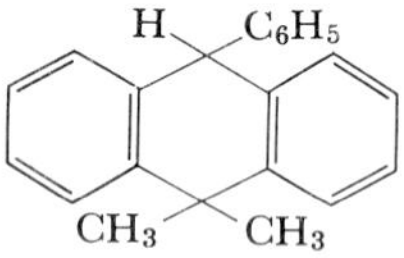 H C_6H_5 CH_3 CH_3	29.0
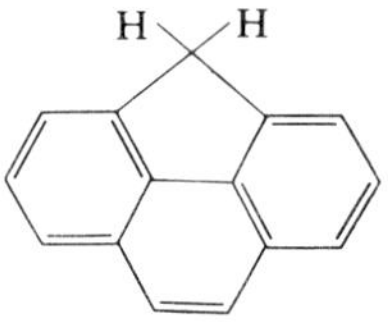 H H	22.6	H p-$C_6H_5C_6H_4C(C_6H_5)_2$	31.2
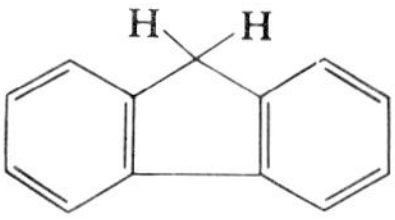 H H	22.9	$(C_6H_5)_3C—H$	32.5
		$(C_6H_5)_2CH_2$	34.1

number of facts suggest that such effects are small. (1) Although the rates of equilibration were highly dependent on solvent, the $K_{obs.}$ was only slightly affected by solvent. (2) The value of $K_{obs.}$ was invariant with total lithium concentration over a 13.7-fold range for ethyl- and propyllithium equilibria. (3) Equilibria involving propyllithium and a mixture of ethyl iodide and isobutyl iodide gave $K_{obs.}$ values similar to those obtained when equilibria were measured for the two systems taken separately. Table III records the values of $K_{obs.}$ obtained.

For purposes of comparison, the authors[7a] converted $K_{obs.}$ to equilibrium constants that refer to a common standard iodide. With phenyl

TABLE III

Equilibrium Constants at − 70° for Reaction

$$RLi + R'I \underset{}{\overset{K_{obs.}}{\rightleftarrows}} RI + R'Li$$

R′	R	Solvent: Ether	Solvent: 40% Ether–60% Pentane
$CH_2{=}CH—$	C_6H_5	258 ± 11	—
C_6H_5	c-C_3H_5	9.55 ± 0.46	—
c-C_3H_5	C_2H_5	—	333 ± 23
C_2H_5	n-C_3H_7	2.38 ± 0.20	2.56 ± 0.10
n-C_3H_7	$(CH_3)_2CHCH_2$	5.11 ± 0.28[a]	5.45 ± 0.22
$(CH_3)_2CHCH_2$	$(CH_3)_3CCH_2$	7.49 ± 0.28[b]	—
$(CH_3)_2CHCH_2$	c-C_4H_7	—	35.3 ± 3.5
$(CH_3)_3CCH_2$	c-C_4H_7	—	5.10 ± 0.11
c-C_4H_7	c-C_5H_9	—	5.82 ± 1.22

[a] In pentane at −23°, $K_{obs.} = 7.77 \pm 0.22$.
[b] In pentane at −23°, $K_{obs.} = 21.59 \pm 1.47$.

iodide as standard, Applequist and O'Brien calculated K_ϕ values, the logarithms of which are recorded in Table IV.

In a second approach to the problem of the acidities of the very weak carbon acids, Salinger and Dessy[7b,c] measured approximate equilibrium

TABLE IV

Values of log K_ϕ *where* K_ϕ *is the Equilibrium Constant for the Reaction*

$$RLi + C_6H_5I \rightleftarrows RI + C_6H_5Li$$

R	log K_ϕ	R	log K_ϕ
$CH_2{=}CH$	−2.41	$(CH_3)_2CHCH_2$	4.6
C_6H_5	0.00	$(CH_3)_3CCH_2$	5.5
c-C_3H_5	0.98	c-C_4H_7	6.1
C_2H_5	3.5	c-C_5H_9	6.9
n-C_3H_7	3.9	—	—

constants for the reactions of various dialkyl-, dialkenyl-, and diarylmagnesium and mercury compounds with one another in tetrahydrofuran at 25°. In Table V are recorded the relative values of the equilibrium constants for the hydrocarbon group–metal redistribution reactions, the logarithms of these relative values, and values of $-\log K_\phi$ taken from Table IV.

TABLE V

Relative Values of Logarithms of Equilibrium Constants for Hydrocarbon Group–Metal Redistribution Reactions in Tetrahydrofuran at 25° Compared with $-\log K_\phi$ *Values from Table IV*

$$\text{R—Mg—} + \text{R}'\text{—Hg—} \overset{K}{\rightleftarrows} \text{R}'\text{—Mg—} + \text{R—Hg—}$$

R	Relative values of K	Relative values of log K	Values of $-\log K_\phi$
$(CH_3)_2CH$	$\sim 10^{-6}$	-6	—
C_2H_5	$\sim 10^{-4}$	-4	-3.5
CH_3	10^{-2}	-2	—
c-C_3H_5	0.13	-0.9	-1.0
$CH_2{=}CH$	0.30	-0.5	2.4
C_6H_5	1.0	0	0
$CH_2{=}CH{-}CH_2$	1.70	0.2	—
$C_6H_5CH_2$	3.70	0.4	—

The last two columns of Table V allow a comparison between the Applequist and Dessy approaches to the problem of determination of the relative stabilities of the anions derived from the weak carbon acids. In the Applequist scale, $\Delta \log K$ for the ethyl and phenyl groups is -4, which compares with -3.5 on the Dessy scale. The $\Delta \log K$ on each scale between the cyclopropyl and phenyl groups is about -1. The similarity in these sets of $\Delta \log K$ values suggests that the same correspondence probably exists for the $\Delta \log K$ and the ΔpK_a values for the corresponding carbon acids. All of the $\Delta \log K$ values on each scale appear reasonable except that between ethylene and benzene on the Applequist scale. The value of 2.4 for $\Delta \log K$ is far greater than the -0.5 on the Dessy scale, and the latter value is much more reasonable. In the absence of other data, these provide the best quantitative comparisons between carbanion stability and structure for highly reactive carbanions.

CORRELATIONS BETWEEN THERMODYNAMIC AND KINETIC ACIDITY

The rates at which protons are transferred from carbon (kinetic acidity) can be measured somewhat more easily than equilibrium constants. The most commonly used method involves base-catalyzed hydrogen isotope exchange between the carbon acid and oxygen or nitrogen acids in the medium. For many of the stronger carbon acids, carbanion formation is rate determining in bromination of ketones, nitroalkanes, and similar compounds. The disappearance of bromine can be easily followed, and used to measure rate of anion formation. The question now arises as to the relationship between kinetic and thermodynamic acidity, and this is treated in this section.

The connection between thermodynamic and kinetic acidity has been examined much more thoroughly for the oxygen than the carbon acids. The Brønsted equation (Equation (10)) provides for a linear relationship between $\log k_A$ (rate constant for kinetic processes involving proton transfer) and $\log K_A$ (K_A is the dissociation constant for the acid). Parameters α_a and G_A are constants characteristic of the particular series of reactions; α is positive and smaller than unity[8] when k_A and K_A are measured in the same solvent-base system.

(10) $$\log k_A = \alpha_a \log K_A + \log G_A$$

Good Brønsted linear relationships have been observed for oxygen acids only when the structures of the acids are very closely related, and when the acidity range spans only a few pK_a units. Differences in steric factors (steric inhibition of resonance and of solvation) and resonance effects from one acid to a second do not allow a wide variety of oxygen acid types to be placed on the same Brønsted plot.[8] These limitations on the applicability of the Brønsted equation to the oxygen acids are expected to be even more severe for the carbon acids, because of the greater opportunity for structural variation in the latter class of acid.

The most comprehensive attempt at correlation of thermodynamic and kinetic acidity of the carbon acids was made by Pearson and Dillon.[9] These authors collected and collated data pertaining to the relation between k_1 and K_a for the proton transfer from carbon acids to water at 25° (Equation (11)). These substances were all carbon acids by virtue

(11) $$-\overset{|}{\underset{|}{C}}-H + H_2O \underset{k_{-1}}{\overset{k_1}{\rightleftharpoons}} -\overset{|}{\underset{|}{C}}^- + H_3O^+ \qquad K_a = k_1/k_{-1}$$

of the presence of at least one acidifying substituent, and usually more than one. The substituents included the ketonic carbonyl group, the

[8] (a) J. E. Leffler and E. Grunwald, "Rates and Equilibria of Organic Reactions," Wiley, New York, 1963, p. 235; (b) R. P. Bell, "The Proton in Chemistry," Cornell Univ. Press, Ithaca, New York, 1959, pp. 87, 144, 159, 163.

[9] R. G. Pearson and R. L. Dillon, *J. Am. Chem. Soc.*, **75**, 2439 (1953).

carbethoxy, nitro, halo, cyano, and trifluoromethyl groups, as well as phenyl, methyl, and hydrogen. The values of K_a (dissociation equilibrium constant) ranged over about 14 powers of 10, and the values of k_1 (rate of dissociation) over about 9 powers of 10. The compounds are listed in Table VI.

TABLE VI

Rate and Equilibrium Data for Carbon Acids in Water at 25°

Cmpd. no.	Compound	K_a [a]	k_1 (sec.$^{-1}$)	k_{-1} (liters/mole sec.$^{-1}$)
1	$CH_2(NO_2)_2$	2.7×10^{-4}	0.83	3.1×10^{3}
2	$CH_3COCH_2COCF_3$	2×10^{-5}	1.5×10^{-2}	7.5×10^{2}
3	$CH_3COCH_2NO_2$	8.0×10^{-6}	3.7×10^{-2}	3.8×10^{3}
4	$C_2H_5O_2CCH_2NO_2$	1.5×10^{-6}	6.3×10^{-3}	4.1×10^{3}
5	(thiophene ring, S)—$COCH_2COCF_3$	8.0×10^{-7}	1.0×10^{-2}	1.2×10^{4}
6	$C_6H_5COCH_2COCF_3$	1.5×10^{-7}	8.3×10^{-3}	5.5×10^{4}
7	$CH_3COCHBrCOCH_3$	1×10^{-7}	2.3×10^{-2}	1.7×10^{5}
8	$C_2H_5NO_2$	2.5×10^{-9}	3.7×10^{-8}	1.5×10^{1}
9	$CH_3COCH_2COCH_3$	1.0×10^{-9}	1.7×10^{-2}	1.7×10^{7}
10	CH_3NO_2	6.1×10^{-11}	4.3×10^{-8}	6.8×10^{2}
11	(cyclopentanone ring, C=O)—$CO_2C_2H_5$	3.0×10^{-11}	2.3×10^{-3}	8.3×10^{7}
12	$CH_3COCH_2CO_2C_2H_5$	2.1×10^{-11}	1.2×10^{-3}	5.8×10^{7}
13	$CH_3COCHCH_3COCH_3$	1.0×10^{-11}	8.3×10^{-5}	8.3×10^{6}
14	$CH_2(CN)_2$	6.5×10^{-12}	1.5×10^{-2}	2.3×10^{9}
15	(cyclohexanone ring, C=O)—$CO_2C_2H_5$	3×10^{-12}	9.7×10^{-6}	3.3×10^{6}
16	$CH_3COCHC_2H_5CO_2C_2H_5$	2×10^{-13}	7.5×10^{-6}	3.8×10^{7}
17	$CH_2(CO_2C_2H_5)_2$	5×10^{-14}	2.5×10^{-5}	5×10^{8}
18	$CH_3COCHCl_2$	10^{-15}	7.3×10^{-7}	6.7×10^{8}
19	$C_2H_5CH(CO_2C_2H_5)_2$	10^{-15}	3.3×10^{-7}	3.3×10^{8}
20	CH_3COCH_2Cl	3×10^{-17}	5.5×10^{-8}	1.7×10^{9}
21	CH_3COCH_3	10^{-20}	4.7×10^{-10}	5×10^{10}

[a] Gross dissociation constant uncorrected for enol content.

A plot of $\log k_1$ against $-\log K_a$ for these compounds (see Figure 1) provided a straight line with a slope of 0.6.[9] The correlation is very rough, and considerable scatter is evident. The simple nitro compounds are conspicuously slow in ionizing, considering their acid strengths, as are the substituted nitro compounds and the trifluoromethyl diketones. In the carbanions derived from these carbon acids, charge is largely localized on the more electronegative atoms. As a consequence, both proton abstraction and donation at carbon is slow compared to carbon acids whose carbanions have charge more concentrated on carbon. On

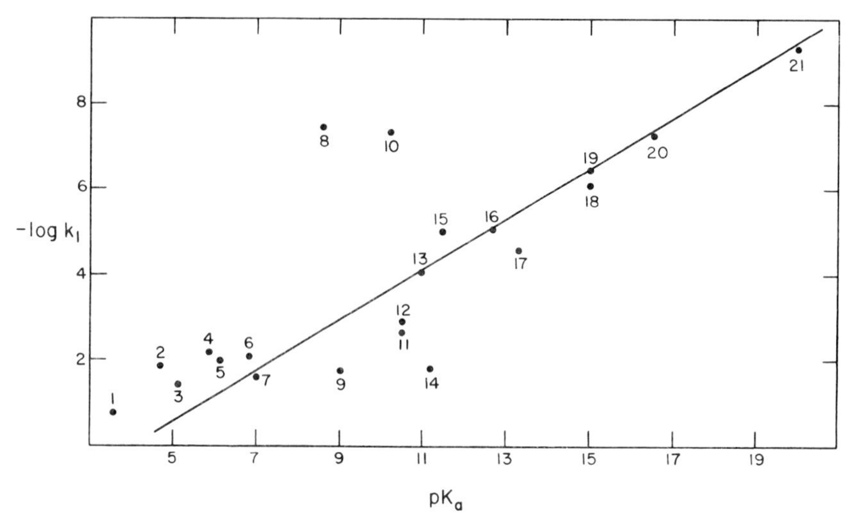

FIG. 1. Plot of negative of the logarithm of the rate of ionization vs. the pK_a of carbon acids of Table VI.

the other hand, the cyano compounds ionize faster, and their carbanions are captured faster than their dissociation constant values would suggest. In these anions, charge is probably less delocalized than in the carbonyl-stabilized carbanions, which largely determine the position of the straight line.

The rates of dissociation of some of the weaker carbon acids were estimated from their *base-catalyzed* rates of hydrogen-deuterium exchange.[9] The assumption was made that the ratio k_{H_2O}/k_{OH^-} was the same for these compounds as for acetone. With this estimated rate and the plot of Figure 1, the K_a's of these carbon acids were estimated, and are tabulated in Table VII.

With these measurements and estimates, the groups can be arranged in the following order as to their ability to acidify carbon–hydrogen

TABLE VII

Estimated Rate and Equilibrium Data for Weak Carbon Acids in Water at 25°

Compound	K_a	k_1 (sec.$^{-1}$)
$CH_3SO_2CH_3$	10^{-23}	3.3×10^{-8}
CH_3CO_2H	10^{-24}	3.3×10^{-9}
CH_3CN	10^{-25}	6.7×10^{-10}
CH_3CONH_2	10^{-25}	3.3×10^{-8}

bonds: $NO_2 > CO > SO_2 > CO_2H > CO_2R > CN \simeq CONH_2 > X > H > R$. This order is different from that obtained for the ability of the same substituents to acidify the oxygen–hydrogen bond: $SO_2 > NO_2 > CO_2H \simeq CN > CO > X$.[9]

The effects on carbon acidity of accumulating several of the same acidifying substituents on the same carbon atom are not additive. The departure from additivity seems greatest for the most strongly acidifying substituent, the nitro group. The inability of all atoms to occupy the same plane in the anion (for steric reasons) seems to be responsible. The relevant data are recorded in Table VIII.

TABLE VIII

Substituent Accumulative Effect on Carbon Acidity[9]

	pK_a		pK_a
CH_3NO_2	11	CH_3COCH_3	20
$CH_2(NO_2)_2$	4	$CH_2(COCH_3)_2$	9
$CH(NO_2)_3$	0	$CH(COCH_3)_3$	6
$CH_3SO_2CH_3$	23	CH_3CN	25
$CH_2(SO_2CH_3)_2$	14	$CH_2(CN)_2$	12
$CH(SO_2CH_3)_3$	0	$CH(CN)_3$	0

The second-order rate constants (k_{-1}) for carbanion capture by hydronium ion were calculated, and are recorded in Table VI. For the weakest acid, acetone ($pK_a = 20$), the recombination rate constant is 5×10^{10} liters/mole sec., and the values of the rate constants for the stronger acids are lower. The rate constant for acetone is in the diffusion control range of about 10^{11} liters/mole sec. for ion recombination,[10] and

[10] L. Onsager, *J. Chem. Phys.*, **2**, 999 (1934).

is similar in value to many of the rate constants for proton transfers from positively charged acids to negatively charged bases measured by Eigen.[11]

The above data indicate that, at best, a linear free energy relationship between thermodynamic and kinetic acidity of the carbon acids is poor for the carbon acids ranging in pK_a from 4 to 20. The complete lack of correlation among carbon, oxygen, and nitrogen acids has been emphasized by Bell.[8b] The four acids of Table IX have about the same pK_a in water (9 to 10), but their dissociation rates vary by about 8 powers of 10.

TABLE IX

Comparison of pK_a's and Dissociation Rate Constants (k_1) of Carbon, Nitrogen, and Oxygen Acids[8b]

Acid	pK_a	$\log k_1$
Nitromethane	10.2	−7.4
Benzoylacetone	9.6	−2.1
Trimethylammonium ion	9.8	+1.1
Boric acid	9.1	+1.0

A second interesting correlation between rates and equilibria involving carbon acids was made by Dessy *et al.*[12] These authors measured the rates of hydrogen-deuterium exchange of cyclopentadiene, acetophenone, phenylacetylene, and fluorene with deuterium oxide, catalyzed by 1 *M* triethylamine in dimethyl formamide. The rate constants were corrected for statistical factors (divided by the number of exchangeable hydrogens), and ratios of rate constants relative to that of phenylacetylene were calculated.

The pK_a of cyclopentadiene was estimated[12] to be about 14–15 from the observation that sodium methoxide in methanol metalates cyclopentadiene, whereas sodium phenoxide does not. Metalation (or at least carbanion formation) was detected by carbonation of the salt to give cyclopentadienecarboxylic acid. This pK_a, taken together with those of McEwen (see previous section), provided some of the data for the correlation of Table X.[12] The remainder was taken from the work of Shatenshtein *et al.*,[13] who have studied the rates of exchange of large

[11] M. Eigen, *Angew. Chem. (Intern. Ed. Engl.)*, **3,** 8 (1964).

[12] R. E. Dessy, Y. Okuzumi, and A. Chen, *J. Am. Chem. Soc.*, **84,** 2899 (1962).

[13] (a) A. I. Shatenshtein, *Dokl. Akad. Nauk SSSR*, **60,** 1029 ff. (1949); (b) A. I. Shatenshtein, *Advan. Phys. Org. Chem.*, **1,** 161 (1963).

numbers of organic compounds with liquid deuterated ammonia. The compounds selected as a basis for the latter correlation were indene, fluorene, triphenylmethane, and diphenylmethane. Table X records the relevant data.

TABLE X

Data for Kinetic–Thermodynamic Acidity Correlation[12]

Compound	pK_a [a]	k (sec.$^{-1}$) at 44° in $(CH_3)_2NCHO$ [b]	k (sec.$^{-1}$) at 120° in ND_3 [c]
Cyclopentadiene	14–15	1.1×10^{-2}	—
Acetophenone	19	2.2×10^{-4}	4×10^{-3}
Phenylacetylene	21	6.6×10^{-5}	—
Indene	21	—	4×10^{-1}
Fluorene	25	2.2×10^{-6}	2×10^{-2}
Triphenylmethane	33	—	2×10^{-7}
Diphenylmethane	35	—	7×10^{-9}

[a] McEwen[2a] and Morton.[2b] [b] 1 *M* in triethylamine.[12] [c] Shatenshtein.[13]

Plots of log k against pK_a are linear for the data of both sets of authors,[12] the slope being about 0.4. The kinetic data for acetophenone in liquid ammonia were not included since it seems probable that imine was formed, and the rate was slower than it otherwise would have been.

This linear correlation may well reflect a constancy for the rate of proton capture by different carbanions in the same proton-donating medium, this rate being diffusion controlled or close to it. For Equation (12), if k_{-1} is a constant, then $K_a = \text{constant} \times k_1$, and the basis for the linearity of the plots becomes clear. The fact that the slopes of the lines were not unity could reflect the fact that K_a and k_1 were not measured in the same media.

$$(12) \qquad -\overset{|}{\underset{|}{C}}-H + \bar{B} \underset{k_{-1}}{\overset{k_1}{\rightleftharpoons}} -\overset{|}{\underset{|}{C}}^{-} + HB \qquad K_a = k_1/k_{-1}$$

A condition for diffusion control of proton capture rates by resonance-stabilized carbanions is that the proton donor is a much stronger acid than the carbon acid generated.[14] This condition might apply to all compounds but cyclopentadiene in the Dessy series, but could not in the Shatenshtein series. Ammonia has a pK_a estimated to be 36 on the McEwen scale,[14b] comparable to that of diphenylmethane.

[14] (a) M. Eigen, *Angew. Chem.* (*Intern. Ed. Engl.*), **3**, 18 (1964); (b) N. S. Wooding and W. C. Higginson, *J. Chem. Soc.*, p. 774 (1952).

Another type of correlation has been developed by Streitwieser.[15] For planar unsaturated hydrocarbons that contain carbon–hydrogen bonds acidified by the unsaturated system, he assumed that the thermodynamic acidity is proportional to the difference in π-delocalization energy (ΔE_{π_i}) between the carbanion ($(E_\pi)_{A_i^-}$) and the parent hydrocarbon ($(E_\pi)_{AH_i}$) (Equation (13)). He further assumed that simple molecular orbital (MO) theory applies; that the value of β (resonance integral) of Equation (13) is an adjustable parameter but is the same from system to system; that the σ-bond (C—H) and solvation energies are comparable from system to system; and that the strain in each hydrocarbon is comparable with that in the derived anion. Equation (14)

$$\Delta E_{\pi_i} = (E_\pi)_{A_i^-} - (E_\pi)_{AH_i} = 2\alpha + \Delta M_i \beta \tag{13}$$

α = Coulomb integral of MO theory

β = Resonance integral of MO theory

relates the pK_a's of the hydrocarbons to the calculable ΔM_i, with a as the intercept and b the empirical slope.

$$pK_i = a + b\Delta M_i \tag{14}$$

For the three planar carbon acids, toluene, fluorene, and indene, a plot of pK_a's (McEwen's values)[2] vs. calculated ΔM's gave a straight line whose slope (b) and intercept (a) were evaluated to give Equation (15). An intercept of about 48 was obtained, which might be considered as the pK_a of methane, since all of the systems are substituted methanes, and $\Delta M = 0$ for methane. With these values of a and b, and values for ΔM that could be calculated for other planar hydrocarbons, values of pK_a for the other planar hydrocarbons could be estimated. These are listed in Table XI, along with those experimental values that are available. The entry of pK_a (exp.) for cycloheptatriene is taken from Dauben and Rifi,[16] and the pK_a (exp.) for cyclopentadiene was taken from Dessy *et al.*[12]

$$pK_a = 48 - 15.5\Delta M \tag{15}$$

In a later study, Streitwieser *et al.*[17] measured the rates of lithium cyclohexylamide-catalyzed exchange between cyclohexylamine (as solvent) and deuterated arylmethanes. They also calculated the values

[15] A. Streitwieser, Jr., *Tetrahedron Letters*, **6**, 23 (1960).
[16] H. J. Dauben, Jr. and M. R. Rifi, *J. Am. Chem. Soc.*, **85**, 3042 (1963).
[17] (a) A. Streitwieser, Jr. and W. C. Langworthy, *J. Am. Chem. Soc.*, **85**, 1757 (1963); (b) A. Streitwieser, W. C. Langworthy, and J. I. Brauman, *ibid.*, **85**, 1761 (1963).

of ΔM_i of Equation (16) to see if a correlation existed between these values and their rates of isotopic exchange. A linear correlation of the type found in Equation (17) was observed for the four compounds of Table XII, where k_0 was the rate constant for exchange of deuterated toluene, and the k_i's were the rate constants for the other hydrocarbons at 49.9°.

$$(16) \quad \Delta E_\pi(\mathrm{ArCH_2D}) = E_\pi(\mathrm{ArCH_2^-}) - E_\pi(\mathrm{ArH}) = 2\alpha + \Delta M_i \beta$$

$$(17) \quad \log(k_i/k_0) = a + b\Delta M_i$$

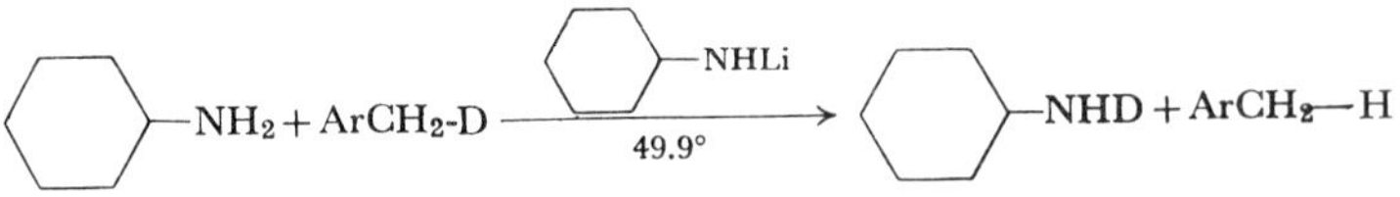

C_6H_{11}-NH_2 + $ArCH_2$-D $\xrightarrow[49.9°]{C_6H_{11}\text{-NHLi}}$ C_6H_{11}-NHD + $ArCH_2$—H

TABLE XI

Calculated pK_a*'s for Planar Hydrocarbons Based on MO Calculated Delocalization Energies*

Compound	ΔM	pK_a (exp.)	pK_a (calc.)
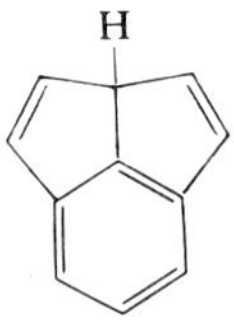	2.411	—	11
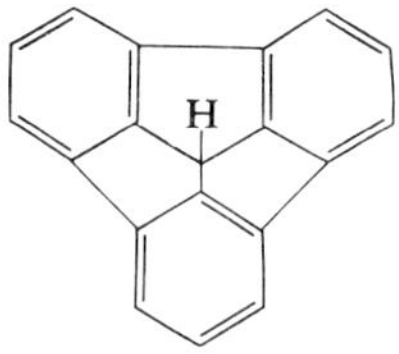	2.225	—	14
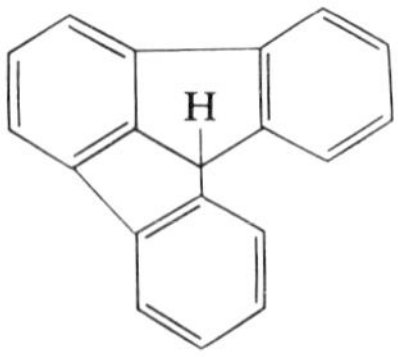	2.115	~11	15
	2.000	~15	17

TABLE XI (*cont.*)

Compound	ΔM	pK_a (exp.)	pK_a (calc.)
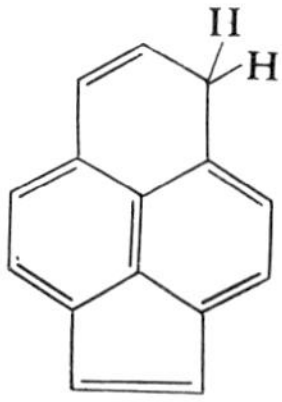	1.964	—	18
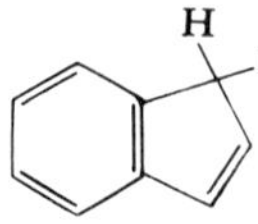	1.747	21	—
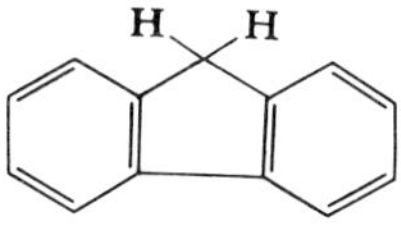	1.523	25	—
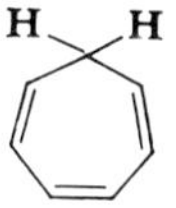	1.110	36	31
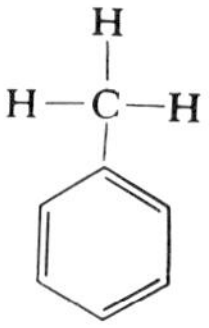	0.721	37	—
CH_4	0.000	—	~48

TABLE XII

Anionic Reactivities of Arylmethanes

Ar in $ArCH_2D$	k_i/k_0 [a]	ΔM_i
Phenyl	1.0	0.721
2-Naphthyl	7.4	0.744
3-Phenanthryl	14	0.754
2-Anthracyl	31	0.769

[a] k_0 is for $C_6H_5CH_2$—D.

In a study that bridges the gap in the kinetic acidities of the very weak and much stronger carbon acids, Streitwieser and co-workers[18] measured the rates of hydrogen-tritium exchange catalyzed by lithium or cesium cyclohexylamide between a series of hydrocarbons and cyclohexylamine. Table XIII records the rates of exchange of the hydrocarbons relative to that for benzene-*t*.

TABLE XIII

Rates of Lithium or Cesium Cyclohexylamide-Catalyzed Exchange Between Tritiated Hydrocarbons and Cyclohexylamine Relative to that of Benzene-t

	$LiNHC_6H_{11}$-*c*		$CsNHC_6H_{11}$-*c*	
Compound	25°	50°	25°	50°
Triphenylmethane-α-*t*	1.2×10^5	—	—	—
Diphenylmethane-α-*t*	2.9×10^4	—	—	—
Toluene-α-*t*	1.1×10^2	—	—	—
Benzene-*t*	1.0	—	—	—
Cumene-α-*t*	—	0.84	1.34	—
Toluene-3-*t*	—	0.54	0.57	—
Toluene-4-*t*	—	0.46	0.48	—
2-Phenylbutane-2-*t*	—	0.31	0.39	0.43
Triptycene-α-*t*	—	—	0.24	—
Toluene-2-*t*	—	0.12	0.20	—
Cyclopropane-*t*	—	—	—	10^{-3} [a]
Cyclobutane-*t*	—	—	—	10^{-6} [a]
Cyclohexane-*t*	—	—	—	1.1×10^{-8}

[a] Private communication from A. Streitwieser, Jr.; preliminary values.

Several interesting relationships are visible in the data of Table XIII.[18] Substitution of cesium for lithium ion has only minor effects on the rate factors, and recalls the fact that the ultraviolet spectra of the lithium and cesium carbon salts were very similar.[5] If the 25° difference in temperature and the difference in metal cation are neglected, the rate factors cover a range of 13 powers of 10, the most kinetically acidic hydrocarbon of the list being triphenylmethane, and the least, cyclohexane. The two methyl groups of cumene reduce the rate of exchange from that of toluene by about 2 powers of 10. The rate factor of triphenylmethane exceeds that of triptycene by a factor of 5×10^5. Since the

[18] A. Streitwieser, Jr., R. A. Caldwell, and M. R. Granger, *J. Am. Chem. Soc.*, **86**, 3578 (1964).

benzene rings of triptycene anion are unable to delocalize charge due to the restrictions of the bicyclic ring system, it seems that some of the acidifying power that phenyl has on α-hydrogen is due to an inductive effect.[18]

In the McEwen acidity scale,[2a] ΔpK_a for cumene and triphenylmethane was about 4.5 units. If the correspondence between the McEwen and Streitwieser scales (see Table III) extends to cumene, then on the Streitwieser scale cumene would have a pK_a of about 37. Furthermore, with an increase in value of pK_a of 4.5 units, the relative rates of exchange decreased by about 5 powers of 10. Thus, an increase of 1 pK_a unit corresponds to a little over one power of 10 decrease in rate. If this relationship should hold over the range from triphenylmethane to cyclohexane, then benzene should have a pK_a of about 37, and cyclohexane a pK_a of 45 on the Streitwieser scale. Use of the Applequist equilibrium data[7] for estimating the difference in pK_a between benzene and cyclohexane gives a ΔpK_a of about 7, or a pK_a for cyclopentane of 44 on the Streitwieser scale. The near correspondence of the pK_a estimates of cyclohexane and cyclopentane based on the two different approaches lends credibility to the reasoning. Application of these methods to the data of Streitwieser, Applequist, and Dessy that dealt with the saturated

TABLE XIV

McEwen-Streitwieser-Applequist-Dessy (MSAD) pK_a *Scale*

Compound	pK_a	Compound	pK_a
Fluoradene	11	Ethylene[b]	36.5
Cyclopentadiene	15	Benzene	37
9-Phenylfluorene	18.5	Cumene (α-position)	37
Indene	18.5	Triptycene (α-position)	38
Phenylacetylene	18.5	Cyclopropane	39
Fluorene	22.9	Methane	40
Acetylene[a]	25	Ethane	42
1,3,3-Triphenylpropene	26.5	Cyclobutane	43
Triphenylmethane	32.5	Neopentane	44
Toluene (α-position)	35	Propane (*s*-position)	44
Propene (α-position)	35.5	Cyclopentane	44
Cycloheptatriene	36	Cyclohexane	45

[a] Based on Wooding and Higginson's value[14b] reconciled to the MSAD scale.

[b] S. I. Miller and W. G. Lee (*J. Am. Chem. Soc.*, **81**, 6313 (1959)) estimated the upper limits to the pK_a's for the 1,2-dihaloethylenes to be 34–36.

hydrocarbons leads to estimates that exhibit an encouraging pattern of internal consistency. This amalgamated pK_a scale, based on 9-phenylfluorene (pK_a = 18.5), will be referred to as the McEwen-Streitwieser-Applequist-Dessy, or MSAD, scale and is found in Table XIV. Further work will undoubtedly lead to modifications of this scale, particularly at the upper end.

Methane has a pK_a of 40 on the MSAD scale. Two other estimates of the pK_a of methane exist. Pearson and Dillon[9] recorded a value of 40 in connection with the accumulative effects on pK_a values of successive substitutions of nitro, acetyl, sulfonyl, or cyano groups for hydrogens of methane. Bell[8b] refined an argument originally due to Schwarzenbach,[19] who assumed that Equation (18) seemed reasonable. Implicit in this equation is the assumption that all bond energies are the same. If corrections are made for the differences in bond energies,[8b] and the pK_a of ammonia is taken to be 35,[8b] then methane would appear to have a value of about 58.

$$\text{(18)} \qquad pK_{CH_4} - pK_{NH_4^+} = pK_{NH_3} - pK_{OH_3^+}$$

KINETIC ACIDITIES OF WEAK HYDROCARBON ACIDS

The pioneering work of Shatenshtein and co-workers[20] in potassium amide-catalyzed hydrogen isotopic exchange between deuterated ammonia and both saturated and unsaturated hydrocarbons establishes that even the former can be regarded as acids. In Table XV are recorded the results of exchange experiments with isopentane, hexane, cyclopentane, and cyclohexane.[21] The limited solubility of the hydrocarbons in the solvent required that these reactions be carried out heterogeneously. However, the results clearly demonstrate that saturated hydrocarbons can behave as proton donors, and suggest that the degree of substitution at carbon by other carbon affects the acidity of the attached hydrogens. The latter conclusion is reinforced by the results of the follow-

[19] G. Schwarzenbach, *Z. Physik. Chem.*, **176A**, 133 (1936).

[20] A. I. Shatenshtein, *Advan. Phys. Org. Chem.*, **1**, 153–201 (1963).

[21] (a) N. M. Dykhno and A. I. Shatenshtein, *Zh. Fiz. Khim.*, **28**, 14 (1954); (b) A. I. Shatenshtein and L. N. Vasil'eva, *Dokl. Akad. Nauk SSSR*, **94**, 115 (1954); (c) A. I. Shatenshtein, L. N. Vasil'eva, N. M. Dykhno, and E. A. Izrailevich, *ibid.*, **85**, 381 (1952); (d) A. I. Shatenshtein, N. M. Dykhno, L. N. Vasil'eva, and M. Faivush, *ibid.*, **85**, 381 (1952); (e) A. I. Shatenshtein, E. A. Yakovleva, M. I. Rikhter, M. Lukina, M. Yu, and B. A. Kazanskiĭ, *Izv. Akad. Nauk SSSR, Otdel. Khim. Nauk*, p. 1805 (1959); (f) A. I. Shatenshtein, *Advan. Phys. Org. Chem.*, **1**, 175 (1963).

ing experiment carried out at 120° for 100 hr. in a 1 N solution of potassium amide in liquid ammonia. Treatment of isobutane labeled with deuterium in the methyl positions resulted in 50% deuterium loss, whereas treatment of isobutane labeled at the methine position produced no exchange. Clearly methyl hydrogens are more acidic than methine hydrogens, and the data of Table XV suggest that methyl hydrogens are more acidic than methylene hydrogens.

TABLE XV

Hydrogen–Deuterium Exchange of Saturated Hydrocarbons with Deuterated Liquid Ammonia at 120°

Hydrocarbon	KNH_2 conc. (M)	Time (hr.)	No. H's exchanged
Isopentane	1.0	330	3.3
n-Heptane	1.0	500	3.0
Cyclopentane	3.0	1300	3.4
Cyclohexane	0.8	180	0.7

Hydrogen itself has been shown to be many orders of magnitude more kinetically acidic than any of the above-mentioned saturated hydrocarbons.[22] Thus, in liquid ammonia–potassium amide solution at −53°, dissolved deuterium gas forms hydrogen deuteride with a rate constant of 81 liters/mole sec. This rate was comparable to the rate of conversion of parahydrogen to orthohydrogen, which appears to involve an anionic mechanism. At 50° the rate of this conversion was estimated to occur about 14 powers of 10 faster in liquid ammonia–potassium amide than in water–potassium hydroxide.[22]

Although more experimental data are desirable, those available suggest the following rank of kinetic acidities: $H_2 > CH_4 > CH_3CH_3 > (CH_2)_6 > (CH_3)_3CH$ (methine hydrogen). This order is what is predicted on the basis of the electron-releasing capacity of the carbon–carbon bond as compared to the carbon–hydrogen bond. Carbanion stability, which correlates with carbon acidity, seems to decrease as hydrogen attached to the carbanion is successively substituted with alkyl groups. This effect, as expected, is in a direction opposite to that observed for carbonium ion stability.

[22] (a) W. K. Wilmarth, J. C. Dayton, and J. M. Flournoy, *J. Am. Chem. Soc.*, **75**, 4549 (1953); (b) W. K. Wilmarth and J. C. Dayton, *ibid.*, **75**, 4553 (1953); (c) J. M. Flournoy and W. K. Wilmarth, *ibid.*, **83**, 2257 (1961).

In further experiments, Shatenshtein *et al.*[21e,23] made a rough survey of the relative rates of hydrogen–deuterium exchange between deuterated ammonia–potassium amide and a number of hydrocarbons, some of which contained three-membered rings. The experimental conditions and results are given in Table XVI.

TABLE XVI

Results of Deuterated Potassium Amide-Catalyzed Exchange Between Hydrocarbons and Deuterated Liquid Ammonia

	Number of hydrogens exchanged				
	At 25°			At 120°	
	0.05 M,[a] 6 hr.	1.0 M,[a] 8 hr.	1.0 M,[a] 240 hr.	1.0 M,[a] 6 hr.	1.0 M,[a] 200 hr.
$CH_3-CH(CH_3)-CH_2-CH_3$	—	—	0	—	2
$(H_2C)_2CH-CH_2-CH_3$ (cyclopropyl)	—	0.3	4	5	6
$CH_3-CH(CH_3)-C_6H_5$	—	—	6	—	12
$(H_2C)_2CH-C_6H_5$ (cyclopropyl)	5	10	10	—	—
$(H_2C)_2CH-CH{=}CH_2$ (cyclopropyl)	2	8	8	—	—
$H_2C-CH(-CH_2-CH_3)$, H_2C-CH_2 (cyclobutyl)	—	—	—	0.1	2

[a] Molarities of KND_2.

These data suggest that the order of acidity of the various hydrogens is the one tabulated (hydrogens are circled). The rank of some of the hydrogens is doubtful, since assignment of relative acidity depends on a comparison of the numbers of hydrogens exchanged (Table XVI) and

[23] A. I. Shatenshtein, *Advan. Phys. Org. Chem.*, **1**, 176 (1963).

the numbers of each kind of hydrogen in the molecule. However, the following firm conclusions can be drawn.

(1) Cyclopropane hydrogens are much more acidic than the hydrogens of their alicyclic counterparts.

$$\text{(H}_2\text{)C}\diagdown\!\!\!|\!\!\!\diagup\text{C(H)—C}_6\text{H}_5 \gtrsim \text{(H}_2\text{)C}\diagdown\!\!\!|\!\!\!\diagup\text{C(H)—C(H)=C(H}_2\text{)} \sim \text{H}_2\text{C}\diagdown\!\!\!|\!\!\!\diagup\text{CH—C}_6\text{(H}_5\text{)}$$

$$\text{(H}_2\text{)C}\diagdown\!\!\!|\!\!\!\diagup\text{CH—CH}_2\text{—CH}_3 \sim \text{CH}_3\text{—C(CH}_3\text{)(H)—C}_6\text{(H}_5\text{)} > \text{H}_2\text{C}\diagdown\!\!\!|\!\!\!\diagup\text{C(H)—CH}_2\text{—CH}_3 >$$

$$\text{C(H}_3\text{)—CH(C(H}_3\text{))—C}_6\text{H}_5 > \text{C(H}_3\text{)—C(C(H}_3\text{))(H)—C(H}_2\text{)—C(H}_3\text{)} \sim \text{(H}_2\text{)C—C(H}_2\text{)—(H}_2\text{)C—C(H}_2\text{)—C(H}_2\text{)—C(H}_3\text{)}$$

(2) Aryl or vinyl groups attached to the cyclopropane ring acidify the hydrogens of the cyclopropane ring.

(3) The hydrogens of an alkylcyclopropane ring are comparable in acidity to the ring hydrogens of an alkylbenzene ring.

(4) A phenyl substituted in the 2-position of an alkane acidifies the hydrogens of both the 1- and 2-positions, the latter much more than the former.

(5) The hydrogens of a cyclobutane ring appear to be comparable in acidity to those of its open-chain counterpart.

The increased acidity of cyclopropane ring hydrogens over those of their open-chain analogs is probably associated with the greater *s*-character of the carbon–hydrogen bond in cyclopropane rings[24] ($sp^{2.28}$-s sigma bond). The greater the *s*-character of the orbital containing the two electrons of a carbanion, the more stable that anion, since the charge of the nucleus is less shielded in a $2s$ than in a $2p$ orbital (see Chapter II).

Another informative series is provided by the rates of exchange of ethylene, propene, isobutene, trimethylethylene, and tetramethylethylene with deuterated liquid ammonia catalyzed by potassium amide.[21b,25] Five of the hydrogens of propene exchange rapidly compared to the sixth hydrogen, or the hydrogens of ethylene. These results indicate that the allylic hydrogens exchange much more readily than

[24] (a) L. L. Ingraham, *in* "Steric Effects in Organic Chemistry" (M. S. Newman, ed.), Wiley, New York, 1956, p. 518; (b) C. A. Coulson and W. E. Moffitt, *J. Chem. Phys.*, **15**, 151 (1947); (c) A. D. Walsh, *Trans. Faraday Soc.*, **45**, 179 (1949).

[25] A. I. Shatenshtein, *Advan. Phys. Org. Chem.*, **1**, 178 (1963).

vinylic hydrogens, and that in the propene system, the terminal hydrogens exchange by multiple rearrangement, as is formulated below:

$$CH_2{=}CH{-}CH_3 + KND_2 \longrightarrow$$

$$\{CH_2{=}CH{-}\bar{C}H_2 \longleftrightarrow \bar{C}H_2{-}CH{=}CH_2\} \xrightarrow{ND_3}$$

$$CH_2{=}CH{-}CH_2D \xrightarrow{KND_2}$$

$$\{CH_2{=}CH{-}\bar{C}HD \longleftrightarrow \bar{C}H_2{-}CH{=}CHD\} \xrightarrow{ND_3}$$

$$\{CH_2{=}CH{-}CHD_2 + CH_2D{-}CH{=}CHD\} \xrightarrow{KND_3} \text{etc.}$$

Substitution of the vinyl hydrogens of propene with methyl groups decreases the rate of exchange of the allylic hydrogens, the decrease being greater, the greater the number of methyl groups. The same conclusions regarding allylic vs. vinylic hydrogen acidity, cyclopropane vs. alkane acidity, and the effect of methyl groups had also been reached by Morton and co-workers on the basis of metalation and carbonation experiments.[26]

Two elegant types of experiments highlight the conclusion that multiple allylic rearrangements make vinylic hydrogens exchangeable as allylic hydrogens. (1) At 120° for 100 hr., only 7 out of 16 hydrogens of compound I exchanged with deuterated liquid ammonia–potassium amide.[21c,27] Compounds such as 1-hexene, cyclohexene, and propylcyclohexene exchange all of their hydrogens under conditions that leave saturated hydrocarbons untouched (e.g., 1.0 *N* potassium amide in deuterated liquid ammonia at 100°).[28,21c,21d,26a]

$$\underset{\substack{|\\C(H_3)}}{(H_2)C{=}C}{-}C(H_2){-}\overset{\substack{CH_3\\|}}{\underset{\substack{|\\CH_3}}{C}}{-}CH_3$$

1 (encircled hydrogens exchanged)

A further discussion of the acidity of allylic hydrogens is found in Chapter IV under Allylic Rearrangements.

[26] (a) A. A. Morton and M. L. Brown, *J. Am. Chem. Soc.*, **69**, 160 (1947); (b) A. A. Morton, F. D. Marsh, R. D. Coombs, A. L. Lyons, S. E. Penner, H. E. Ramsden, V. B. Baker, E. L. Little, and R. L. Letsinger, *ibid.*, **72**, 3875 (1950); (c) E. J. Lampher, L. M. Redman, and A. A. Morton, *J. Org. Chem.* **23**, 1370 (1958).

[27] (a) A. I. Shatenshtein, L. N. Vasil'eva, and N. M. Dykhno, *Zh. Fiz. Khim.*, **28**, 193 (1954); (b) A. I. Shatenshtein, *Advan. Phys. Org. Chem.*, **1**, 181 (1963).

[28] (a) A. I. Shatenshtein and E. A. Izrailevich, *Dokl. Akad. Nauk SSSR*, **108**, 294 (1956); (b) A. I. Shatenshtein, *Advan. Phys. Org. Chem.*, **1**, 179 (1963).

Two groups of investigators studied the rates of base-catalyzed exchange of deuterium attached to aromatic nuclei. The Shatenshtein group[29] observed the rate constants (sec.$^{-1}$) listed below the formulas in 0.02 N solutions of potassium amide in liquid ammonia at 25°. Naphthalene exchanges in the α-position about 10 times and in the β-position about four times as fast as does benzene, whereas the three kinds of hydrogens in biphenyl exchange at about the same rate as do the β-hydrogens of naphthalene.

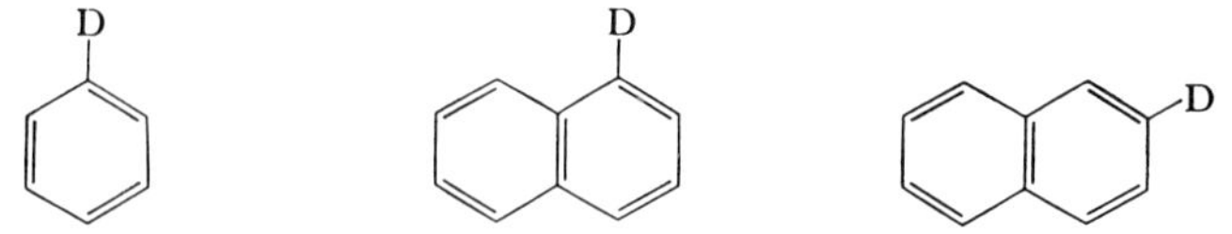

D D D

8.2×10^{-5} 8.0×10^{-4} 3.6×10^{-4}

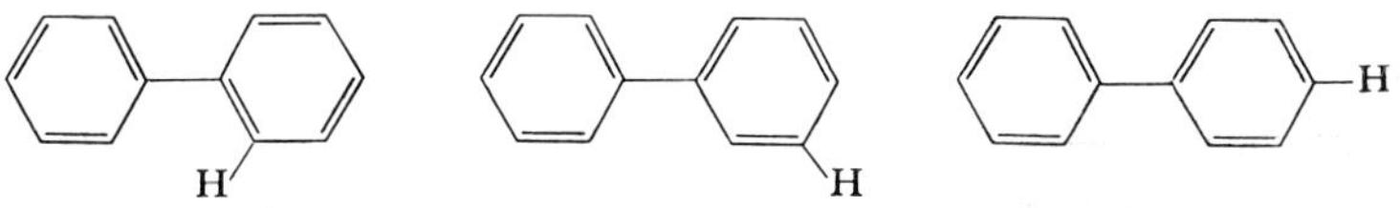

H H H

3.0×10^{-4} 2.6×10^{-4} 2.5×10^{-4}

The second group[30] examined the rates of detritiation of the various positions of a series of polynuclear aromatic compounds. They observed an approximately linear correlation between the logarithm of the rates relative to that of phenyl, and the sum of the reciprocal distances of the carbanionic site from the other π-carbons of the molecule ($\sum 1/r$). The relevant data are listed in Table XVII.

The authors[30] interpreted the correlation in terms of the electron-attracting field effect of the aromatic π-carbons (inductive effect), which stabilizes the aryl anion. They concluded that carbanion-carbene resonance was unimportant since they also observed a greater exchange rate at the 3-biphenylyl ($k_D = 14.4 \times 10^{-5}$ liters/mole sec. at 49.9°) than

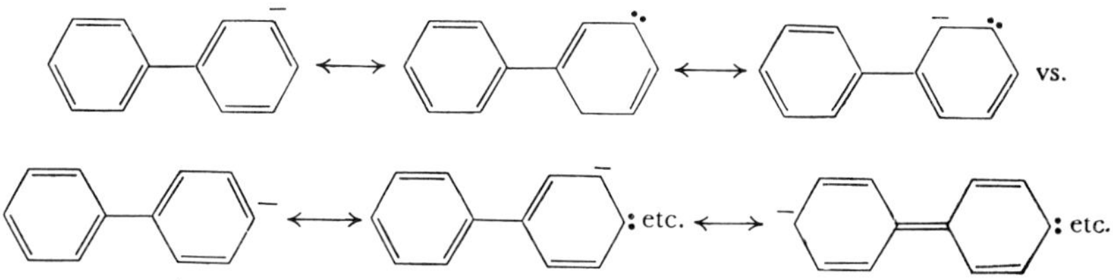

vs.

: etc. : etc.

[29] (a) A. I. Shatenshtein and E. A. Izrailevich, *Zh. Fiz. Khim.*, **28**, 3 (1954); (b) E. N. Yurygina, P. P. Alikhanov, E. A. Izrailevich, P. N. Manochkina, and A. I. Shatenshtein, *Zh. Fiz. Khim.*, **34**, 587 (1960); (c) A. I. Shatenshtein, *Advan. Phys. Org. Chem.*, **1**, 182 (1963).

[30] A. Streitwieser, Jr. and R. G. Lawler, *J. Am. Chem. Soc.*, **85**, 2854 (1963).

TABLE XVII

Rates of Exchange of Tritiated Aromatic Hydrocarbons with Cyclohexylamine–Potassium Cyclohexylamide at 49.9° [30]

Ar-*t*	$k \times 10^{-5}$ liters/mole sec.$^{-1}$	$\Sigma\ 1/r$
Phenyl-*t*	2.3	3.655
2-Naphthyl-*t*	10.5 [a]	4.932
1-Naphthyl-*t*	15.9 [a]	5.321
1-Anthracyl-*t*	34	6.308
9-Phenanthryl-*t*	50 [a]	6.599
2-Pyrenyl-*t*	61	6.513
1-Pyrenyl-*t*	75	6.918
4-Pyrenyl-*t*	97	7.165
9-Anthracyl-*t*	143	6.988

[a] Calculated from the rate of dedeuteration, assuming that $k_D/k_T = 1.4$.

at the 4-biphenylyl position ($k_D = 8.6 \times 10^{-5}$ liters/mole sec. at 49.9°). Had this kind of resonance been important, the relative values would have been reversed.

Substitution of methyl groups in a benzene ring depresses the rate of exchange of deuterium attached to the ring.[30, 31] The relative rates of

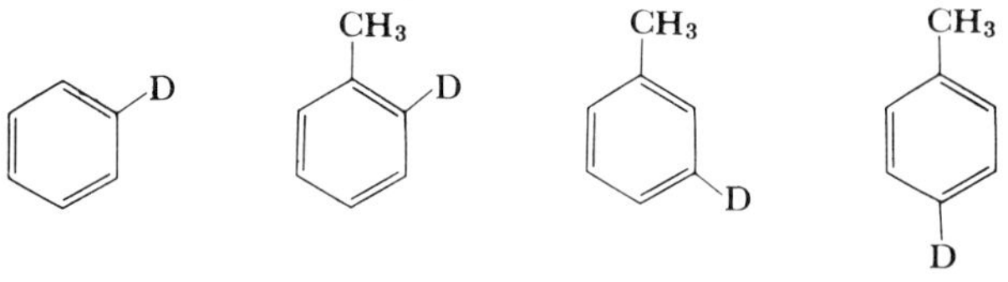

D CH3 D CH3 CH3 D D

Relative rates 258 58 120 105

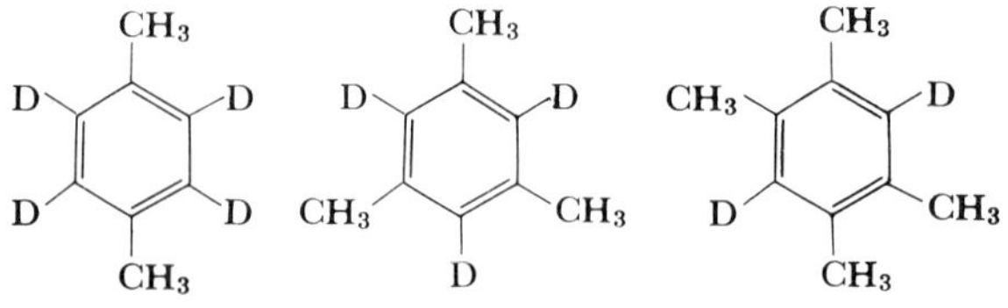

CH3 D D D D CH3 CH3 D D CH3 CH3 D CH3 CH3 D D CH3 CH3

Relative rates 26 4 1

[31] (a) A. I. Shatenshtein and E. A. Izrailevich, *Zh. Obshch. Khim.*, **32**, 1930 (1962); (b) A. I. Shatenshtein, *Tetrahedron*, **18**, 95 (1962); (c) A. I. Shatenshtein, *Advan. Phys. Org. Chem.*, **1**, 184 (1963).

dedeuteration in 0.2 *N* potassium amide in liquid ammonia at 25° are indicated. The electron-releasing effects of the methyl groups, coupled with their ability sterically to inhibit solvation of the derived carbanion produced, are probably responsible for these trends.

Confirmation of the relative rates of exchange of aryl hydrogens (or their isotopes) of benzene and of the *m*- and *p*-positions of toluene is found in the work of Streitwieser and Lawler,[30] who used as solvent cyclohexylamine–potassium cyclohexylamide at 49.9°. These authors observed factors of 1.9 and 2.0, respectively.[30] These factors compare to those of 2.1 and 2.5 obtained by Shatenshtein *et al.*[31]

Three groups of investigators have examined the rates of base-catalyzed hydrogen isotope exchange at the α-position of phenyl-substituted alkanes.[32, 33, 34] Toluene, ethylbenzene, and cumene were common to all three of the studies, *o*-, *m*-, and *p*-xylene to two,[33, 34] and the higher methylated benzenes were studied by one of the groups.[34] The Shatenshtein group worked with liquid ammonia–potassium amide (0.02 *N*), the Streitwieser group with cyclohexylamine–potassium cyclohexylamide (0.06 *N*), and the Schriesheim group with dimethyl sulfoxide–potassium *tert*-butoxide (0.60 *N*). Table XVIII records the rate data in each solvent–base mixture relative to that of toluene.

The agreement is qualitatively very good considering the potential complications (see Chapter III). The rate order, $C_6H_5CH_3 >$ $C_6H_5CH_2CH_3 > C_6H_5CH(CH_3)_2$, is expected on the basis of the electron-releasing character of a methyl group. The effect of methyl substitution at the *m*-position of the benzene ring is felt less strongly than at the α-position, but the effect is smaller than expected. The direct inductive effect, steric inhibition of resonance, and solvation should all be more important at the α- than at the *m*-position. Most surprising of all is the fact that methyl substitution at the *p*-position is more rate retarding than at the α-position in dimethyl sulfoxide as solvent, whereas in cyclohexylamine substitution in the α-position is the more rate retarding, but only by a factor of 2.7. These facts suggest that possibly the rates of

[32] (a) N. M. Dykhno and A. I. Shatenshtein, *Zh. Fiz. Khim.*, **28**, 11 (1954); (b) A. I. Shatenshtein and E. A. Izrailevich, *Zh. Fiz. Khim.*, **32**, 2711 (1958); (c) A. I. Shatenshtein, L. N. Vasil'eva, N. M. Dykhno, and E. A. Izrailevich, *Dokl. Akad. Nauk SSSR*, **85**, 381 (1952); (d) A. I. Shatenshtein, *Advan. Phys. Org. Chem.*, **1**, 183 (1963); (e) A. I. Shatenshtein, I. O. Shapiro, F. S. Jakuhin, G. G. Isajewa, and Yu. I. Ranneva, *Kinetika i Kataliz*, **5**, 752 (1964).

[33] (a) A. Streitwieser, Jr. and D. E. Van Sickle, *J. Am. Chem. Soc.*, **84**, 249 (1962); (b) A. Streitwieser, Jr. and H. F. Koch, *ibid.*, **86**, 404 (1964).

[34] J. E. Hofmann, R. J. Muller, and A. Schriesheim, *J. Am. Chem. Soc.*, **85**, 3002 (1963).

TABLE XVIII

Relative Rates of Hydrogen Isotopic Exchange of Arylalkanes at the α-Position (per Hydrogen Atom)

Compound	NH_3–NH_2K at 10°[a]	c-$C_6H_{11}NH_2$–c-$C_6H_{11}NHK$ at 50°[b]	CH_3SOCH_3–*tert*-BuOK at 30°[c]
$C_6H_5CH(CH_3)_2$	0.029	0.008	0.023
$C_6H_5CH_2CH_3$	0.14	0.116	0.22
$C_6H_5CH_3$	1	1	1
m-$CH_3C_6H_4CH_3$	0.67	0.60	0.51
p-$CH_3C_6H_4CH_3$	0.25	0.31	0.033
o-$CH_3C_6H_4CH_3$	—	0.60	1.4

[a] Shatenshtein and co-workers.[32]
[b] Streitwieser and co-workers.[33]
[c] Hofmann *et al.*[34]

exchange do not reflect rates of ionization, because k_{-1} might be greater than k_2 in Equation (19). Since k_1 is undoubtedly much lower valued than k_{-1} or k_2, and the steady state approximation applies, $k_{obs.} = k_1 k_2/(k_{-1}+k_2)$. If $k_{-1} \gg k_2$, then $k_{obs.} = k_2 k_1/k_{-1} = k_2 K$ where $K = k_1/k_{-1}$. Thus the observed rate constant is the product of an equilibrium constant and a rate constant for a process that does not involve breakage or formation of covalent bonds.

$$\text{(19)} \quad \begin{array}{l} -\overset{|}{\underset{|}{C}}-H + :\bar{B} \underset{k_{-1}}{\overset{k_1}{\rightleftharpoons}} -\overset{|}{\underset{|}{C}}^{-}\cdots HB \xrightarrow[k_2]{DB} \\ -\overset{|}{\underset{|}{C}}^{-}\cdots DB \longrightarrow -\overset{|}{\underset{|}{C}}-D + :\bar{B} \end{array}$$

The observation that low and even negative hydrogen-deuterium isotope effects were observed for base-catalyzed hydrogen-deuterium exchange reactions of certain carbon acids[35a] was interpreted on the

[35] (a) D. J. Cram, D. A. Scott, and W. D. Nielsen, *J. Am. Chem. Soc.*, **83**, 3696 (1961); (b) A. I. Shatenshtein, I. O. Shapiro, F. S. Iakushin, A. A. Isaewa, and Yu I. Ranneva, *Kinetics Catalysis* (*USSR*) *Engl. Transl.*, **5**, 752 (1964); (c) A. Streitwieser, W. C. Langworthy, and D. E. Van Sickle, *J. Am. Chem. Soc.*, **84**, 251 (1962); (d) J. E. Hofmann, A. Schriesheim, and R. E. Nichols, Private Communication.

basis of $k_{-1} \gg k_2$. In such a case, the isotope effect would involve only k_2 and K, values for which should be close to unity or slightly negative. The isotope effect for hydrogen-deuterium exchange in cyclohexylamine–lithium cyclohexylamide of toluene is $k_H/k_D > 10$.[33b, 35c] However, $k_H/k_D \sim 0.6$ for toluene in dimethyl sulfoxide–potassium *tert*-butoxide.[35d] These results suggest that in the amine solvent $k_2 \gg k_{-1}$ and the isotope effect is associated with k_1, but in dimethyl sulfoxide the isotope effect reflects Kk_2. In other words, the rate of exchange in the amine corresponds to the rate of dissociation of the carbon acid, but in dimethyl sulfoxide the rate of dissociation is faster than the rate of exchange. Likewise, the low isotope effect for the isotopic exchange of the α-position of thiophene in dimethyl sulfoxide–lithium *tert*-butoxide ($k_D/k_T \sim 0.8$) points to $k_{-1} \gg k_2$ in this system as well.[35b]

The rates of exchange obtained in cyclohexylamine for toluene and 13 substituted toluenes[33b] give an excellent Hammett $\sigma\rho$-plot, with $\rho = 4.0$. This high value points to great development of charge in the transition state.

The study carried out in dimethyl sulfoxide–potassium *tert*-butoxide[34] included a large number of other alkylated benzenes, and the rates per hydrogen relative to toluene are recorded in Table XIX. Exchange in these compounds was exclusively at the α-position. Under the conditions used for the rate comparisons of Table XIX (0.6 M potassium *tert*-butoxide at 30°), *tert*-butylbenzene gave no isotopic exchange after 334 hr., and benzene gave no isotopic exchange at 50° after 100 hr.

The most surprising relationship visible in these data is the greater rate of exchange of *o*-xylene over that for toluene. Although the explanation may lie in the relative values of k_{-1} and k_2 (Equation (19)), other explanations are also possible.[34] In *o*-xylene, steric strain may be released in passing from the ground to the transition state, and the transition state may even be stabilized by $\text{C—H} \cdots \bar{\text{C}}$ hydrogen bonding. Carbanions are known to form strong hydrogen bonds.[36] Conceivably, a nonplanar bridged hydrogen heterocycle could be an intermediate. The hydrogen could be above the plane of the carbon skeleton and located on an axis perpendicular to and bisecting a line connecting the two methyl carbon atoms.

An additivity of methyl substituent effects is visible in the rate data for the higher methylbenzenes.[34] Substituent factors are defined in Equations (20). If the effects are additive, then Equation (21) should apply for 1,3,5-trimethylbenzene and Equation (22) for 1,2,3,4-tetramethyl-

[36] L. L. Ferstandig, *J. Am. Chem. Soc.*, **84**, 3553 (1962).

TABLE XIX

Relative Rates of Exchange per Benzylic Hydrogen of Alkylated Benzenes with Tritiated Dimethyl Sulfoxide 0.6 M in Potassium tert-Butoxide at 30° [34]

Compound	$k_{\text{compound}}/k_{\text{toluene}}$
Toluene	1.00
o-Xylene	1.40
m-Xylene	0.51
p-Xylene	0.033
Ethylbenzene	0.22
Cumene	0.023
tert-Butylbenzene	0.00
1,2,4-Trimethylbenzene	0.26
1,3,5-Trimethylbenzene	0.22
1,2,3-Trimethylbenzene	1.12
1,2,4,5-Tetramethylbenzene	0.018
1,2,3,5-Tetramethylbenzene	0.12
1,2,3,4-Tetramethylbenzene	0.46
Pentamethylbenzene	0.056
Hexamethylbenzene	0.010
o-Ethyltoluene	0.51
m-Ethyltoluene	0.24
p-Ethyltoluene	0.026
p-Cymene	0.023
p-tert-Butyltoluene	0.033

benzene. Application of these principles to the higher methylbenzenes provides a set of predicted relative rates which are roughly comparable with those observed (see Table XX).

$$f_o = \ln(k_{o\text{-xylene}}/k_{\text{toluene}}) \tag{20}$$

$$f_m = \ln(k_{m\text{-xylene}}/k_{\text{toluene}})$$

$$f_p = \ln(k_{p\text{-xylene}}/k_{\text{toluene}})$$

$$\ln(k_{1,3,5\text{-trimethylbenzene}}/k_{\text{toluene}}) = 2f_m \tag{21}$$

$$\ln(k_{1,2,3,4\text{-tetramethylbenzene}}/k_{\text{toluene}}) = 0.5(f_o+f_m+f_p)+0.5(2f_o+f_m) \tag{22}$$

Another interesting series of comparisons is the relative values of the substituent effects, methyl, ethyl, isopropyl, and *tert*-butyl on the rate of exchange of *m*- and *p*-alkyl-substituted toluenes. In the comparison in

TABLE XX

Predicted and Observed Relative Rates for Hydrogen Exchange in Polymethylated Benzenes[34]

Hydrocarbon	$k_{hydrocarbon}/k_{toluene}$	
	Predicted	Observed
1,2,3-Trimethylbenzene	1.13	1.12
1,2,4-Trimethylbenzene	0.26	0.26
1,3,5-Trimethylbenzene	0.26	0.22
1,2,3,4-Tetramethylbenzene	0.51	0.46
1,2,3,5-Tetramethylbenzene	0.20	0.12
1,2,4,5-Tetramethylbenzene	0.024	0.018
Pentamethylbenzene	0.12	0.056
Hexamethylbenzene	0.017	0.010

Table XXI, the observed values for *p*- and *m*-ethyltoluene and *p*-cymene are accompanied by values that have been corrected for the fact that the observed rates reflected exchange of two different kinds of hydrogens. The assumption was made in the correction that a methyl substituent has the same effect on all other alkyl groups as it does on another methyl group.

TABLE XXI

Variation in Inductive Effect with Variation in Alkyl Groups on Hydrogen Exchange Rates at α-Carbon[34]

Hydrocarbon	$k_{hydrocarbon}/k_{toluene}$
Toluene	1
p-Xylene	0.033
p-Ethyltoluene	0.026 (0.041) [a]
p-Cymene	0.023 (0.03) [a]
p-*tert*-Butyltoluene	0.033
m-Xylene	0.51
m-Ethyltoluene	0.25 (0.33) [a]

[a] Values corrected for fact that two different kinds of hydrogens are undergoing exchange.

The data indicate that in the *p*-position, methyl, ethyl, isopropyl, and *tert*-butyl have approximately the same inductive effects, but that ethyl has a larger rate-retarding effect than methyl by a factor of 2 when the groups occupy the *m*-position.

EFFECT OF MEDIUM ON KINETIC AND EQUILIBRIUM ACIDITY OF CARBON ACIDS

In the foregoing sections, attention has been focused on constitutional effects on the thermodynamic and kinetic acidity of carbon acids. Because of the large solvation energies involved in transforming a covalent carbon–hydrogen bond into a carbanion, one might expect that the rates of carbanion generation and capture are highly dependent on the character of the solvent, as well as on the charge type of the basic catalyst involved. In this section, environmental effects on catalyst activity and on carbanion stability are treated.

Four qualitatively different kinds of solvents for carbon acids and their derived carbanions have been recognized: proton-donating polar solvents such as water and the lower alcohols and polyols, e.g., methanol and ethylene glycol; proton-donating nonpolar solvents such as the butanols, aniline, and similar compounds; non-proton-donating polar solvents such as dimethyl sulfoxide, sulfolane, and dimethyl formamide; non-proton-donating nonpolar solvents such as tetrahydrofuran, ether, dioxane, benzene, or cyclohexane. These divisions are somewhat arbitrary since a continuity of changes in properties is observed in appropriate studies.

Two general classes of bases have been used to generate carbanions from carbon acids. Of the electrically neutral bases, ammonia, tripropylamine, pyridine, and aniline are frequently employed. A large variety of metal and quaternary ammonium bases is available: the metal hydroxides, alkoxides, phenoxides, acetates; the metal alkyls, aryls, and benzyls; the metal hydrides; the metal amides; the quaternary ammonium hydroxides. Most of the latter types of bases involve highly ionic bonds and will be referred to as charged bases.

The activity of each kind of base should vary with changes in solvent, as should carbanion stability. Few systematic studies of base and solvent type on the acidity of the carbon acids have been made. However, enough has been done to indicate that the field is fertile. A few examples follow.

Perhaps the most dramatic effects of medium on the rate of carbanion generation are found in a comparison of alcohols and dimethyl sulfoxide. The initial observation of the high activity of alkoxide salts in dimethyl sulfoxide was made in connection with base-catalyzed cleavages of II.[37] Substitution of dimethyl sulfoxide for *tert*-butyl alcohol as solvent with

[37] D. J. Cram, J. L. Mateos, F. Hauck, A. Langemann, K. R. Kopecky, W. D. Nielsen, and J. Allinger, *J. Am. Chem. Soc.*, **81**, 5774 (1959).

potassium *tert*-butoxide as base resulted in a rate increase large enough to allow the temperature to be lowered by 100–150° to obtain comparable rates. In *tert*-butyl alcohol, the rates of the cleavage reaction were affected in only a minor way by a change from potassium to sodium to lithium *tert*-butoxide. In dimethyl sulfoxide, the activities of the three salts varied enough so that with the potassium base, the reaction occurred orders of magnitude faster than with the sodium salt which, in turn, was a much more effective catalyst than the lithium salt.

$$C_2H_5-C(CH_3)(C_6H_5)-C(OH)(C_6H_5)-CH_3 + \overset{+}{M}\overset{-}{O}C(CH_3)_3 \rightleftharpoons C_2H_5-C(CH_3)(C_6H_5)-C(\overset{-}{O}\ \overset{+}{M})(C_6H_5)-CH_3 \longrightarrow$$

II

$$C_2H_5-\overset{-}{C}(CH_3)(C_6H_5)\ M^+ + C_6H_5-C(=O)-CH_3 \xrightarrow{ROH} C_2H_5-C(CH_3)(C_6H_5)-H$$

The effect was found to apply to a number of base-catalyzed reactions that involved breaking carbon–hydrogen or carbon–carbon bonds as the rate-determining step. For example, the Wolff-Kishner reaction can be conducted at room temperature by adding a hydrazone slowly to a solution of potassium *tert*-butoxide in dimethyl sulfoxide[38]; and bromobenzene reacts with the same solution at room temperature to give a mixture of phenol and *tert*-butoxybenzene.[39] The Cope elimination reaction of amine oxides proceeds at a rate many powers of 10 greater in dimethyl sulfoxide than in water.[40] Base-catalyzed allylic rearrangements of alkenes proceed in dimethyl sulfoxide–potassium *tert*-butoxide many powers of 10 faster than in *tert*-butyl alcohol–potassium *tert*-butoxide,[41] as does the cleavage of non-enolizable ketones,[42a] and base-catalyzed 1,2-elimination reactions.[42b]

A careful study of this effect was made by examination of the kinetics of racemization of optically active 2-methyl-3-phenylpropionitrile

[38] D. J. Cram and M. R. V. Sahyun, *J. Am. Chem. Soc.*, **84**, 1734 (1962).
[39] D. J. Cram, B. Rickborn, and G. R. Knox, *J. Am. Chem. Soc.*, **82**, 6412 (1960).
[40] M. R. V. Sahyun and D. J. Cram, *J. Am. Chem. Soc.*, **85**, 1263 (1963).
[41] (a) A. Schriesheim and C. A. Rowe, Jr., *J. Am. Chem. Soc.*, **84**, 3161 (1962); (b) D. J. Cram and R. T. Uyeda, *ibid.*, **86**, 5466 (1964).
[42] (a) P. G. Gassman and F. V. Zalar, *Chem. Eng. News*, **42**, 44 (April 20, 1964); (b) T. J. Wallace, J. E. Hofmann, and A. Schriesheim, *J. Am. Chem. Soc.*, **85**, 2739 (1963).

(III).[39,43] Under the conditions of the experiments, the rate constants for racemization (k_α) and exchange (k_e) were equal, a fact that clearly established a symmetrical carbanion as an intermediate. With potassium methoxide, the reaction in methanol was almost first order (1.09) in base concentration up to 0.6 *M*. The rate was independent of whether potassium, sodium, or tetramethylammonium methoxide was employed as base. The rate was unaffected by addition of 0.353 *M* potassium iodide to a 0.260 *M* solution of potassium methoxide in methanol. Similar

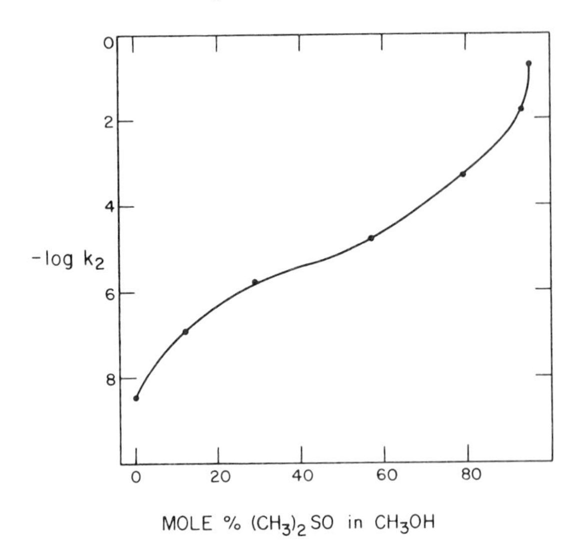

FIG. 2. Plot of mole percent dimethyl sulfoxide in methanol against negative of logarithm of k_2 (liter mole^{-1} second^{-1}) for methoxide-catalyzed racemization of 2-methyl-3-phenylpropionitrile at 25°.

results were observed in ethylene glycol, and in dimethyl sulfoxide–methanol mixtures at low base concentration (0.01 *M* or lower). Under these conditions, it is reasonable to expect that potassium methoxide is completely dissociated, and that methoxide anion is the catalytically active species. This conclusion is compatible with the dielectric constants for the three solvents ($\epsilon = 34$, 35, and 49 for methanol, ethylene glycol, and dimethyl sulfoxide, respectively, at about 20°). In *tert*-butyl alcohol as solvent and potassium *tert*-butoxide as base, variation of the concentration of base from 0.01 to 0.44 *M* gave an increase in rate constant of only 2.5. In this solvent of much lower dielectric constant ($\epsilon = 11$ at

[43] D. J. Cram, B. Rickborn, C. A. Kingsbury, and P. Haberfield, *J. Am. Chem. Soc.*, **83**, 3678 (1961).

about 20°), the base undoubtedly exists as ion pairs that are somewhat aggregated.

$$C_6H_5CH_2-\overset{*}{C}(CH_3)(C{\equiv}N)-H$$

III

The values of the rate constants for base-catalyzed racemization in various solvents relative to that for potassium methoxide in methanol at 25° are shown in Table XXII. Figure 2 records a plot of the mole percent dimethyl sulfoxide in methanol against the log of the second-order rate constant. These results indicate that the rate of carbanion formation can be increased by about 8 powers of 10 by substitution of dimethyl sulfoxide —1.5% by weight in methanol—for pure methanol.

TABLE XXII

Relative Rates of Racemization of 2-Methyl-3-phenylpropionitrile (III) *with Potassium Alkoxides as Bases at 25°*

Solvent, % by weight	$k_{solvent}/k_{methanol}$
100% $HOCH_2CH_2OH$	0.32
100% CH_3OH	1.0
75% CH_3OH–25% $(CH_3)_2SO$	3.2×10^1
50% CH_3OH–50% $(CH_3)_2SO$	1.6×10^2
24% CH_3OH–76% $(CH_3)_2SO$	4.9×10^3
10% CH_3OH–90% $(CH_3)_2SO$	1.3×10^5
3% CH_3OH–97% $(CH_3)_2SO$	1.4×10^6
1.5% CH_3OH–98.5% $(CH_3)_2SO$	5.0×10^7
2% CH_3OH–98% $(CH_2)_4SO_2$	7.8×10^4
$(CH_3)_3COH$, 0.01 M in H_2O	1.4×10^6
$(CH_3)_3COH$, <0.003 M in H_2O	4.2×10^6

The shape of the curve in Figure 2 is interesting. Although the plot is about linear from 35 to 80 mole % dimethyl sulfoxide, at lower and higher dimethyl sulfoxide concentrations the rates rise much more steeply. At least two solvent effects are visible in the data, one at low and one at high dimethyl sulfoxide concentration. Equilibria of the sort shown in Equation (23) undoubtedly exist, and are established much faster than protons are removed from carbon. Of the two forms of the base, A is probably of much lower energy. The transition states for breaking the carbon–hydrogen bond by each form of the base are symbolized by A‡ and B‡. Transition state B‡ is more stable than A‡ since in B‡ a strong hydrogen bond between methanol and dimethyl sulfoxide is being made, whereas in A‡, a strong hydrogen bond between methoxide and methanol is being exchanged for two weaker hydrogen bonds between methanol molecules. The ground state for both processes is the same since at all times A and B are in equilibrium with each other. As dimethyl sulfoxide is added to the medium, the rate increases because more and more material passes through transition state B‡ as the concentration of B increases. At about 35 mole % dimethyl sulfoxide, all of the reaction passes through B‡, even though most of the base is still in the form of A.

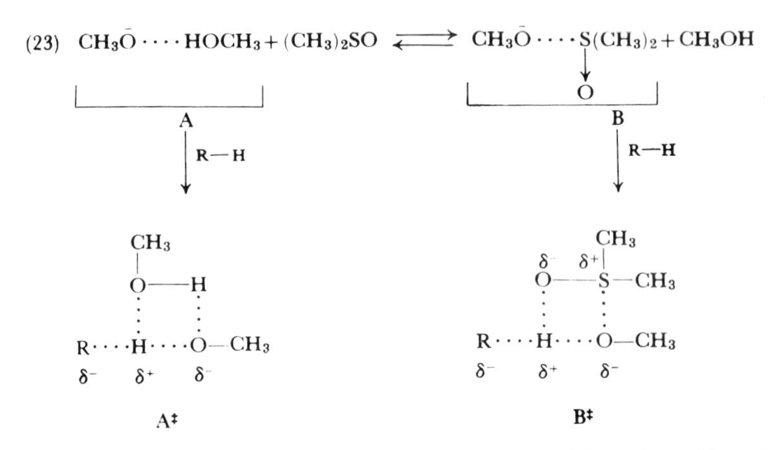

As the concentration of dimethyl sulfoxide is raised from 35 to 80 mole %, the energy of the ground state is gradually raised since the concentration of B increases. Above 80 mole % sulfoxide, the curve rises much more steeply because of a scavenging effect dimethyl sulfoxide has for methanol molecules (Equation (24)), and because of the decrease in concentration of methanol. In pure dimethyl sulfoxide, all of the base is in the form of B, and the rate is very rapid. This hypothesis is oversimplified, since secondary solvation effects of A and B are undoubtedly important. On the other hand, these secondary solvation effects would operate in the same direction as the primary effects symbolized by A, B, A‡, and B‡, and are omitted for simplicity.

$$(24)\quad \underbrace{CH_3\bar{O}\cdots HOCH_3 + 2(CH_3)_2SO}_{A} \rightleftharpoons \underbrace{CH_3\bar{O}\cdots \underset{\underset{O}{\downarrow}}{S}(CH_3)_2 + (CH_3)_2S\rightarrow O\cdots HOCH_3}_{B}$$

Other interesting relationships are visible in Table XXII. Ethylene glycol–potassium ethylene glycoxide gives a rate constant only about a third that of methanol–potassium methoxide. Furthermore, *tert*-butyl alcohol–potassium *tert*-butoxide provides a rate constant about 6 powers of 10 faster than methanol–potassium methoxide. The latter rate increase probably reflects the fact that methoxide anion in methanol is more heavily solvated than potassium *tert*-butoxide in *tert*-butyl alcohol. In passage to the transition state for proton abstraction by a charged base, charge is being dispersed and some solvation energy being overcome. Tetramethylene sulfone (sulfolane) 2 mole % in methanol gave a rate increase over methanol–potassium methoxide of about 5 powers of 10. The same factors appear to operate here as were described for dimethyl sulfoxide, but are less dramatic. The superior ability of dimethyl sulfoxide to hydrogen-bond hydroxyl groups as compared to sulfolane is probably responsible for the difference between the two solvents. Dimethyl sulfoxide acts as a better scavenger for methanol molecules than sulfolane, and leaves the methoxide anion less solvated.

Others[44] had observed that the activity coefficient of hydroxide ion in sulfolane (5 mole % in water) is largely responsible for the remarkable increase in basicity (6 H_- units[45]) of a 0.01 M solution of phenyltrimethylammonium hydroxide in the sulfolane solvent mixture as compared with water. Apparently similar effects play a role in dimethyl sulfoxide and sulfolane.

In another investigation, the rates of racemization and exchange of 1-phenylmethoxyethane-1-*d* were studied in dimethyl sulfoxide–potassium *tert*-butoxide.[46] In this solvent, the rates of racemization and isotopic exchange were equal to each other ($k_e/k_\alpha = 1$). The reaction rate was approximately half order in base from about 0.05 to 0.23 M potassium *tert*-butoxide. This observation is consistent with a mechanism in which the potassium *tert*-butoxide in dimethyl sulfoxide is largely in the form of ion pairs, but dissociated *tert*-butoxide ion is the active form of the base. In other words, in Equation (25), $k_{-1} \gg k_1$, and $k_3 \ggg k_2$. This

$$(25)\qquad (CH_3)_3C\overset{-}{O}\overset{+}{K} \underset{k_{-1}}{\overset{k_1}{\rightleftarrows}} (CH_3)_3C\overset{-}{O} + K^+$$

$$(CH_3)_3C\overset{-}{O}\overset{+}{K} \xrightarrow[k_2]{\overset{*}{R}-H} R^- \qquad\qquad (CH_3)_3C\overset{-}{O} \xrightarrow[k_3]{\overset{*}{R}-H} R^-$$

[44] C. H. Langford and R. L. Burwell, Jr., *J. Am. Chem. Soc.*, **82**, 1503 (1960).
[45] M. A. Paul and F. A. Long, *Chem. Rev.*, **57**, 1 (1957).
[46] D. J. Cram, C. A. Kingsbury, and B. Rickborn, *J. Am. Chem. Soc.*, **83**, 3688 (1961).

interpretation was strengthened by the observation that when a 0.0537 M solution of potassium *tert*-butoxide in dimethyl sulfoxide was made 0.49 M in potassium iodide, the rate decreased by a factor of 22. Furthermore, potassium *tert*-butoxide gave a rate about 2 powers of 10 greater than that obtained with sodium *tert*-butoxide.

Solutions of potassium *tert*-butoxide–0.9 M *tert*-butyl alcohol in dimethyl sulfoxide gave rates of racemization between 6 and 7 powers of 10 greater than those observed in *tert*-butyl alcohol. Since $k_e/k_\alpha \sim 2$–5 in *tert*-butyl alcohol, had isotopic exchange rates been compared, about 6 powers of 10 would have been obtained. Thus, the rate of isotopic exchange catalyzed by potassium *tert*-butoxide–0.9 M *tert*-butyl alcohol in dimethyl sulfoxide is about 10^{13} times that observed in methanol–potassium methoxide.

This vast increase in kinetic activity of alkoxide anions in dimethyl sulfoxide over that in hydroxylic solvents has its thermodynamic counterpart. The H_- function[47] (see Equation (26)) is a property of a medium that measures the ability of that medium to remove a proton under equilibrium conditions from weak acids.

$$H_- = pK_{HA} - \log\frac{[HA]}{[A^-]} = -\log\frac{a_{H^+}f_{A^-}}{f_{HA}} \tag{26}$$

In Equation (26), pK_{HA} is the pK_a of an acid HA, a_{H^+} is the activity of H^+ and f_{A^-} and f_{HA} are activity coefficients of A^- and HA, respectively.

With a series of weak, overlapping indicator acids (substituted anilines and diphenylamines) ranging in pK_a from 12.2 to 18.4, H_- values were determined for a 0.025 M solution of sodium methoxide in various mixtures of methanol–dimethyl sulfoxide. These values ranged from 12.2 in pure methanol to 19.4 in 95 mole percent dimethyl sulfoxide–5 mole percent methanol. A plot of H_- against mole percent methanol resembled in shape the plot (see Figure 2) of $\log k$ for racemization of optically active 2-methyl-3-phenylpropionitrile (III) against mole percent methanol in dimethyl sulfoxide. The plot of $\log k$ against H_- is approximately linear (see Figure 3) with a slope of 0.87. The rate constants cover a range of about 10^6 and the H_- values, a range of 7 units.

Two general mechanisms for the base-promoted racemization reaction are compared for their compatibility with the correlation embodied in Figure 3. Both mechanisms involve an intermediate carbanion, and differ only in the relative values of k_{-1} and k_2 in Equations (27) and (28).

[47] R. Stewart, J. P. O'Donnell, D. J. Cram, and B. Rickborn, *Tetrahedron*, **18**, 317 (1962).

In the first mechanism, $k_{-1} \gg k_2$, and the asymmetrically solvated carbanion is formed in an equilibrium step followed by a rate-affecting

(27) $$\underset{\text{Optically active}}{-{}^{*}\overset{|}{\underset{|}{C}}-H+B^-} \underset{k_{-1}}{\overset{k_1}{\rightleftharpoons}} \underset{\text{Optically active}}{-{}^{*}\overset{|}{\underset{|}{C}}^{-}\cdots HB} \xrightarrow{k_2} \underset{\text{Racemic}}{-\overset{|}{\underset{|}{C}}^{-}\cdots HB} \xrightarrow{k_{-1}} \underset{\text{Racemic}}{-\overset{|}{\underset{|}{C}}-H}$$

(28) $$k_{\text{obs.}} = k_1 \frac{k_2}{(k_{-1}+k_2)}$$

racemization step associated with k_2. In this case, Equation (28) reduces to (29). In the second mechanism, $k_2 \gg k_{-1}$, and the initially formed

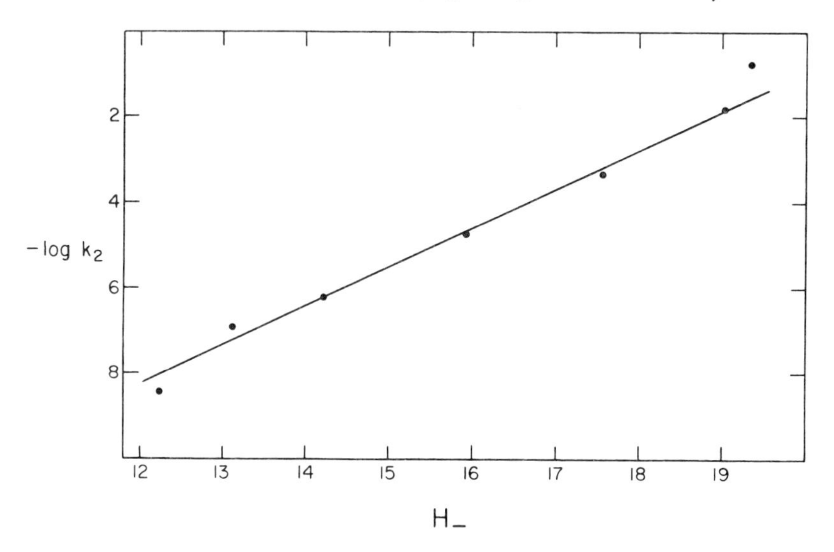

FIG. 3. Plot of H_- function against negative of logarithm of k_2 (liter mole^{-1} second^{-1}) for methoxide-catalyzed racemization of 2-methyl-3-phenylpropionitrile at 25°.

carbanion leads directly to racemic product. Thus, Equation (28) reduces to (30).

(29) $$k_{\text{obs.}} = \frac{k_1}{k_{-1}} \times k_2 = Kk_2$$

(30) $$k_{\text{obs.}} = k_1$$

The correlation of Figure 3 implies that the free energies of the rate-limiting transition states and the anionic form of the indicator acids respond in a similar way to the changes in solvent from pure methanol to dimethyl sulfoxide. The transition states for the rate-limiting steps of the two mechanisms are formulated, along with the anionic form of the indicator acids. In the first mechanism, the transition state involves breaking of only a *hydrogen bond*, whereas in the second mechanism, the transition state involves the breaking of a *covalent bond*. Clearly, the structure of the transition state of the *first mechanism* more nearly resembles the structure of the anionic form of the indicator acids and, therefore, the correlation supports the preequilibrium mechanism (first) for racemization or exchange.

Transition state for pre-equilibrium (1st) mechanism

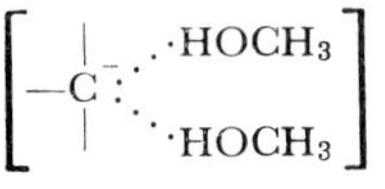

$$\left[-\overset{|}{\underset{|}{C}}{}^{-}\!:\begin{matrix}\cdots HOCH_3 \\ \cdots HOCH_3\end{matrix} \right]$$ Solvent changes $CH_3OH \rightarrow (CH_3)_2SO$

Hydrogen bonds are made and broken

Anionic form of indicator acids

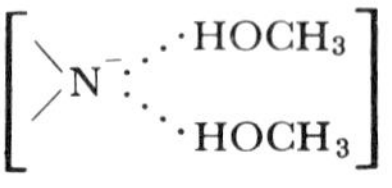

$$\left[>N^{-}\!:\begin{matrix}\cdots HOCH_3 \\ \cdots HOCH_3\end{matrix} \right]$$ Solvent changes $CH_3OH \rightarrow (CH_3)_2SO$

Two pairs of electrons on nitrogen are hydrogen bonded

Transition state for non-pre-equilibrium (2nd) mechanism

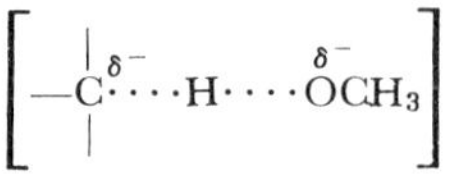

$$\left[-\overset{|}{\underset{|}{C}}{}^{\delta-}\cdots H \cdots \overset{\delta-}{O}CH_3 \right]$$ Solvent changes $CH_3OH \rightarrow (CH_3)_2SO$

A covalent bond is broken

Results of a different study[48] indicate that the H_- values must increase dramatically as the mole percent methanol in dimethyl sulfoxide is reduced below 5 mole percent. These authors prepared solutions of the sodium and potassium salts of dimethyl sulfoxide[49] (dimsylsodium[50] and dimsylpotassium) in dimethyl sulfoxide, and titrated solutions of proton donors in dimethyl sulfoxide with triphenylmethane as indicator. The triphenylmethane anion is deeply colored. This procedure had been previously applied by others[50] for converting alcohols and other weak acids to their sodium salts in dimethyl sulfoxide. Good equivalence points had been observed with *dimsylsodium*[50] in all cases except with glycerol (see Table XXIII). This fact suggests that the pK_a of dimethyl sulfoxide $\gg$ than that of triphenylmethane, and that the pK_a of triphenylmethane $\gg$ than those of the weak acids titrated.

[48] E. C. Steiner and J. M. Gilbert, *J. Am. Chem. Soc.*, **85**, 3054 (1963).
[49] E. J. Corey and M. Chaykovsky, *J. Am. Chem. Soc.*, **84**, 866 (1962).
[50] G. G. Price and M. C. Whiting, *Chem. & Ind. (London)*, p. 775 (1963).

TABLE XXIII

Equivalence Points in Titration of Acids with Dimsylmetallics with Triphenylmethane as Indicator

Acid	Lit.[a] pK_a	Equivalents needed for $(C_6H_5)_3C^-$ color	
		$NaCH_2SOCH_3$[b]	KCH_2SOCH_3[c]
Benzoic acid	4.2	—	1.01
Acetic acid	4.8	1.03	—
Phenol	9.9	1.02	0.99
Formanilide	—	—	1.00
Glycerol	—	1.56	1.00
1-Methoxy-2-propanol	—	—	0.54
Water	15.7	1.00	0.64
Ethanol	18	—	0.31
n-Butyl alcohol	—	0.98	0.45
tert-Butyl alcohol	19	1.06	0.33
Cyclopentadiene	16	0.99	0.97
Indene	21	0.98	0.98
Diphenylamine	23	0.99	1.00
Aniline	27	<0.01	<0.01
Triphenylmethane	33	—	—

[a] Literature: Danner,[51a] Stearns and Wheland,[51b] and Hine and Hine.[51c]
[b] Price and Whiting.[50]
[c] Steiner and Gilbert.[48]

The same results were obtained with *dimsylpotassium* for weak carbon acids, for carboxylic acids, and for phenol. However, for water and the simple alcohols, the color of the triphenylmethide anion was visible after only 0.33 to 0.64 equivalents of *dimsylpotassium* had been added.[48]

When the same alcohols were titrated with metal triphenylmethide in tetrahydrofuran as solvent, the sodium salt gave an end point (color of triphenylmethide anion) after about one equivalent of base had been added, the potassium salt after 0.85 equivalent, and the cesium salt after about 0.50 equivalent. Clearly, the phenomenon is not limited to dimethyl sulfoxide as solvent.

The implications of this result is that the intrinsic acidities of water and the alcohols are comparable to that of triphenylmethane. This conclusion is at sharp variance with the accepted pK_a's of the simple alcohols as

16–19,[51] and of triphenylmethane as 32.5.[2a, 5] The explanation is that hydroxylic compounds drastically reduce the basicity of oxygen anions by hydrogen bonding, and metal cations do the same by incomplete dissociation and by aggregation. These effects are reduced in dimethyl sulfoxide at extremely low concentrations of alcohol, and by cations that provide the largest metal alkoxide dissociation constants (cesium or potassium).

Ledwith and McFarlane[52] prepared solutions of dimsylsodium and dimsylpotassium by dissolving the metals directly in dimethyl sulfoxide. The metal hydroxide produced was removed by filtration, and the dimethyl sulfide by distillation. This solution was then used to measure equilibrium constants between dimethyl sulfoxide and *tert*-butyl alcohol and their respective potassium salts, as well as between triphenylmethane and dimethyl sulfoxide and their respective potassium salts. These authors found triphenylmethane is 8×10^3 times as acidic as dimethyl sulfoxide, which is comparable with the value of 1.3×10^3 observed by Steiner and Gilbert.[48] Furthermore, Ledwith and McFarlane[52] reported *tert*-butyl alcohol was 830 times as acidic as triphenylmethane at 10^{-2} to 10^{-3} *M* *tert*-butyl alcohol concentration, whereas at similar concentrations Steiner and Gilbert[48] obtained comparable acidities for triphenylmethane and *tert*-butyl alcohol. The latter authors prepared their dimsylpotassium from potassium amide, and did not generate potassium hydroxide or potassium methyl mercaptide. Presence of the latter in the dimethyl sulfoxide solutions could be responsible for Ledwith and McFarlane's[52] low value for the acidity of *tert*-butyl alcohol.

In an extension of the earlier work, Steiner and Gilbert[53a] constructed three new acidity scales based on water–dimethyl sulfoxide mixtures, methanol–dimethyl sulfoxide mixtures, and on dimethyl sulfoxide itself as solvent. All three systems are based on the aqueous scale with dilute aqueous solution as the reference state. At the upper end of each scale, potassium bases of the indicator systems were employed. The three scales were interrelated through the reasonable assumption that the pK_a of triphenylmethane is the same in dimethyl sulfoxide, in dimethyl sulfoxide containing 0.11% water (by weight), and in dimethyl sulfoxide containing 0.02% methanol (by weight). The three scales agree within 0.5 pK unit over a range of 12 units. Taken together, the scales cover a *range of* 18

[51] (a) P. S. Danner, *J. Am. Chem. Soc.*, **44**, 2832 (1922); (b) R. S. Stearns and G. W. Wheland, *ibid.*, **69**, 6025 (1947); (c) J. Hine and M. Hine, *ibid.*, **74**, 5266 (1952).

[52] A. Ledwith and N. McFarlane, *Proc. Chem. Soc.*, p. 108 (1964).

[53] (a) E. C. Steiner and J. M. Gilbert, *J. Am. Chem. Soc.*, **87**, 382 (1965); (b) E. C. Steiner and J. M. Gilbert, *ibid.*, **87**, in press (1965).

TABLE XXIV

Acidity Scales in Water–Dimethyl Sulfoxide, in Methanol–Dimethyl Sulfoxide, and in Dimethyl Sulfoxide

Acid	H_2O–$(CH_3)_2SO$ pK_a	CH_3OH–$(CH_3)_2SO$ pK_a	$(CH_3)_2SO$ pK_a
$(CH_3)_2SO$	—	—	31.3
$(C_6H_5)_2CH_2$	—	—	28.6
CH_3	—	27.7	—
4-$C_6H_5C_6H_4CH_2C_6H_5$	—	27.2	—
$(C_6H_5)_3CH$	27.2	27.2	27.2
H, H, O	—	27.1	—
4-$C_6H_5C_6H_4CH(C_6H_5)_2$	—	25.3	—
C_6H_5, H, O	24.3	24.2	24.2
(4-$C_6H_5C_6H_4)_3CH$	—	22.8	—
H, H	20.5	20.5	20.5
4-$NO_2C_6H_4NH_2$	18.4	18.5	18.6
H, H	—	18.2	—
2,4-$(NO_2)_2C_6H_3NH_2$	15.0	14.5	14.7
2,4-$(NO_2)_2C_6H_3NHC_6H_5$	13.8	13.2	—

p*K units*, and therefore provide an excellent means of relating the acidities of many carbon and other acids. Table XXIV contains the data pertaining to these scales.

In the three solvent systems, the dielectric constants are relatively high, and the salts of the acids are probably dissociated. This, coupled with the highly diffuse charge in the anions, accounts for the consistency of the pK_a values from solvent to solvent. However, the pK_a values of the weaker carbon acids (diphenylmethane, triphenylmethane, and biphenyldiphenylmethane) are 5.6 ± 0.3 pK units lower than on the Streitwieser scale (see Table II). This difference is probably due to the fact that the dielectric constant of cyclohexylamine is much lower than that of dimethyl sulfoxide, water, and methanol, and that the salts of the carbon acids exist as ion pairs in cyclohexylamine.

The relative acidities of dimethyl sulfoxide and triphenylmethane were determined by a third method.[49] An equilibrium mixture of dimsylsodium and triphenylmethane in dimethyl sulfoxide was quenched by addition of deuterium oxide. The amount of deuterium in the recovered triphenylmethane was used to calculate the relative acidities of the two substances. A value of K(dimethyl sulfoxide)/K(triphenylmethane) = 21 was obtained.

That this method provides spurious results was demonstrated by Ritchie,[54] who showed that *during* quenching of salts of carbon acids in dimethyl sulfoxide with hydroxylic acids, proton transfers occur between dimethyl sulfoxide and the carbanion at rates competitive with those between the added hydroxylic acids and the carbanion. Thus, the final isotopic composition of the isolated carbon acid in no way reflects the equilibrium composition of the mixture quenched.

In one series of experiments,[54] one equivalent of triphenylmethane was added to 2 equivalents of dimsylsodium in dimethyl sulfoxide. To the resulting mixture was added a large excess of various deuterated acids, and the isolated triphenylmethane was analyzed for deuterium. Table XXV records the results. Ritchie interprets these and other results as reflecting diffusion control of the rates of proton transfer between carbon acids and carbanions as well as between oxygen acids and carbanions in dimethyl sulfoxide.

Another potential source of difficulty in interpreting the results of quenching experiments arises from the possibility that rates of mixing may limit the rates of proton transfers. As the first amount of added deuterated acid diffuses into the basic solution, deuterated triphenylmethane is produced. However, neutralization is not yet complete, and proton and deuteron transfers continue between triphenylmethane, deuterated triphenylmethane, and dimethyl sulfoxide on the one hand, and triphenylmethylsodium and dimsylsodium on the other, until enough of the added stronger acid has diffused into the area to neutralize completely the stronger bases.

[54] C. D. Ritchie, *J. Am. Chem. Soc.*, **86**, 4488 (1964).

TABLE XXV

Isotopic Composition of Triphenylmethane Obtained by Addition of Different Deuterated Acids to Triphenylmethylsodium in Dimethyl Sulfoxide

Acid added	% $(C_6H_5)_3CH$	% $(C_6H_5)_3CD$
Large excess D_2O	46	49
9 Equiv. CD_3NO_2	15	85
9 Equiv. CD_3NO_2–6 Equiv. C_6H_5OH	44	55
9 Equiv. CD_3NO_2–4 Equiv. *p*-$CH_3C_6H_4SO_3H$	57	43

An example of this type of phenomenon was encountered in the following experiments.[55] Solutions of potassium *tert*-butoxide in *tert*-butyl alcohol rapidly racemize carbon acid IV by carbanion formation.[56] This reaction is not surprising since the pK_a of *tert*-butyl alcohol is approximately 19, and that of IV has been estimated to be about 21.[56] Addition with normal mixing of a 0.34 *M* solution of potassium *tert*-butoxide in *tert*-butyl alcohol to a *tert*-butyl alcohol solution 0.1 *M* in trimethylacetic acid and 0.2 *M* in IV resulted in 70% racemization of IV (sodium trimethylacetate does not racemize IV). The total amount of base added was less than half that needed to neutralize the carboxylic acid present. In a second experiment, a 0.1 *M* solution of potassium *tert*-butoxide in *tert*-butyl alcohol was added to a well-stirred solution of *tert*-butyl alcohol 0.1 *M* in IV and 0.1 *M* in acetic acid. Again, about half the amount of base was added that was needed to neutralize the carboxylic acid present. In this experiment, recovered IV had not racemized.

CH_3 H

$CON(CH_3)_2$

IV

Apparently, in the first experiment local concentrations of base exceeded local concentrations of acid and racemization occurred. In the second, the solutions were dilute and the stirring was good enough to avoid this condition.

[55] D. J. Cram and L. Gosser, unpublished work.

[56] D. J. Cram and L. Gosser, *J. Am. Chem. Soc.*, **85**, 3890 (1963).

CHAPTER II

Carbanion Structure and Mechanism of Stabilization

Preconceptions about the structure of carbanions unstabilized by substituents depend on analogies with the structures of ammonia or amines. Ammonia is isoelectronic with methyl anion, and the fact that ammonia possesses a rapidly inverting pyramidal structure[1] suggests that the methyl anion is also pyramidal, but that the rate constant for inversion is extremely high. Such a configuration places the unshared pair of electrons and the negative charge of carbanions in an sp^3 orbital, which is 25% $2s$. An alternative structure for the methyl anion places the electron pair and negative charge in a p-orbital, and the three hydrogens and carbon in a plane, with the bonds composed by sp^2-s orbital overlap. In a third structure, carbon remains unhybridized, the hydrogens are bonded through the p-orbitals of carbon, and the electron pair is in the $2s$-orbital. Of these three possibilities, sp^3 is the most probable, and saturated carbanions should possess a pyramidal configuration.

$$H_3N\text{:} \rightleftharpoons \text{:}NH_3 \qquad H_3C\text{:}^- \rightleftharpoons {}^-\text{:}CH_3$$

Through an elegant application of nuclear magnetic resonance technique, Saunders and Yamada[2] measured at 25° in aqueous hydrochloric acid of various concentrations the inversion rate of dibenzylmethylamine, as well as the rate of proton transfer from conjugate acid to the free amine. The rate constant for amine inversion proved to be

[1] (a) J. F. Kincaid and F. C. Henriques, Jr., *J. Am. Chem. Soc.*, **62**, 1472 (1940); (b) W. Gordy, W. V. Smith, and R. F. Trambarillo, "Microwave Spectroscopy," Wiley, New York, 1953.

[2] M. Saunders and F. Yamada, *J. Am. Chem. Soc.*, **85**, 1882 (1963).

$2 \pm 1 \times 10^5$ sec.$^{-1}$, whereas at 0.35 *M* amine hydrochloride concentration in the presence of 2 *N* hydrochloric acid, the rate constant for proton transfer came to $6 \pm 3 \times 10^8$ liters/mole sec. Good evidence was obtained that salt could lose its proton to give amine and be reprotonated without inversion. This experiment suggests that the same possibility exists for carbon acids and carbanions in the proper systems, and correlates with the results of study of the stereochemistry of base-catalyzed hydrogen–deuterium exchange discussed in Chapter III.

Most carbanions are stabilized by substituents or by the unshared pair of electrons occupying orbitals very high in 2*s*-character. This chapter is concerned with carbanion structure and the mechanism of stabilization. The following types of stabilization are discussed in turn: *s*-orbital effects, conjugative effects, inductive effects, homoconjugative effects, aromatization effects, negative hyperconjugative effects, and *d*-orbital overlap effects.

s-CHARACTER EFFECTS

In Chapter I, the results of studies were discussed which indicated that $HC{\equiv}CH > CH_2{=}CH_2 > (CH_2)_3 > CH_3CH_3$ in kinetic and thermodynamic acidity. In Table I are listed the four compounds, the hybridizations of the carbon attached to hydrogen, the percent *s*-character which carbon contributes to the C—H bond, and an estimate of the pK_a of the substances. The kinetic acidity of acetylene approximates that of phenylacetylene,[3] which has a pK_a of 21 on the McEwen[4a] and of 18.5 on the McEwen-Streitwieser-Applequist-Dessy[4] scale (MSAD scale; see Table XIV, Chapter I). However, acetylene has a pK_a between that of fluorene and aniline,[3b] and therefore comes to approximately 25 on the MSAD scale.

The bond hydridization of the *exo*-bond of cyclopropane is the result of calculations based on a variety of physical data.[5]

A near linear correlation between carbon acid strength and the

[3] (a) R. E. Dessy, Y. Okuzumi, and A. Chen, *J. Am. Chem. Soc.*, **84**, 2899 (1962); (b) N. S. Wooding and W. C. E. Higginson, *J. Chem. Soc.*, p. 774 (1952).

[4] (a) W. K. McEwen, *J. Am. Chem. Soc.*, **58**, 1124 (1936); (b) A. Streitwieser, Jr., J. I. Brauman, J. H. Hammons, and A. H. Pudjaatmaka, *ibid.*, **87**, 384 (1965); (c) D. E. Applequist and D. F. O'Brien, *ibid.*, **85**, 743 (1963); (d) R. M. Salinger and R. E. Dessy, *Tetrahedron Letters*, **11**, 729 (1963); (e) R. E. Dessy, private communication.

[5] (a) J. E. Kilpatrick and R. Spitzer, *J. Chem. Phys.*, **14**, 463 (1946); (b) C. A. Coulson and W. E. Moffitt, *ibid.*, **15**, 151 (1947); (c) A. D. Walsh, *Trans. Faraday Soc.*, **45**, 179 (1949).

TABLE I

Correlation of Percent s-Character Contributed by Carbon to Carbon–Hydrogen Bond and Estimated Acidity of Compound Containing That Bond

Compound	Bond hybridization	% *s*-Character	Estimated pK_a
HC≡CH	sp	50	25
$H_2C{=}CH_2$	sp^2	33	36.5
CH_2 / H_2C—CH_2 (cyclopropane)	$sp^{2.28}$	30	39
CH_3—CH_3	sp^3	25	42

percent *s*-character contributed by carbon to the carbon–hydrogen bond is evident from Figure 1, in which the estimated pK_a's of the four compounds are plotted against the percent *s*-character contributed to the

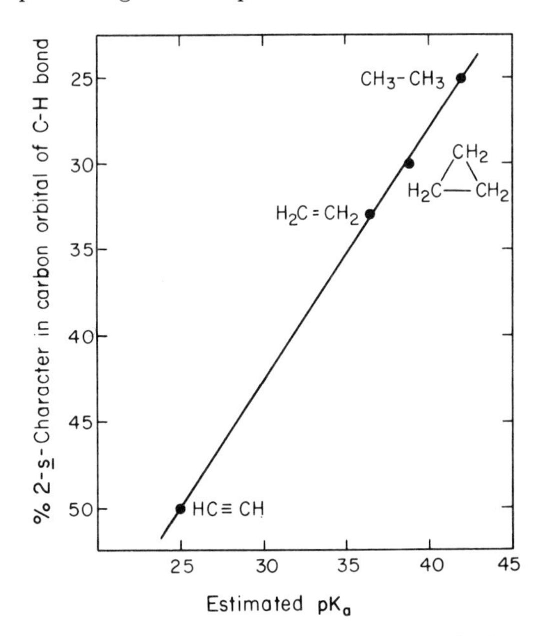

Fig. 1. Plot of estimated pK_a of hydrocarbons vs. percent 2*s*-character in carbon orbital of C—H bond.

carbon–hydrogen bond by carbon. Clearly, the greater the *s*-character of the carbon–hydrogen bond, the stronger the carbon acid. Electrons in 2*s*-orbitals are closer to the nucleus than electrons in 2*p*-orbitals, and therefore carbanions should be the more stable the higher the 2*s*-character of the orbital they occupy.[5c, 6]

Increased *s*-character in a carbon–hydrogen bond results whenever the other bonds to carbon are restricted by ring systems which reduce C—C—C bond angles. A number of investigations other than those discussed in Chapter I have dealt with the acidity of highly strained systems. Thus the circled hydrogens in compounds I,[7a] II,[7c] and III[7b] have been observed to undergo base-catalyzed hydrogen–deuterium exchange,[7a] or ready metalation.[7b, c] The amount of *s*-character in the exocyclic orbitals of carbons 1 and 2 of tricyclic compound IV has been determined from the ^{13}C—H nuclear spin-spin coupling constants in the nuclear magnetic resonance spectrum of IV to be 40 and 29%, respectively.[8] These values, coupled with the correlation of Figure 1, suggest pK_a's of approximately 32 and 39 for these hydrogens, respectively. The same technique applied to C-1 of III produced a value of 44% *s*-character, which suggests a pK_a of about 29 (Figure 1) for the substance. In qualitative agreement with theory, the circled hydrogen of III underwent potassium *tert*-butoxide-catalyzed hydrogen–deuterium exchange with *tert*-butyl alcohol about 10^4 times faster than the circled hydrogen of IV.

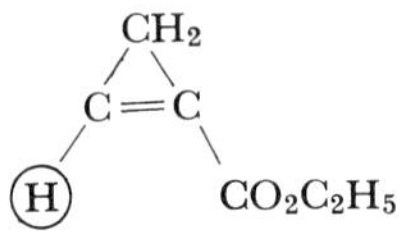

CH_2, C=C, (H), $CO_2C_2H_5$

I

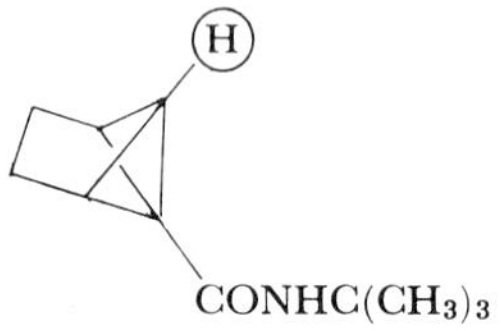

(H), $CONHC(CH_3)_3$

II

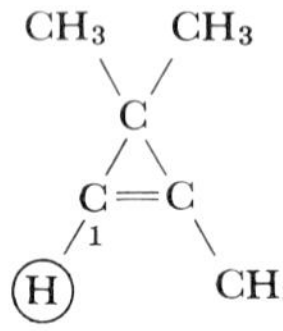

CH_3 CH_3, C, C=C, 1, (H), CH_3

$pK_a \sim 29$

III

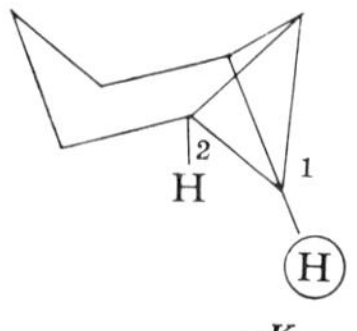

2, H, 1, (H)

$pK_a \sim 32$

IV

[6] (a) C. A. Coulson and W. E. Moffitt, *Phil. Mag.* [7], **40**, 1 (1949); (b) H. A. Bent, *Chem. Rev.*, **61**, 275 (1961).

[7] (a) K. B. Wiberg, R. K. Barnes, and J. Albin, *J. Am. Chem. Soc.*, **79**, 4994 (1957); (b) G. L. Closs and L. E. Closs, *ibid.*, **85**, 99 (1963); (c) J. Meinwald, C. Swithenbank, and A. Lewis, *ibid.*, **85**, 1880 (1963).

[8] G. L. Closs and L. E. Closs, *J. Am. Chem. Soc.*, **85**, 2022 (1963).

Results of a systematic study of the rates of triethylamine-catalyzed hydrogen–deuterium exchange between deuterium oxide and a series of cyclic ketones in dimethyl formamide are listed in Table II.[9] The carbonyl group provides most of the acidifying power for these compounds, but the data are correlated qualitatively by the amount of *s*-character of the carbon–hydrogen bond. Thus for the cyclanones, cyclobutanone > cyclopentanone > cyclohexanone > cycloheptanone > dipropyl ketone in rate. For the phenyl cyclyl ketones, cyclopropyl > cyclobutyl > cyclopentyl > cyclohexyl in rate. The complex steric effects make this interpretation tenuous, particularly because of the small differences involved.

TABLE II

Relative Rates of Triethylamine(1 M)-Catalyzed Hydrogen Deuterium Exchange Between Cyclic Ketones and Deuterium Oxide (5 M) in Dimethyl Formamide at 40° [9]

Ketone	Rel. rate[a]	Ketone	Rel. rate[a]
Dipropyl ketone	1	Isobutyrophenone	1
Cycloheptanone	2.4	Cyclohexyl phenyl ketone	0.7
Cyclohexanone	12	Cyclopentyl phenyl ketone	4.1
Cyclopentanone	85	Cyclobutyl phenyl ketone	12
Cyclobutanone	290	Cyclopropyl phenyl ketone	14

[a] Corrected for statistical factors where necessary.

That considerable cancellation of steric factors is probably important here is suggested by a second investigation.[10] The rates of neutralization of nitrocyclanes with hydroxide ion were studied in 50% water–50% dioxane at 0°. The rates relative to that of 3-nitropentane were as follows: nitrocyclopropane, $\ll 1$; nitrocyclobutane, 96; nitrocyclopentane, 27; nitrocyclohexane, 7.5; nitrocycloheptane, 13.5; nitrocyclooctane, 10. In this study, nitrocyclopropane was the least acidic of the compounds, and by a large factor, in spite of the fact that its C—H bond possesses the greatest *s*-character. These data indicate that in the absence of strongly acidifying substituents, the *s*-character criterion adequately accounts for ranking of carbon acids, but that in the presence of strongly acidifying

[9] H. Schechter, M. J. Cullis, R. E. Dessy, Y. Okuzumi, and A. Chen, *J. Am. Chem. Soc.*, **84**, 2905 (1962).

[10] H. Stone, P. W. K. Flanagan, F. G. Traynham, and H. Schechter, *Abstr. Papers Presented before Am. Chem. Soc. at Atlantic City, September, 1956*, p. 82–0.

substituents, more powerful effects dominate the acidity orders. Possibly the *p*-orbitals of the nitro group overlap the orbital occupied by the electron pair in proportion to the amount of *p*-character in the orbital of the pair. Thus, carbanion-stabilization effects due to *s*-character and due to delocalization of charge are not additive.

Carbanions stabilized by *s*-character effects should be pyramidal when tertiary, and the rate of inversion reduced when they are incorporated in three-membered rings, much as has been observed with the ethyleneimine analogs.[11] Attachment of electron-delocalizing substituents to the amino group of ethyleneimine increases the rate of inversion,[11] and the same probably applies to the carbanion analog. This point is discussed further in Chapter III.

CONJUGATIVE EFFECTS

Carbanions stabilized by substituents capable of delocalizing the electron pair onto more electronegative elements are expected to approach sp^2-p or trigonal hybridization. Such a configuration maximizes overlap between the *p*-orbital of carbon and those of the substituent, and allows negative charge to reside largely on the heteroatom. The nitro, carbonyl, and cyano carbanions are probably planar or near planar, at least when free of restrictions of ring systems.

$H_2\bar{C}{-}\overset{+}{N}(=O)O \longleftrightarrow H_2C{=}\overset{+}{N}(\bar{O})\bar{O}$ Nitromethane anion

$H_2\bar{C}{-}C(a){=}O \longleftrightarrow H_2C{=}C(a){-}\bar{O}$ Acetyl anion

$H_2\bar{C}{-}C{\equiv}N \longleftrightarrow H_2C{=}C{=}\bar{N}$ Acetonitrile anion

Deformations from the planar state in response to steric, solvation, or ring constraint effects will completely destroy the conjugative effect only in the extreme case in which a dihedral angle of 90° is enforced between the plane of the carbanion and that of the carbonyl or nitro group.

Sass and co-workers[12] determined the crystal structure of ammonium

[11] (a) A. T. Bottini and J. D. Roberts, *J. Am. Chem. Soc.*, **78**, 5126 (1956); (b) A. T. Bottini and J. D. Roberts, *ibid.*, **80**, 5203 (1958).

[12] C. Bugg, R. Desiderato, and R. L. Sass, *J. Am. Chem. Soc.*, **86**, 3157 (1964).

tricyanomethide and pyridinium dicyanomethylide. The tricyanomethide anion was almost planar and trigonal. The C—C—C bond angles were about 119.5°, and the central carbon atom was 0.13 Å above the plane occupied by the three nitrogen atoms. Each C—C≡N unit made an angle of 3° with respect to its projection in this plane. In pyridinium dicyanomethylide, the pyridine ring and the attached carbon were coplanar, and the C—C—C bond angle of the side chain was 119°. The two nitrogens of the cyano groups lay 0.13 Å above the plane of the ring.

Tricyanomethide ion

Pyridinium dicyanomethylide

A good example of the loss of acidity of a carbon acid due to steric inhibition of resonance of its conjugate base is found in the work of Bartlett and Woods.[13] These authors found that the bicyclic diketone formulated was much less acidic than its open-chain analog. The pair of electrons of the bridgehead anion of the bicyclic system are held in an sp^3 orbital whose angular disposition makes overlap with the p-orbitals of the carbonyl group minimal.

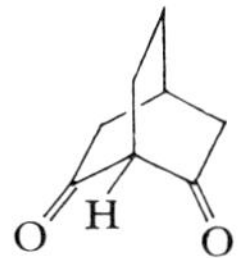

Bicyclic β-diketone, very weak carbon acid

Cyclic β-diketone, moderately strong carbon acid

With hydrocarbon substituents capable of conjugating with carbanions such as the vinyl, ethynyl, and phenyl groups, the hybridization is less clear. Delocalization effects favor planar carbanions, but s-character effects, unsymmetrical solvation or ion-pairing effects, or steric effects may deform the carbanion from a normal, planar configuration.

The symmetry properties of the allylic anion in the absence of a

[13] P. D. Bartlett and G. F. Woods, *J. Am. Chem. Soc.*, **62**, 2933 (1940).

counter ion require equal distribution of charge at the two ends of the chain, and in the molecular orbital which extends above and below the plane of the anion. An unsymmetrical distribution of substituents would disturb this balance. In the absence of cations, the charge distribution in and the structure of the propynyl (or allenyl) anion are less clear. To the extent that charge was distributed on the less substituted end of the molecule, the anion would be nonplanar, since allene itself is nonplanar. If charge were largely localized at the more substituted end, the carbon at that end would tend to become pyramidal, and the anion as a whole nonplanar. Probably the geometry of the anion represents a compromise between these extreme structures. The less substituted end of the molecule probably carries the greater amount of charge, since the electron pair could occupy an orbital containing more *s*-character when concentrated there. Total concentration at the more substituted end gives a pyramidal configuration with the two electrons in an orbital with 25% *s*-character. Total concentration at the less substituted end puts the two electrons in an orbital with 33% *s*-character.

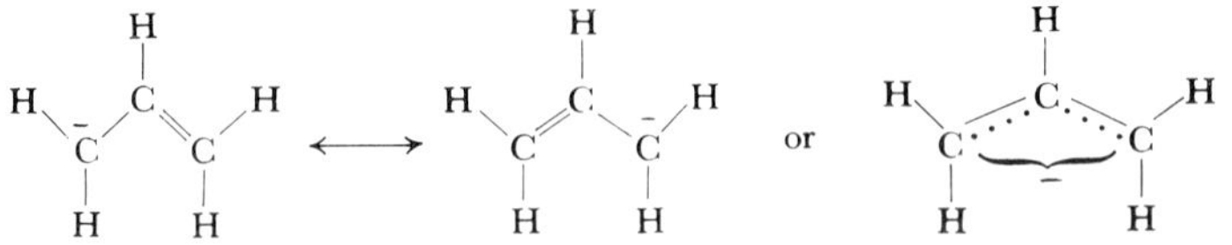

Allylic anion

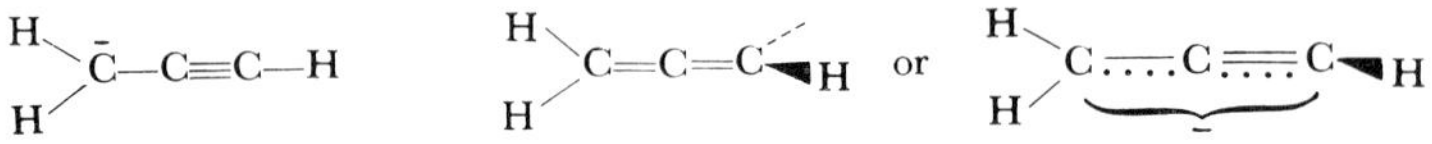

Two extreme structures for the propynyl anion

Compromise structure

The π-electron densities at the various positions of benzyllithium, diphenylmethyllithium, and triphenylmethyllithium have been calculated using nuclear magnetic resonance spectroscopy.[14] The spectra, and therefore the charge distributions, of triphenylmethylsodium and triphenylmethyllithium were almost the same in tetrahydrofuran, and the spectrum of the latter gave only minor changes when the solvent was changed from tetrahydrofuran to hexamethylphosphoramide to dimethyl sulfoxide. Thus, the triphenylmethyl organometallic compounds must be essentially completely ionized, although not dissociated.

The charge distribution for the various positions is listed in Table III. These data are in qualitative agreement with the results of self-consistent-field molecular orbital calculations for the ions. Only 38% of

[14] V. R. Sandel and H. H. Freedman, *J. Am. Chem. Soc.*, **85**, 2328 (1963).

TABLE III

Charge Distribution in Phenyl-Substituted Methyllithium Compounds, Values Being Given in Units of Absolute Value of the Charge of an Electron[14]

Compound	α-Position	*o*-Position	*m*-Position	*p*-Position
$C_6H_5CH_2Li$	0.38	0.12	0.10	0.18
$(C_6H_5)_2CHLi$	0.08	0.08	0.07	0.16
$(C_6H_5)_3CLi$	0.13	0.00	0.08	0.13

the charge of benzyllithium is found at the α-position, and considerably less for the other two organometallic compounds. The low charge concentration at the *o*-positions may reflect electrostatic repulsions associated with the proximity of the *o*-positions in the most probable (propeller) conformation. The *p*-positions are considerably richer in charge than the *o*-positions, and in the case of triphenylmethyllithium, the *m*-position carries more charge than does the *o*-position. The fact that 62% of the charge is in the nucleus suggests that the α-carbon of benzyllithium possesses a configuration somewhere between planar and pyramidal, and that the inversion rate is extremely high.

Systems such as the cycloheptatrienide anion[15] are probably planar or near planar, although no data bearing on the question are yet available.

INDUCTIVE EFFECTS

No differentiation is made here among inductive, field, and electrostatic effects, except when situations arise that call for such a distinction. Otherwise they will be collectively referred to as the inductive effect.

About the only common substituents that stabilize carbanions by a pure inductive effect are the quaternary ammonium and fluoride groups. Although attempts to determine the acidity of the α-hydrogens of quaternary ammonium systems have frequently led to substitution, elimination, or rearrangement reactions, data are available concerning tetramethylammonium iodide. The compound is metalated by phenyllithium[16a] but not by benzylsodium.[16b] Clearly the salt is a stronger acid than benzene, but the lack of reaction with benzylsodium may reflect only a very slow proton transfer. The salt undergoes sodium

[15] (a) H. J. Dauben, Jr. and M. R. Rifi, *J. Amer. Chem. Soc.*, **85**, 3042 (1963); (b) W. von E. Doering and P. P. Gaspar, *ibid.*, **85**, 3043 (1963).

[16] (a) G. Wittig and M. H. Wetterling, *Ann.*, **557**, 193 (1947); (b) W. Schlenk and J. Holtz, *Ber.*, **50**, 274 (1917).

deuteroxide-catalyzed (0.31 *M*) exchange with deuterium oxide at 83° with a rate constant of 9.4×10^{-10} sec.$^{-1}$.[17] Exchange of tetramethylammonium deuteroxide in *tert*-butyl alcohol-*O-d* at 50° occurs with an estimated rate of about 10^{-7} sec.$^{-1}$ at 0.3 *M* concentration, about 2 powers of 10 slower than 3-phenyl-1-butene undergoes proton abstraction.[18] These data collectively suggest that the tetramethylammonium ion has a pK_a comparable with that of triphenylmethane, or about 33 on the MSAD scale (Table XIV, Chapter I).

The pK_a of trifluoromethane has been estimated[19] to be about 31 on the McEwen scale,[4a] but this value is probably too high in view of the more accurate work of Streitwieser[4b] (see discussion later in this Chapter). The three fluorine atoms attached to carbon in trifluoromethane are somewhat more acidifying than the formal charge of a tetraalkylammonium group.

No data are available that reveal the configuration of carbanions stabilized solely by the inductive effect. If the electrons occupy a *p*-orbital in trimethylammonium ylid, the two charges are probably slightly further apart than if the electrons are in an sp^3-orbital. However, if the electron pair was in an *s*-orbital, the charges would be closer to one another. Thus the possibility exists that the electron pair is in an orbital even richer in *s*-character than sp^3.

Some experimental data support this possibility. The bond angles A—N—A in compounds such as $:NA_3$ decrease as the substituent A becomes more electronegative.[6b] For example, the C—N—C bond angle in trimethylamine is 109°, the H—N—H angle in ammonia is 106° 46′, and the F—N—F angle in nitrogen trifluoride is 102° 30′. These data suggest that as the electronegativity of the substituent increases, the central atom diverts increasing amounts of *s*-character to the orbital occupied by the unshared electron pair.[6b] As the substituent withdraws electrons from the sigma bond that bond gets more *p*-character, since the *p*-orbitals are more extended than the *s*-orbitals. Just how these effects combine to minimize the energy of trimethylammonium ylid or trifluoromethyl anion is a matter for conjecture.

The inductive effect can operate in a system in which a pyramidal configuration for the carbanion is enforced by its incorporation at the bridgehead of the triptycene system.[20] In terms of the MSAD acidity

[17] W. von E. Doering and A. K. Hoffmann, *J. Am. Chem. Soc.*, **77**, 521 (1955).
[18] D. J. Cram and R. T. Uyeda, *J. Am. Chem. Soc.*, **86**, 5466 (1964).
[19] S. Andreades, *J. Am. Chem. Soc.*, **86**, 2003 (1964).
[20] A. Streitwieser, Jr., R. A. Caldwell, and M. R. Granger, *J. Am. Chem. Soc.*, **86**, 3578 (1964).

scale (Table XIV, Chapter I), triphenylmethane has an estimated pK_a of 32.5, triptycene of 38, and methane of 40. Because of the enforced orientation of the orbitals of the triptycyl anion, little charge delocalization can occur. Thus the three phenyl groups exert an inductive effect great enough in triptycene to account for about 2 pK_a units on the MSAD scale. The experiment further suggests that the inductive effects of the phenyl groups in triphenylmethane make a substantial contribution to the acidifying properties of these groups.[20]

Results that have a bearing on the relative importance of the inductive and resonance effects of phenyl on carbanion stability were obtained in the [1.*n*]paracyclophane system.[21] The relative rates of potassium *tert*-butoxide-catalyzed hydrogen–deuterium exchange with *tert*-butyl alcohol-*O*-*d* of compound V and compounds VI were determined and are listed in Table IV. Very little exchange occurred at other than the

TABLE IV

Relative Rates at 195° of Potassium tert-Butoxide-Catalyzed Hydrogen–Deuterium Exchange Between tert-Butyl Alcohol-O-d and the [1.n]Paracyclophanes and an Open-Chain Model (Diarylmethyl Hydrogens)

Compound		Relative rate
$(p\text{-}C_2H_5C_6H_4)_2C(H_2)$ V		1
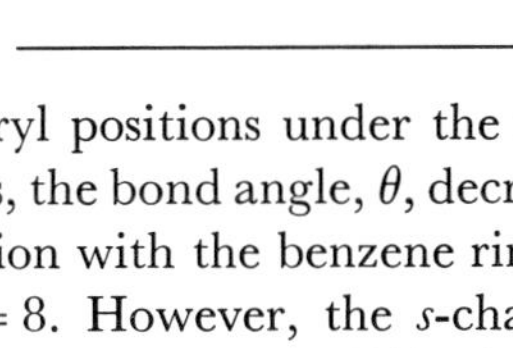 VI: $(CH_2)_m$ bridged diphenylmethane, θ C(H_2)	$m = 12$	2
	$m = 11$	2.3
	$m = 10$	0.2
	$m = 9$	0.06
	$m = 8$	0.002

benzhydryl positions under the reaction condition. As the value of m decreases, the bond angle, θ, decreases, and the carbanion is taken out of conjugation with the benzene rings, probably almost completely in VI with $m = 8$. However, the *s*-character of the C—H bonds increases, which should tend to acidify the compound. The drop in the rates of exchange of about 3 powers of 10 in going from $m = 12$ to $m = 8$ reflects a

[21] D. J. Cram and L. A. Singer, *J. Am. Chem. Soc.*, **85**, 1084 (1963).

superposition of these effects on one another. Certainly the inductive effect of the phenyl groups is playing a major role in keeping the difference in kinetic acidity as low as it is.

Most carbanion-stabilizing groups act through a mixture of electronic effects, in most cases through a mixture of inductive and conjugative effects. Attempts to separate the effects center on the use of the Hammett substitution constants, σ_m and σ_p.[22] Several authors have attempted these separations, notably Taft *et al.*,[23a, b] and Cohn and Jones.[23c] The latter authors dealt with substituent constants (σ^-) based on the pK_a's of *m*- and *p*-substituted phenols in water at 25°. The σ_p^- parameters were dissected into resonance (σ_R^-) and inductive (σ_I^-) components by use of the assumptions of Equations (1) and (2). As usual, the more positive the σ^- value, the more anion stabilizing the substituent. The substituents, σ_I^- and σ_R^- values, and the ratios of $\sigma_{Rp}^-/\sigma_{Ip}^-$ are recorded in Table V.

$$\sigma_R^- = \sigma_p^- - \tfrac{2}{3}\sigma_m^- \tag{1}$$

$$\sigma_I^- = \sigma_m^- \tag{2}$$

TABLE V

Relative Importance of Inductive and Resonance Effects on the Ability of Substituents to Acidify Phenol in Water at 25°[23c]

Substituent	σ_p^-	σ_m^-	σ_{Rp}^-	σ_{Ip}^-	$\sigma_{Rp}^-/\sigma_{Ip}^-$
CH_3O	−0.13	0.12	—	—	—
CO_2^-	0.24	−0.02	—	—	—
Br	0.25	0.40	—	—	—
SO_3^-	0.40	0.28	0.21	0.19	1.1
$CONH_2$	0.61	0.28	0.42	0.19	2.2
$CO_2C_2H_5$	0.64	0.37	0.39	0.25	1.6
$N(CH_3)_3^+$	0.77	0.85	—	—	—
$COCH_3$	0.84	0.33	0.62	0.22	2.8
CN	0.88	0.59	0.48	0.40	1.2
CHO	1.04	0.48	0.72	0.32	2.3
NO_2	1.24	0.69	0.78	0.46	1.7
SO_2CH_3[45a]	0.98	0.70	0.51	0.47	1.1
$SOCH_3$[45b]	0.73	0.53	0.38	0.35	1.1
SO_2CF_3[46]	1.36	0.92	0.75	0.61	1.2

[22] H. H. Jaffé, *Chem. Rev.*, **53**, 191 (1953).

[23] (a) R. W. Taft, Jr. and I. C. Lewis, *J. Am. Chem. Soc.*, **80**, 2436 (1958); (b) R. W. Taft, E. Price, I. R. Fox, I. C. Lewis, K. K. Andersen, and G. T. Davis, *ibid.*, **85**, 709 (1963); (c) L. A. Cohn and W. M. Jones, *J. Am. Chem. Soc.*, **85**, 3397, 3402 (1963).

The values of $\sigma_{Rp}^{-}/\sigma_{Ip}^{-}$ for groups that conjugate well ($CONH_2$, $CO_2C_2H_5$, $COCH_3$, CN, CHO, and NO_2) range from 1.2 for the cyano to 2.8 for the acetyl group. Had the σ constants been based on pK_a values of substituted toluenes, or been determined in a non-hydrogen bonding solvent, different contributions for the inductive and resonance effects would undoubtedly have been observed. The striking feature about the data is that the inductive effect makes a sizeable contribution to the total substituent effect even in those groups that conjugate best with negative charge. The same was observed with the phenyl group in triphenylmethane.[20]

A striking example of the lack of additivity of inductive and conjugative effects is found in the fact that the four α-hydrogens of *m*-methylbenzal fluoride undergo potassium *tert*-butoxide catalyzed hydrogen-deuterium exchange with *tert*-butyl alcohol-*O*-*d* at comparable rates.[24] Had the two effects been complementary, exchange at the carbon attached to the two fluorine atoms and the aryl group would have occurred much faster than at the other α-position. Possibly in the difluoromethyl anion the inductive effect of the two fluorine atoms increased the *p*-character of the carbon atom's contribution to the carbon fluorine bonds, and the electron pair and charge of the anion were left in an orbital unusually rich in *s*-character. Such an orbital would have poor overlap potential with the π-electrons of the attached benzene ring, and thus the inductive and delocalization effects could not both fully exert their stabilizing influence on the same anion.

A number of kinetic studies of base-catalyzed isotopic exchange rates provide additional insight with respect to inductive effects, and to their interplay with conjugative effects. Dessy *et al.*[3] studied the rates of triethylamine(1 *M*)-catalyzed isotopic exchange between a group of mono-substituted acetylenes and 5 *M* deuterium oxide in dimethylformamide. The relative rates are recorded in Table VI.

Hydrogen as a substituent is comparable to that of phenyl, in spite of the electron-withdrawing inductive effect of the phenyl group. This suggests that a small conjugative effect operates in the opposite direction, which tends to distribute positive charge into the benzene ring. The

$$C_6H_5-C{\equiv}C-H \longleftrightarrow {}^{+}C_6H_5{=}C{=}\bar{C}-H$$

similarity in kinetic acidity of acetylene and phenylacetylene contrasts with the difference of about 6 pK_a units in their thermodynamic acidity

[24] D. J. Cram and J. P. Lorand, unpublished results.

TABLE VI

Relative Rates of Trimethylamine(1 M)-Catalyzed Isotopic Exchange Between Mono-Substituted Acetylenes and 5 M Deuterium Oxide in Dimethylformamide at 40°[3]

Substituent	Relative rate	Substituent	Relative rate
C_6H_5	1.0	*o*-$CF_3C_6H_4$	3.5
H	0.73	*m*-$CF_3C_6H_4$	5.2
C_4H_9	0.058	*p*-$CF_3C_6H_4$	3.3
CH_3O	2.0	*m*-ClC_6H_4	28
$(C_6H_5)_3Si$	68	*p*-ClC_6H_4	4.8
o-FC_6H_4	4.7	*p*-BrC_6H_4	2.1
m-FC_6H_4	7.7	*p*-$CH_3OC_6H_4$	1.0
p-FC_6H_4	1.5	*p*-$HC{\equiv}CC_6H_4$	13.8

on the MSAD scale. None of the groups in the *p*-position of the substituents of Table VI distribute negative charge by simple conjugation, and therefore the magnitude of such an effect is obscure. The fact that the *p*-methoxyphenyl group acts the same as the phenyl indicates that the electron-releasing effect of the methoxy group just about cancels any extra inductive effect it provides. The same conclusion is reached from the fact that methoxy as a substituent attached directly to the acetylene gives a rate of exchange about three times that of acetylene itself. Comparison of the relative rates of the *m*- and *p*-halophenylacetylenes also indicates that conjugative effects of the electron pairs on the halogens are superimposed on their inductive effects, the latter factor being the more important. Both the data and theory eliminate conjugative effects between aryl and the negative charge of the acetylide ion. Other investigators reached the same conclusion.[25] The rate-retarding effect of butyl compared with hydrogen (factor of 13) is also expected on the basis of the electron-releasing inductive effect of alkyl groups. These rate comparisons may be complicated by a lack of identity between the observed rate constants for isotopic exchange and the rate constants for ionization of the carbon-hydrogen bonds (see Chapters I and III).

Shatenshtein and co-workers[26a,b] and Roberts and co-workers[26c] have studied the potassium amide-catalyzed exchange of *o*-, *m*-, and

[25] H. B. Charman, D. R. Vinard, and M. Kreevoy, *J. Am. Chem. Soc.*, **84**, 347 (1962).

[26] (a) A. I. Shatenshtein, *Advan. Phys. Org. Chem.*, **1**, 187 (1963); (b) A. I. Shatenshtein and E. A. Izrailevich, *Zh. Obshch. Khim.*, **32**, 1930 (1962) and references listed therein; (c) G. H. Hall, R. Piccolini, and J. D. Roberts, *J. Am. Chem. Soc.*, **77**, 4540 (1955).

p-deuterium of mono-substituted benzene compounds with liquid ammonia. The rates relative to benzene itself (corrected for statistical factors) are listed in Table VII along with σ_I[23a] (sigma, inductive) values.

Although the rate factors for the *o*-substituents varied by 7 powers of 10, a plot of σ_I against log (rate factor for exchange of deuterium *ortho* to substituent) gave a reasonably straight line.[26] This linear free energy correlation suggests that the kinetic acidity at the position *ortho* to substituents is dominated by inductive effects. Other factors appear to play important roles when the substituents are *meta* or *para* to the deuterium undergoing exchange. For example, deuterium exchanges 10^2 times faster when *para* to the trifluoromethyl than to the fluorine group, although fluorine is inductively more carbanion stabilizing by 0.1 σ_I unit. This comparison coupled with others in Table VII suggests that

TABLE VII

Rates of Potassium Amide-Catalyzed Hydrogen Isotope Exchange Between Liquid Ammonia and o-, m-, and p-Deuterium of Mono-Substituted Benzene Compounds Relative to That of Deuterated Benzene (Corrected for Statistical Factors) [26a]

Substituent	*o*	*m*	*p*	σ_I
CH_3	0.2	0.4	0.4	−0.05
H	1	1	1	0.00
$(CH_3)_2N$	1.4	0.2	0.07	0.1
C_6H_5	4.7	3.3	2.9	0.1
$C_6H_5(CH_3)N$	33	2.9	1.3	—
CH_3O	5×10^2	1	0.05	0.25
C_6H_5O	2×10^4	50	4	0.38
CF_3	10^5	10^4	10^4	0.41
F	10^6	10^3	10^2	0.52

conjugative effects are superimposed on inductive effects, and become relatively more important the further the substituent is from the carbanion generated. For example, although the methoxy group has

$$CH_3O\text{—}C_6H_4\text{—}D \longleftrightarrow CH_3\overset{+}{O}{=}C_6H_4{=}\overset{-}{}\text{—}D$$

$\sigma_I = 0.25$, and should enhance the rate of exchange when in the *p*-position by at least a power of 10 if only an inductive effect is operative, the *p*-methoxy group depresses the rate by a factor of 20. Resonance of the type formulated for anisole seems to lower the acidity of the *p*-deuterium somewhat.

Roberts and co-workers[26c] carried out similar exchange experiments, and pointed to the inductive effect of the functional groups as exerting the main control over the reaction rates.

A fine example of a blend of inductive and conjugative effects on kinetic acidity is observed in the ease with which quinaldine undergoes hydrogen–deuterium exchange.[27] When it is heated with various deuterated alcohols at 120°, exchange occurs at the methyl group without any added base. The rates of this exchange correlate reasonably well with the acidities of the alcohols involved (as determined by an indicator method) relative to the acidity of 2-propanol.[28] Table VIII lists the relevant data.

TABLE VIII

Rates of Exchange of the Methyl Hydrogens of Quinaldine with Various Deuterated Alcohols at 120°[27]

Alcohol	K/K_0[a]	$k(\text{sec.}^{-1}) \times 10^7$
$(CH_3)_2CHOH(D)$	1	7
$C_2H_5OH(D)$	12.5	20
OH_2	15	—
$CH_3OH(D)$	50	30
$HO(CH_2)_3OH$	150	—
$HCONH_2$	162	—
$(D)HOCH_2CH_2OH(D)$	540	90
$HOCH_2CHOHCH_2OH$	2200	—
$C(CH_2OH)_4$	5500	—

[a] $K = [A^-]/[HA][i\text{-Pr}\bar{O}]$ for alcohol in *i*-PrOH, whereas K_0 applies to *i*-PrOH itself.

The fact that the alcohols provide a higher rate of exchange than does ammonia suggests a mechanism for the exchange which involves ion pair formation prior to cleavage of the C—H bond.[29]

A special case of operation of what is probably the inductive effect is found in the acidity of ferrocene, whose nucleus can be metalated and

[27] (a) A. I. Shatenshtein, *Advan. Phys. Org. Chem.*, **1**, 169 (1963); (b) A. I. Shatenshtein and E. N. Zuyagintseva, *Dokl. Akad. Nauk SSSR*, **117**, 852 (1957).

[28] J. Hine and M. Hine, *J. Am. Chem. Soc.*, **74**, 5268 (1952).

[29] W. Gordy and S. C. Stanford, *J. Chem. Phys.*, **8**, 170 (1940); **9**, 204 (1941).

even dimetalated with butyllithium under conditions that leave benzene intact.[30]

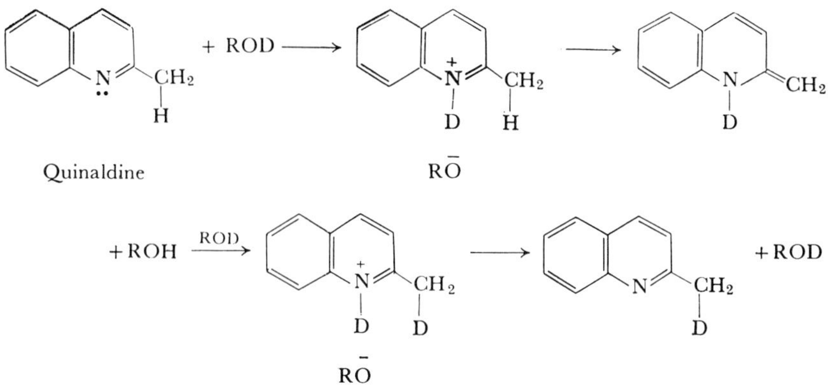

+ ROD ⟶ ⟶

Quinaldine $R\bar{O}$

+ROH $\xrightarrow{ROD}$ ⟶ +ROD

$R\bar{O}$

HOMOCONJUGATIVE EFFECTS

A possible example of homoconjugation in carbanion stabilization is found in the enhanced kinetic acidity of the methyl groups of compounds such as ethylbenzene or 1,1,1-triphenylethane as compared with those of saturated alkanes.[31] For example, the methyl hydrogens of ethylbenzene underwent complete exchange at 120° in 50 hr. with deuterated ammonia which was 0.7 *M* in potassium amide (deuterated). Under similar conditions, heptane exchanged less than half of its methyl hydrogens after 500 hr. In spite of poor solubility, 1,1,1-triphenylethane exchanged its methyl hydrogens much faster than did *tert*-butylbenzene.

Although little doubt exists that β-aryl groups acidify hydrocarbons, the character of the effect is not clear. Inductive effects fall off rapidly with distance, but may be important enough when operating from the β-position to explain the results. An alternative explanation is that homoconjugative effects are operative, and that neighboring aryl

[30] (a) R. A. Benkeser, D. Goggin, and G. Schroll, *J. Am. Chem. Soc.*, **76**, 4025 (1954); (b) A. N. Nesmeyanov, E. G. Perevalova, R. V. Golovnya, and O. A. Nesmeyanova, *Dokl. Akad. Nauk SSSR*, **97**, 459 (1954); (c) D. W. Mayo, P. D. Shaw, and M. Rausch, *Chem. & Ind. (London)*, p. 1388 (1957); (d) R. A. Benkeser and J. L. Bach, *J. Am. Chem. Soc.*, **86**, 890 (1964).

[31] (a) A. I. Shatenshtein, *Advan. Phys. Org. Chem.*, **1**, 185 (1963); (b) A. I. Shatenshtein and E. A. Izrailevich, *Zh. Fiz. Khim.*, **32**, 2711 (1958); (c) A. I. Shatenshtein and E. A. Izrailevich, *Zh. Obshch. Khim.*, **28**, 2939 (1958).

participates in the proton removal. If such a mechanism applies, a "phenanion" should be generated, which is expected to produce rearranged product. Rearranged product was not reported. Rearrangements similar to that formulated are discussed in Chapter VI.

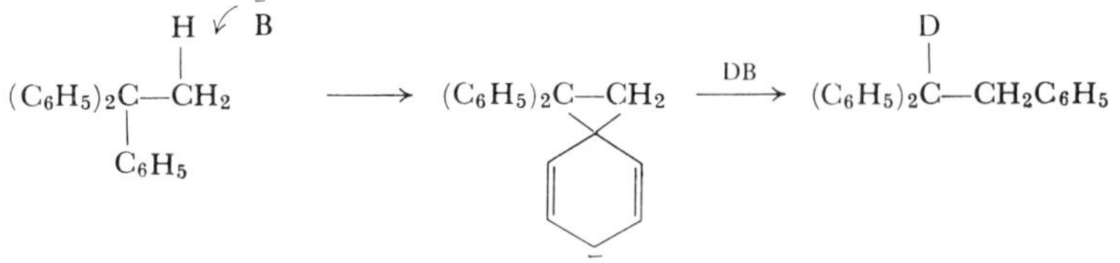

$$(C_6H_5)_2C(C_6H_5)-CH_2-H \;\; \overset{-}{B} \longrightarrow (C_6H_5)_2C-CH_2 \text{(phenanion)} \xrightarrow{DB} (C_6H_5)_2CD-CH_2C_6H_5$$

1,1,1-Triphenylethane — A "phenanion" — 1,1,2-Triphenylethane-1-*d*

Phenyl groups also play some acidifying role with respect to the methyl hydrogens of anisole, dimethylaniline, and diphenylmethylamine.[32] The relative rates of potassium amide-catalyzed exchange between these compounds deuterated in the methyl group and ammonia are listed, along with that of toluene similarly deuterated. Unfortunately, comparisons of these rates with those of their saturated counterparts are not available. The inductive effect of the O—C and N—C bonds are probably playing the major role in acidifying the hydrogens, but the phenyl groups are also undoubtedly contributing, possibly in several ways: homoconjugatively, inductively, and by delocalization of the electrons on oxygen or nitrogen into the benzene ring.

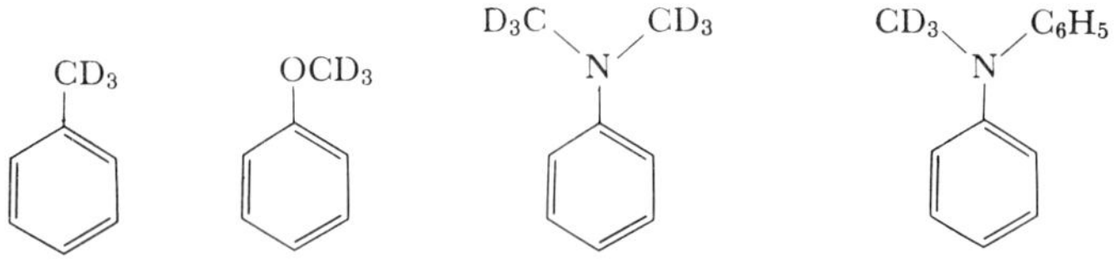

$C_6H_5CD_3$	$C_6H_5OCD_3$	$C_6H_5N(CD_3)_2$	$C_6H_5N(CD_3)C_6H_5$
5.5×10^3	20	1	2

$$^{-}C_6H_5{=}\overset{+}{O}-\overset{-}{C}D_2$$

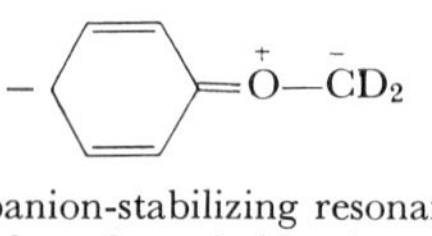

Carbanion-stabilizing resonance form for anisole anion

[32] (a) A. I. Shatenshtein, *Advan. Phys. Org. Chem.*, **1**, 186 (1963); (b) A. I. Shatenshtein, Yu. I. Ranneva, and T. T. Kovalenko, *Zh. Obshch. Khim.*, **32**, 967 (1962) and references quoted here.

An authentic example of homoconjugation was found in the formation of a homoenolate anion when optically active camphenilone (VII) racemized and was deuterated when treated with potassium *tert*-butoxide in *tert*-butyl alcohol-*O*-*d* at 185°.[33] The two processes proceeded at the same rate, a fact which required a homoenolate ion as intermediate. The stereochemistry is discussed in Chapter III.

VII

Optically active

Homoenolate anion, contains plane of symmetry

DB

Racemic VII, deuterated in 6-position

Another possible example of homoconjugation is found in the observation that di-*tert*-butylketone undergoes hydrogen–deuterium exchange when treated with *tert*-butyl alcohol-*O*-*d* which was 1 *M* in potassium *tert*-butoxide at 230°.[34]

AROMATIZATION EFFECTS

One of the successes of simple molecular orbital theory is the prediction that of the completely conjugated planar monocyclic polyolefins, those that possess $4n+2$ π-electrons ($n = 0, 1, 2, 3$, etc.) will be particularly stable because they have completely filled bonding molecular orbitals with substantial electron delocalization energies.[35] The smaller carbocyclic unsaturated anions that fulfill the $4n+2$ condition are the cyclobutadiene dianion, the cyclopentadiene anion, the cyclooctatetraene dianion, and the cyclononatetraene anion. Because of the presumed stability of these anions, their conjugate carbon acids should have pK_a's considerably lower valued than their open-chain counterparts. On the other hand, cycloheptatriene should possess an acid strength comparable with that of heptatrienes. The fact that the differ-

[33] A. Nickon and J. L. Lambert, *J. Am. Chem. Soc.*, **84**, 4604 (1962).

[34] D. J. Cram and R. O. Smith, unpublished results.

[35] A. Streitwieser, Jr., "Molecular Orbital Theory for Organic Chemists," Wiley, New York, 1961, p. 256.

ence in pK_a of cyclopentadiene and cycloheptatriene is about 21 units attests to the qualitative agreement between theory and experiment.

Indene and fluorene, whose anions are isoelectronic with naphthalene and anthracene (or phenanthrene), respectively, also exhibit enhanced

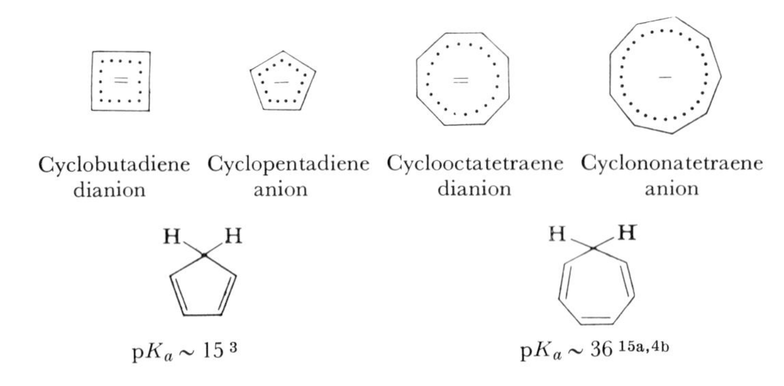

acidities, but of a lower order,[4b] than cyclopentadiene, the pK_a's being reduced by about 3.5 pK_a units for each benzo group.

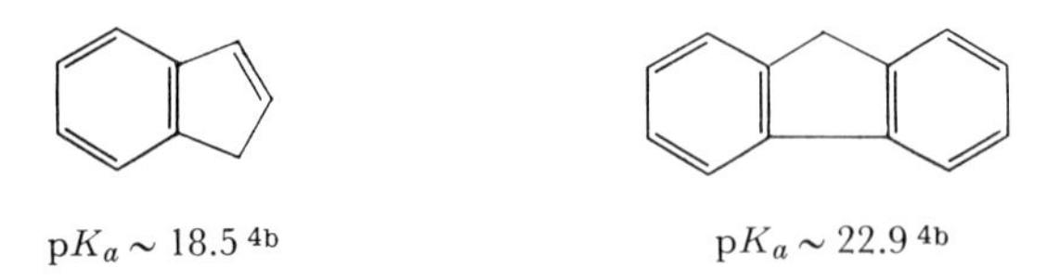

The cyclooctatetraene dianion was prepared by treatment of cyclooctatetraene with alkali metals in ether or liquid ammonia.[36] Carbonation of these salts produced diacids, a fact that established that salt had been prepared.

The dipotassium salt has been isolated in a crystalline state,[37] and the nuclear magnetic resonance spectra of this salt and the dilithium salt were determined in tetrahydrofuran. Both salts exhibited a single sharp peak insignificantly displaced from the resonance of cyclooctatraene itself. The equilibria between cyclooctatetraene, its dianion, and radical

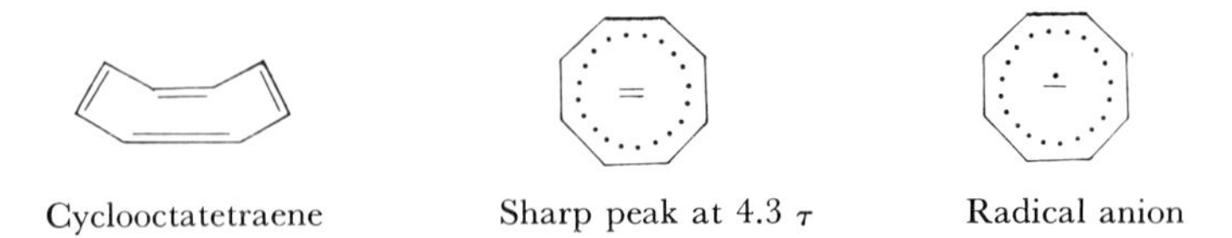

[36] (a) W. Reppe, O. Schlichting, K. Klager, and T. Toepel, *Ann.*, **560**, 1 (1948); (b) A. C. Cope and F. A. Hochstein, *J. Am. Chem. Soc.*, **72**, 2515 (1950).

[37] T. J. Katz, *J. Am. Chem. Soc.*, **82**, 3784, 3785 (1960).

anion were demonstrated to favor strongly the dianion. In compounds such as naphthalene, equilibria strongly favor radical anions.[38]

These observations have been interpreted as consistent only with a planar and essentially aromatic dianion, which contains 10 electrons ($n = 2$ in the $4n+2$ rule).

Salts of the cyclononatetraenide anion have been synthesized by two methods.[39] In the first,[39a] dipotassium cyclooctatetraenide was treated with chloroform in tetrahydrofuran to give compound VIII, which when mixed with potassium or lithium in tetrahydrofuran produced the desired salts. The nuclear magnetic resonance spectra of the potassium salt gave a single sharp peak at 2.96 and of the lithium salt one at 3.15 τ (tetrahydrofuran). These data indicate that the ring system is planar and aromatic.[39a]

In the second preparation of the salt,[39b] compound VIII was synthesized by treatment of cyclooctatetraene with methylene chloride and methyllithium, and VIII gave the salt when treated with lithium. The

$2K^+ + CHCl_3 \longrightarrow$ VIII (H, Cl) $\xrightarrow[\text{Li}]{\text{K or}}$ $\overset{+}{K}$ or $\overset{+}{Li}$

VIII

Potassium or lithium cyclononatetraenide

tetraethylammonium cyclononatetraenide was prepared by treatment of the lithium salt with tetraethylammonium chloride. The lithium salt gave a single sharp peak in the nuclear magnetic resonance spectrum at 3.28, and the quaternary ammonium salt at 3.18 τ. The authors concluded the anion was planar, and possessed a ninefold rotational axis.[39b] The lithium salt did not undergo detectable electron exchange with cyclooctatetraene, but did produce lithium cyclopentadienide when treated with cyclopentadiene. This experiment demonstrates that cyclopentadiene is a stronger acid than cyclononatetraene.

Another exhibition of the stability associated with the $4n+2$ electron system is found in the stability of the pentalene dianion.[40] When treated with butyllithium, dihydropentalene gave dilithium pentalenide, a stable white salt, which in its nuclear magnetic resonance spectrum

[38] (a) G. J. Hoijtink, E. De Boer, P. H. van der Meij, and W. P. Weijland, *Rec. Trav. Chim.*, **75**, 487 (1956); (b) N. S. Hush and J. Blackledge, *J. Chem. Phys.*, **23**, 514 (1955).

[39] (a) T. J. Katz and P. J. Garratt, *J. Am. Chem. Soc.*, **85**, 2852 (1963); (b) E. A. LaLancette and R. E. Benson, *ibid.*, **85**, 2853 (1963).

[40] T. J. Katz and M. Rosenberger, *J. Am. Chem. Soc.*, **84**, 865 (1962).

exhibited a triplet centered at 4.27 τ (intensity 2), and a doublet centered at 5.02 τ (intensity 4). This spectrum identifies the salt as a planar pseudo-aromatic system.

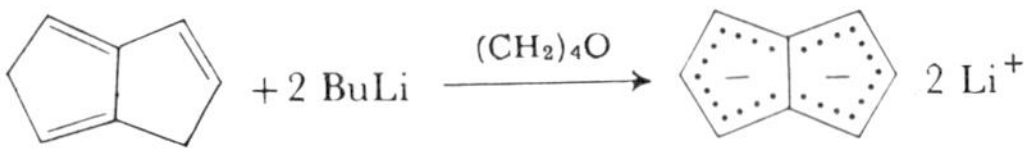

$$\text{Dihydropentalene} + 2\,\text{BuLi} \xrightarrow{(CH_2)_4O} \text{Lithium pentalenide}\ \ 2\,Li^+$$

Dihydropentalene   Lithium pentalenide

NEGATIVE HYPERCONJUGATION EFFECTS

Since carbonium ions are stabilized by hyperconjugation that involves carbon–hydrogen bonds, the possibility exists that carbanions can be stabilized by "negative hyperconjugation" that involves carbon–fluorine or carbon–oxygen bonds. Evidence that the trifluoromethyl group exerts a greater electron-withdrawing effect in the *p*- than in the *m*-position of benzoic acid or anilinium ion is found in the early work of Roberts *et al.*[41a] These and other workers[41b] obtained $\sigma_m = 0.41$, $\sigma_p = 0.53$, $\sigma_m^- = 0.49$, and $\sigma_p^- = 0.65$ and interpreted the conjugating ability of the trifluoromethyl group in terms of no-bond resonance structures of the types formulated. If it is assumed[23c] that Equations (1) and (2) apply, then $\sigma_{R_p}^-/\sigma_{I_p}^- = 0.65$, which makes the conjugative contribution to the overall electrical effect of the trifluoromethyl group somewhat less than the conjugative contribution of the other strongly electron-withdrawing groups, but nevertheless, substantial (see Table V).

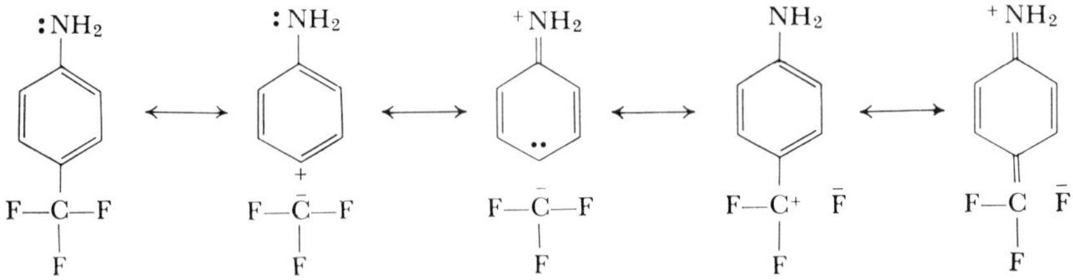

$$:NH_2\text{–}C_6H_4\text{–}CF_3 \longleftrightarrow :NH_2\text{–}\overset{+}{C_6H_4}\text{–}\bar{C}F_3 \longleftrightarrow {}^{+}NH_2{=}C_6H_4(:)\text{–}\bar{C}F_3 \longleftrightarrow NH_2\text{–}C_6H_4\text{–}C^{+}F_2\ \bar{F} \longleftrightarrow {}^{+}NH_2{=}C_6H_4{=}CF_2\ \bar{F}$$

Sheppard,[41c] in a broad study of the substituent effects of fluorine-containing groups has obtained $\sigma_m^- = 0.52$ and $\sigma_p^- = 0.67$ for the perfluoroisopropyl group as applied to the anilinium ion–aniline equilibrium. The values are not far from those of the trifluoromethyl group,

41 (a) J. D. Roberts, R. L. Webb, and E. A. McElhill, *J. Am. Chem. Soc.*, **72**, 408 (1950); (b) W. A. Sheppard, *ibid.*, **84**, 3072 (1962); (c) W. H. Sheppard, private communication.

and indicate that the perfluoroisopropyl group possesses some conjugating properties. These can be represented in terms of a series of no-bond resonance structures which contribute to the resonance hybrid.

$$:NH_2\text{-}C_6H_4\text{-}CF(CF_3)_2 \longleftrightarrow :NH_2\text{-}C_6H_4^{+}\text{-}\bar{C}F(CF_3)_2 \longleftrightarrow {}^{+}NH_2{=}C_6H_4^{\bar{\cdot\cdot}}\text{-}\bar{C}F(CF_3)_2 \longleftrightarrow :NH_2\text{-}C_6H_4\text{-}\overset{+}{C}F(CF_3)\ \bar{C}F_3 \longleftrightarrow {}^{+}NH_2{=}C_6H_4{=}CF(CF_3)\ \bar{C}F_3$$

Hine[42] correlated a large amount of data in terms of the "double bond–no-bond resonance" concept, which is best visualized in terms of the resonance structures for carbon tetrafluoride or tetramethoxymethane. Support for this hypothesis was found in heats of combustion and reaction, in equilibrium constants involving the halomethanes, and in comparisons of rate constants for hydrolysis of mono-, di-, and triethoxyalkanes.

$$F\text{—}CF_2\text{—}F \longleftrightarrow \bar{F}\quad F_2C{=}\overset{+}{F} \quad \text{etc.} \qquad CH_3O\text{—}C(OCH_3)_2\text{—}OCH_3 \longleftrightarrow CH_3\bar{O}\quad (CH_3O)_2C{=}\overset{+}{O}CH_3 \quad \text{etc.}$$

Some of the strongest evidence for negative hyperconjugation as an important effect in stabilizing carbanions results from the work of Andreades.[19] Monohydrofluorocarbons (IX–XII) were submitted to sodium methoxide-catalyzed hydrogen–deuterium exchange in methanol-*O-d*, and the rates determined. Fortunately, elimination reactions were slower than the exchange reactions. Table IX records the results.

The dramatic differences in rates in passage from primary to tertiary fluorocarbon acids probably largely reflect differences in fluorocarbanion stabilities, with *tert* > *sec* > *prim*. These results indicate that nine β-fluorine atoms are far more carbanion stabilizing than three α-fluorine atoms. Negative hyperconjugative stabilization of carbanions provides a good explanation for this effect. The larger the number of β-fluorines, the larger the number of equivalent resonance structures that can be drawn, and the more stabilized is the carbanion. If the inductive effect is described in terms of no-bond resonance structures which contribute to a

[42] J. Hine, *J. Am. Chem. Soc.*, **85**, 3239 (1963).

TABLE IX

Relative Rates of Sodium Methoxide-Catalyzed Hydrogen–Deuterium Exchange between Methanol-O-d and Monohydrofluorocarbons at — 29° [19]

Compound	Compound No.	No. equivalent anionic resonance structures	Relative rate	pK_a estimates [a]
CF_3—H	IX	0	1.0	28
$CF_3(CF_2)_5CF_2$—H	X	2	6	27
$(CF_3)_2CF$—H	XI	6	2×10^5	18
$(CF_3)_3C$—H	XII	9	10^9	11

[a] Streitwieser scale.

resonance hybrid, then inductive effects that operate from the β-position become identical with hyperconjugative effects.

9 Equivalent structures   9 Equivalent structures

Estimates of the pK_a's of compounds IX–XII were made as follows.[19] Since compound XII underwent hydrogen isotope exchange in ethanol-*O-d* at 80°, it was estimated to have a pK_a of about 11.[43] The rate of exchange of 9-tritiofluorene was determined in sodium methoxide–methanol and found to be 200 times that of compound X. A plot of $\log k$ vs. pK_a for fluorene ($pK_a \sim 25$)[4a] and fluoradene ($pK_a \sim 11$)[43] was made, and the pK_a's of compounds IX, X, and XI estimated by use of this plot. Had the Streitwieser scale been applied,[4b] pK_a values of about 28 and 27 would have been obtained for compounds IX and X, and XI would have been around 18. These are recorded in Table IX.

The question arises as to the type of hybridization at carbons in the carbanions derived from compounds X–XII. The tendency of the charge to occupy orbitals as rich as possible in *s*-character tends to make the carbanions pyramidal. It is not clear which type (electrons in *p*- or sp^3-orbitals) of hybridization at carbon gives the best overlap with the electrons of the carbon–fluorine bond. Steric repulsions between the

[43] H. Rapoport and G. Smolinsky, *J. Am. Chem. Soc.*, **82**, 934 (1960).

groups attached to the carbanion certainly favor a planar configuration for the anion.

d-ORBITAL EFFECTS

Considerable experimental evidence has accumulated which indicates that carbanions are stabilized by overlap of the orbital containing the electron pair and the 3*d*-orbitals of attached second-row elements. Identification of this effect is usually complicated by the presence of large stabilizing inductive effects. The kinds of approaches which allow separation of these effects will be discussed first, followed by a brief review of the results of application of molecular orbital theory to the problems of angular and hybridization dependence for *d*-orbital carbanion stabilization by the sulfone group. Experimental data which bear on these questions will then be examined.

One approach to the separation of *d*-orbital and inductive effects is best exemplified by the work of Doering and Hoffmann,[17] who determined the rates and activation parameters for the deuteroxide-catalyzed exchange of tetramethylammonium, tetramethylphosphonium, and trimethylsulfonium ions with deuterium oxide. Table X records the results.

TABLE X

Relative Rates and Activation Parameters for Deuteroxide-Catalyzed Hydrogen–Deuterium Exchange between Ammonium, Phosphonium, and Sulfonium Ions and Deuterium Oxide [17]

Compound	Relative rates at 62°	$\Delta H^{\ddagger}$ (Kcal./mole)	$\Delta S^{\ddagger}$ (e.u.)	C—X bond dist. (Å)
$(CH_3)_4N^+$	1	32.2 ± 0.6	-15 ± 2	1.47
$(CH_3)_4P^+$	2.4×10^6	25.6 ± 0.2	$+4 \pm 1$	1.87
$(CH_3)_3S^+$	2.0×10^7	22.4 ± 0.5	-1 ± 2	1.81

These results amply demonstrate that *d*-orbital effects strongly stabilize the transition state for carbanion formation. Comparison of the rates for the ammonium and phosphonium ions indicates that the latter is 6 powers of 10 faster than the former, and that the heats of activation favor exchange of the phosphonium ion by 6.6 Kcal./mole. The carbanion-stabilizing inductive (electrostatic) effect should be much stronger for the ammonium than for the phosphonium ion, since the C—P bond is 27% longer than the C—N bond, and coulombic interactions fall off rapidly with distance. Thus the difference of 6.6 Kcal./mole in heat of activation for the two ions would be considerably greater in the absence of the inductive effect. Clearly the differences in rates and heats of activation are associated with carbanion stabilization by *d*-orbital effects.[17]

The large difference in rate between the ammonium and phosphonium exchanges is due in about equal measure to differences in heat and in entropy of activation (Table X). The formal charge of the starting materials strongly binds and orients solvent molecules. In the transition states, probably the carbon–hydrogen bond is broken almost completely. Charge is generated in the transition state for the ammonium case, solvent becomes more thoroughly oriented and bound, and the entropy is large and negative. In the phosphonium case, the transition state is partially dipolar, partially covalent due to *d*-orbital overlap effects. Thus relatively smaller changes in solvent binding and orientation occur, and the entropy changes are small and positive. As expected, the sulfonium case resembles the phosphonium, the entropy change being around zero.[17]

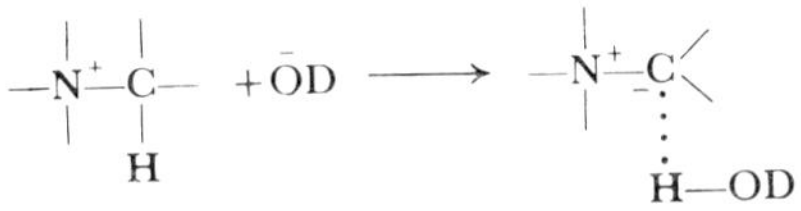

$$-\overset{+}{N}-\underset{H}{C}- \; + \bar{O}D \longrightarrow -\overset{+}{N}-\bar{C}\langle \cdots H-OD$$

Starting state — Dipolar transition state

$$-\overset{+}{P}-\underset{H}{C}- \; + \bar{O}D \longrightarrow -\overset{+}{P}-\bar{C}\langle \cdots H-OD \longleftrightarrow -P{=}C\langle \cdots HOD$$

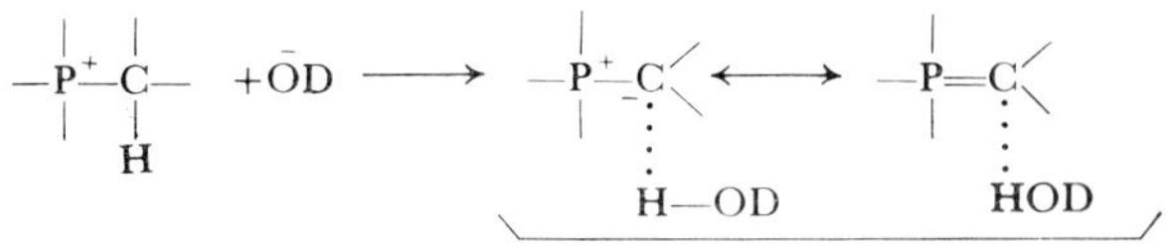

Less dipolar transition state

A second example[44] of the identification of 3-*d*-orbital carbanion-stabilizing effects involves a comparison of the rates at which compounds XIII and XIV undergo potassium alkoxide-catalyzed tritium–hydrogen exchange with alcohols. The two sulfur groups acidified the carbon acid

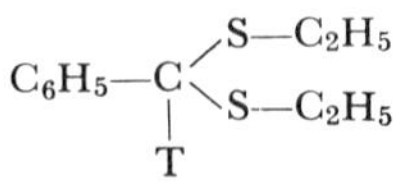

$$C_6H_5-C(T)(S-C_2H_5)_2 \qquad C_6H_5-C(T)(O-C_2H_5)_2$$

XIII — Relative rate, 10^6

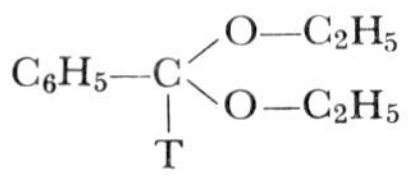

XIV — Relative rate, 1

[44] (a) S. Oae, W. Tagaki, and A. Ohno, *Tetrahedron*, **20**, 417, 427 (1964); (b) D. S. Tarbell and W. E. Lovett, *J. Am. Chem. Soc.*, **78**, 2259 (1956).

more than the two oxygens did by enough to make exchange of XIII occur 6 powers of 10 faster than XIV. Had the greater inductive effect of the oxygens not been operative, this number could have been much greater.

Similarly, Tarbell and Lovett[44b] observed that allyl hexyl sulfide was isomerized to hexyl propenyl sulfide under conditions (78°, 3.7 *M* sodium ethoxide in ethanol) that left allyl hexyl ether unchanged.

A second approach to the identification of *d*-orbital overlap effects is found in the work of Bordwell and Cooper.[45] These authors measured the pK_a values of the *p*- and *m*-methylsulfonyl- and methylsulfinyl-benzoic acids and phenols. The phenoxide anion is strongly conjugating but the carboxylate anion is not. Comparison of the differences between σ_p constants (*p*-sigma substituent constants of the Hammett variety) for the phenols and benzoic acids was employed as a measure of the conjugative ability of the methylsulfonyl and methylsulfinyl groups. These differences are listed.

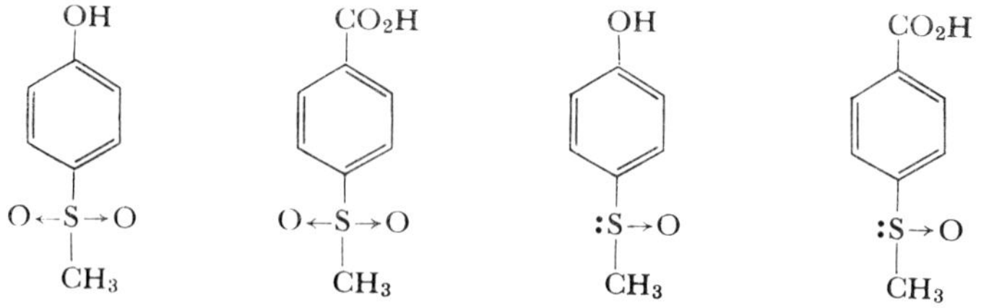

OH	CO_2H	OH	CO_2H
O←S→O, CH_3	O←S→O, CH_3	:S→O, CH_3	:S→O, CH_3
$\sigma = 0.98$	$\sigma = 0.72$	$\sigma = 0.73$	$\sigma = 0.48$
$\Delta\sigma = 0.26$		$\Delta\sigma = 0.25$	

In Table V, the values of $\sigma^-_{Rp}/\sigma^-_{Ip}$ (ratio of resonance to inductive effect contributions to σ for *p*-substituted phenols) are listed, and are found to be equal to 1.1–1.2 for the four 3*d*-orbital-containing groups, sulfinyl (CH_3SO),[45b] sulfonyl (CH_3SO_2),[45a] trifluoromethylsulfonyl (CF_3SO_2),[46] and sulfonate ($\bar{O}_3S$).[23c] Thus the inductive and *d*-orbital overlap contributions to acidification of the phenols are about equal for all four groups.

The ratio of σ_R to σ_I was found to be unity in a study[47] of substituent constants in compounds of general structure XV. This evidence indicates again that the *d*-orbital and inductive effects seem to depend directly

[45] (a) F. G. Bordwell and G. D. Cooper, *J. Am. Chem. Soc.*, **74**, 1058 (1952); (b) F. G. Bordwell and P. J. Boutan, *J. Am. Chem. Soc.*, **79**, 717 (1957).

[46] W. A. Sheppard, *J. Am. Chem. Soc.*, **85**, 1314 (1963).

[47] (a) C. Y. Meyers, B. Cremonini, and L. Maioli, *J. Am. Chem. Soc.*, **86**, 2944 (1964).

on one another. Other evidence for outer shell expansion in sulfur has been summarized elsewhere.[48]

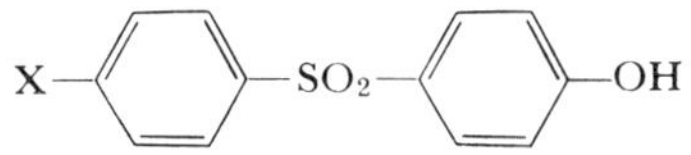

$$X-C_6H_4-SO_2-C_6H_4-OH$$

XV ($X = NH_2$, OH, OCH_3, CH_3, and NO_2)

The character of bonding involving 3*d*-orbitals of elements such as phosphorus and sulfur and unshared pairs of electrons on atoms sigma bonded to these elements has been a subject of considerable theoretical interest.[49] Questions of particular importance are those concerned with (1) the conformational requirement for carbanion stabilization by *d*-orbital effects of attached substituents; and (2) the configurational requirement at carbon for operation of such *d*-orbital effects.

Koch and Moffitt,[49b] on the basis of molecular orbital calculations, concluded that the best overlap between a 2*p*-orbital of a carbon atom attached to a sulfone group and the 3*d*-orbitals of the sulfur atom occurs in the two conformations drawn. Lipscomb and co-workers[49e] determined the crystal structure of compound XVI, and observed that the molecule possessed a "Case II" conformation. Hybridization at nitrogen was intermediate between sp^2 and sp^3, being $sp^{2.23}$, and the S—N bond distance was 1.62 Å, compared to a distance of 1.74 Å calculated for a S—N single bond. The C—N—C bond angles were 113°, the C—N—S bond angles were 119°, the N—S—N bond angle was 113°, and the O—S—O bond angle was 120°. The same investigators[49e] then made LC atomic orbital–molecular orbital calculations on a model of XVI in which the methyl groups were replaced by fluorines (XVII); the hybridization at nitrogen and the overall conformation of the molecule were retained. This "Case II" conformation was compared with a "Case I" conformation in which the hybridization at nitrogen was kept the same and the F_2N groups were rotated 90° about the N—S bond. The difference in total orbital energies indicated that the "Case II" conformation was the more stable, and that this stability arises from *d*-orbital interactions. In the "Case I" conformation, the lone pairs on

[48] G. Cilento, *Chem. Rev.*, **60**, 146 (1960).

[49] (a) G. E. Kimball, *J. Chem. Phys.*, **8**, 188 (1940); (b) H. P. Koch and W. E. Moffitt, *Trans. Faraday Soc.*, **47**, 7 (1951); (c) D. P. Craig, A. Maccoll, R. S. Nyholm, L. E. Orgel, and L. E. Sutton, *J. Chem. Soc.*, p. 332 (1954); (d) H. H. Jaffé, *J. Phys. Chem.*, **58**, 185 (1954); (e) T. Jordan, H. W. Smith, L. L. Lohr, Jr., and W. N. Lipscomb, *J. Am. Chem. Soc.*, **85**, 846 (1963); (f) D. W. J. Cruickshank, *J. Chem. Soc.*, p. 5486 (1961).

nitrogen have to compete more with the lone pairs of the oxygen atoms for *d*-orbitals than in the "Case II" conformation.

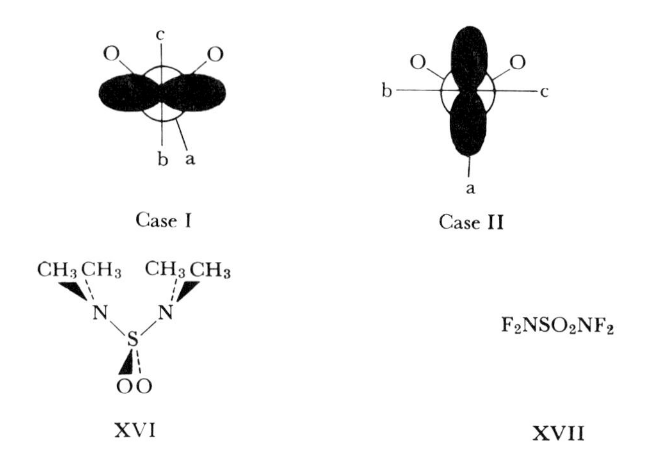

Case I Case II

XVI XVII

Serious questions arise as to how good a model a molecule of XVI in a crystal lattice is for a sulfonylcarbanion in solution. Even more problematic is whether the calculations on XVII have any bearing on the questions of the conformation and configuration of a sulfonylcarbanion in solution. The charge on the carbanion gives rise to electrostatic repulsions between the lone pair of electrons of the carbanion and the partial negative charges of the oxygens. These repulsions are minimized with pyramidal hybridization at carbon and a "Case II" conformation. These repulsions do not exist in either XVI or XVII since the molecules are not charged. Sulfonylcarbanions in solution in polar solvents have large solvation energies which should have an effect on both the conformation and configuration of the species. These solvation energies are absent in a crystal of XVI and in a calculation involving XVII.

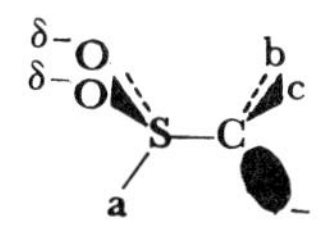

Conformation and configuration of
a sulfonyl carbanion which minimizes
electrostatic repulsions

More direct experimental evidence concerning the configuration and conformation of carbanions stabilized by *d*-orbital overlap effects comes from comparisons of the acidities of open-chain carbon acids and those confined by ring systems. Doering and Levy[50] compared the acidities of the open-chain (XVIII) and bicyclic (XIX) trisulfonylmethanes.

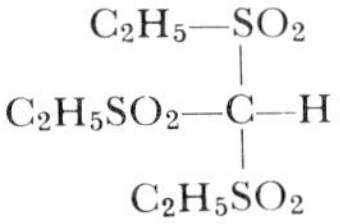

$$\begin{array}{c} C_2H_5{-}SO_2 \\ | \\ C_2H_5SO_2{-}C{-}H \\ | \\ C_2H_5SO_2 \end{array}$$

XVIII

SO_2, H, O_2S, SO_2, CH_3 (bicyclic)

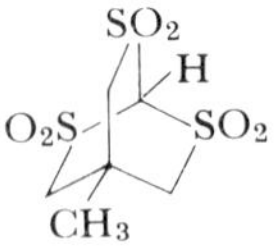

XIX

$pK_a = 3.30$

Unfortunately the pK_a of XVIII could not be determined, but this open-chain compound was found to be a stronger acid than the bicyclic sulfone (XIX). This result indicates that the adaptable geometry of the open-chain carbanion makes that carbanion more stable than its bicyclic counterpart. The bicyclic anion possesses an enforced conformation and configuration that maximize pole-dipole electrostatic repulsions, whereas the open-chain anion can assume both a conformation and a configuration that minimize these repulsions. This experiment sheds little light on the question of whether the sulfonylcarbanion is more stable in a pyramidal or planar configuration. Clearly, conformation must be specified before the question can be answered.

In a second study, Zimmerman and Thyagarajan[51] metalated a one-to-one mixture of sulfones XX and XXI with less than an equivalent of phenyllithium, phenylsodium, or phenylpotassium. The resulting mixture was then quenched with D_2O, and the equilibrium constant (K) for Equation (3) was calculated from the relative amounts of deuterium in the recovered sulfones XX and XXI. A value of $K = 1.4 \pm 0.4$ was obtained when phenyllithium was employed. Unfortunately, this

$$(3)\quad \underset{\text{XX}}{C_6H_5SO_2{-}\overset{|}{\underset{H}{C}}(CH_2)_2} + C_6H_5SO_2{-}\bar{C}(CH_3)_2 \underset{k_2}{\overset{k_1}{\rightleftharpoons}} C_6H_5SO_2{-}\bar{C}(CH_2)_2 + \underset{\text{XXI}}{C_6H_5SO_2\underset{H}{C}(CH_3)_2}$$

$$C_6H_5SO_2{-}\bar{C}(CH_3)_2 \xrightarrow{k_3,\ D_2O} C_6H_5SO_2{-}\underset{D}{C}(CH_3)_2$$

$$C_6H_5SO_2{-}\bar{C}(CH_2)_2 \xrightarrow{k_4,\ D_2O} C_6H_5SO_2{-}\underset{D}{C}(CH_2)_2$$

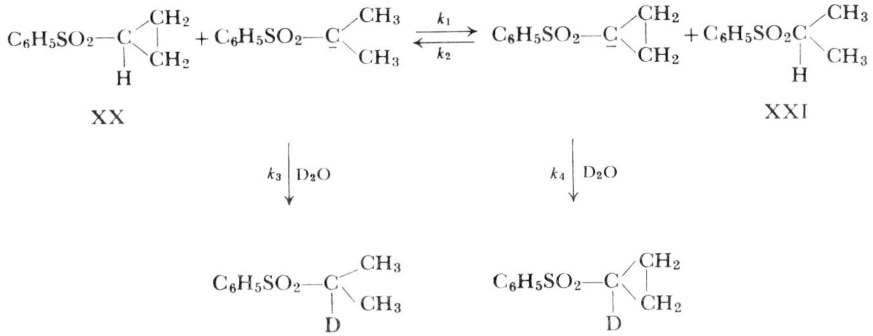

[50] W. von E. Doering and L. K. Levy, *J. Am. Chem. Soc.*, **77**, 509 (1955).
[51] H. E. Zimmerman and B. S. Thyagarajan, *J. Am. Chem. Soc.*, **82**, 2505 (1960).

method of determining the equilibrium constant is open to question, since the method when applied to triphenylmethane and dimethyl sulfoxide[52] led to erroneous results[53] (see Chapter I for discussion). The difficulty that possibly intervenes is that during the quench, proton and deuterium transfers occur between the two carbon acids and their conjugate bases before the deuterium oxide becomes uniformly distributed throughout the medium. This arises because k_1 and k_2 are of values comparable with k_3 and k_4.

The pK_a's of a series of open-chain and cyclic phenolic sulfones were compared in a study by Corey and co-workers.[54] In compound XXII,

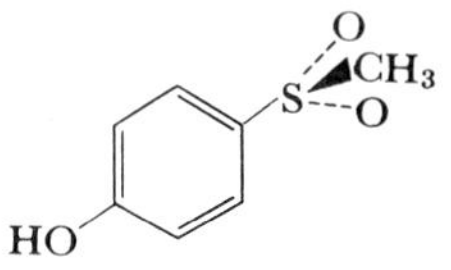

XXII

pK_a = 7.72

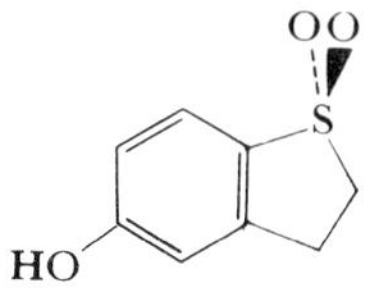

XXIII

pK_a = 7.81
pK_a = 7.68 (corr.)

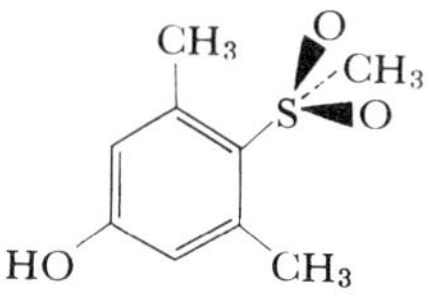

XXIV

pK_a = 8.13
pK_a = 7.87 (corr.)

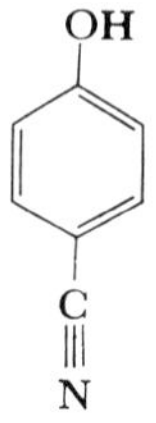

XXV

pK_a = 7.95

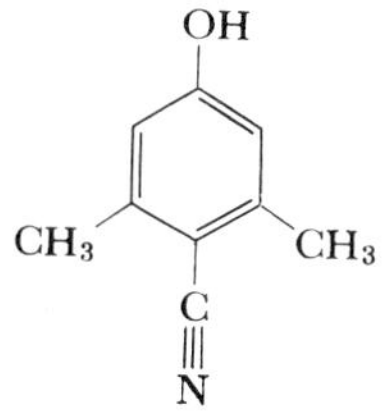

XXVI

pK_a = 8.21

the sulfone group is free to assume a conformation of lowest energy, which on the basis of Lipscomb and co-workers' calculations,[49e] should be the conformation drawn ("Case II" conformation). In XXIII a "Case I" conformation is enforced by the ring system. Comparison of the pK_a's of XXV and XXVI shows that substitution of two *m*-methyl groups on *p*-cyanophenol raises the pK_a by 0.26 units, or by 0.13 units per methyl group. If the pK_a of XXIII is corrected for the inductive

[52] E. J. Corey and M. Chaykovsky, *J. Am. Chem. Soc.*, **84**, 866 (1962).

[53] (a) E. C. Steiner and J. M. Gilbert, *J. Am. Chem. Soc.*, **85**, 3054 (1963); (b) C. D. Ritchie, *ibid.*, **86**, 4488 (1964).

[54] E. J. Corey, H. Konig, and T. H. Lowry, *Tetrahedron Letters*, **12**, 515 (1962).

effect of the *m*-methylene group by subtracting 0.13 pK_a units, one obtains an adjusted pK_a of 7.68, which is very close to the 7.72 value for the pK_a of XXII. Thus the enforced "Case I" conformation of XXIII does not seem to affect the ability of the *d*-orbitals of sulfur to stabilize negative charge compared with a system that is free to assume a "Case II" conformation.

Similarly if the pK_a of XXIV is corrected for the presence of the two *m*-methyl groups, an adjusted pK_a of 7.87 is obtained, which compares with the adjusted value of 7.68 for XXIII. The sterically most feasible conformation of XXIV is of the "Case II" variety and the enforced conformation of XXIII is of the "Case I" type. Yet the adjusted pK_a of the latter compound is higher valued than that for the former. Clearly, the acidities of these compounds show little conformational dependence.

In the phenols XXII–XXIV, overlap was between 2*p*- and 3*d*-orbitals in the anion of the phenol, and the question of hybridization at carbon did not arise. In the series of cyclic disulfonylmethanes formulated in Table XI,[54] the variation of pK_a with ring size can be best

TABLE XI

Variation of pK_a *of Cyclic Disulfonylmethane with Ring Size in Water at* 25°[54]

Compound	No. atoms in ring ($n+3$)	pK_a
cyclo-[SO_2–$(CH_2)_n$–SO_2–CH_2]	5	13.9
	6	12.61
	7	11.75
	8	10.99
$(CH_3SO_2)_2CH_2$	—	12.50

explained on the basis of electrostatic and solvation effects. The open-chain model, dimethylsulfonylmethane, has a pK_a of 12.5, and the anion probably has the conformation and configuration pictured in formula XXVII, which minimizes combined electrostatic and steric repulsions. This requires a pyramidal carbanion. The five-membered ring sulfone anion with its enforced conformation probably has a small electrostatic driving force to assume a trigonal configuration. The greater electrostatic repulsions in XXVIII as compared to XXVII make the conjugate acid of XXVII the stronger acid of the two. In the six-membered ring sulfone anion, electrostatic repulsions are again minimized with a pyramidal carbanion, but the ring enforced conformation is not as favorable as in the open-chain compound. However, the fact that the sulfone substituents are "tied back" in the cycle and are not in the open-chain compound tends to provide less steric inhibition of solvation in XXIX than in XXVII. As a result, the pK_a's of the six-membered ring sulfone

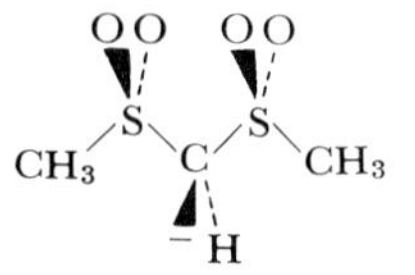

XXVII (sp^3-hybridization of C^-)

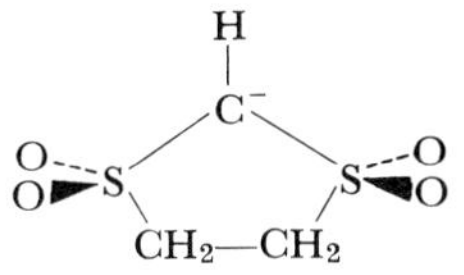

XXVIII (sp^2-hybridization of C^-)

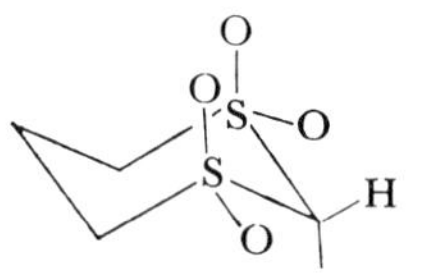

XXIX (sp^3-hybridization of C^-)

and open-chain model approach one another. As the rings get bigger, the longer chains provide conformations for pyramidal carbanions which grow more favorable electrostatically, and the acidity increases. The reason the seven- and eight-membered rings are stronger acids than the open-chain model is that their derived anions are "tied back" and better solvated than are their open-chain counterparts.

Data that bear on the above explanation have been obtained by Breslow and Mohacsi.[55] These authors prepared and measured the pK_a's of compounds XXX–XXXIII, and noted an interesting inversion in the acidity relationships between the cyclic and noncyclic compounds. When the carbanion was generated α- to a carbethoxyl group, the cyclic compound was the more acidic (XXXIII > XXXII in acidity), but in the absence of the carbethoxyl group, the reverse was true (XXX > XXXI in acidity). The carbanion α to the carbethoxyl group occupies a p-orbital because of overlap with the p-orbitals of the carbonyl group. With this configuration, there is little change in electrostatic repulsions with changes in conformation. Compound XXXIII is more acidic than

[55] R. Breslow and E. Mohacsi, *J. Am. Chem. Soc.*, **83**, 4100 (1961).

XXXII because the cyclic anion, being tied back, has less steric inhibition of solvation than its noncyclic counterpart. In the anion of XXX,

$(C_6H_5SO_2)_2CH_2$ XXX $pK_a = 11.5$

(4-$CH_3C_6H_3$)$(SO_2)_2CH_2$ (cyclic) XXXI $pK_a > 13$

$(C_6H_5SO_2)_2CHCO_2C_2H_5$ XXXII $pK_a = 4.60$

(4-$CH_3C_6H_3$)$(SO_2)_2CHCO_2C_2H_5$ (cyclic) XXXIII $pK_a = 3.35$

the change is in an sp^3 orbital, and considerable electrostatic advantage is associated with the system being able to assume the best conformation. As a result, XXX is more acidic than XXXI, in which the advantage is lost because of the ring system.

In what appears to be an anomalous result, the same authors[56] reported that the acidities of compounds XXXIV and XXXV were almost equal. This result was based on a metalation and quench competition experiment similar to that which gave spurious results in another case[52] (see Chapter I for discussion), and is open to question.

$C_6H_5SO_2CH_2SC_6H_5$ XXXIV

$C_6H_4(SO_2)(S)CH_2$ (cyclic) XXXV

These data taken together are evidence for the following conclusions concerning the relationships of conformation, configuration, and α-sulfonylcarbanion stability. (1) When the pair of electrons of a carbanion occupies a *p*-orbital, its stability relative to its conjugate acid has little dependence on conformation. (2) When the pair of electrons of a carbanion occupies an sp^3 orbital, its stability relative to its conjugate acid has a large dependence on conformation, that conformation being preferred which minimizes electrostatic repulsions. (3) When restrictions of a single ring system force a conformation on the carbanion which is

[56] R. Breslow and E. Mohacsi, *J. Am. Chem. Soc.*, **84**, 684 (1962).

electrostatically poor for a pyramidal configuration, the carbanion assumes a planar configuration. (4) A carbanion at the bridgehead of a bicyclic system is held in a pyramidal configuration, even though the conformation is electrostatically poor, and as a result, the carbanion is destabilized relative to the analogous open-chain anion. (5) An open-chain carbanion assumes a pyramidal configuration and a conformation that minimizes electrostatic repulsions. This configuration maximizes the *s*-character of the orbital that contains the lone pair of electrons of the carbanion. (6) Carbanions of ring systems possess less steric inhibition of solvation than do their open-chain counterparts. (7) Carbanion stabilization by 3*d*-orbitals has conformational and configurational dependences that are less important than electrostatic and solvation effects.

Additional results that have a bearing on the question of the configuration of α-sulfonylcarbanions involved use of the *N*-methanesulfonylaziridine (XXXVI) as a model.[57] The nuclear magnetic resonance spectra (60 Mc./sec.) of XXXVI and *N*-phenylaziridine (XXXVII) in carbon disulfide were compared from −80° to 25°. The protons of the aziridine ring of XXXVI gave a sharp band at 25°. Below about −30° this band starts to broaden and splits into two bands at about −40°. At still lower temperatures an A_2B_2 pattern appears such as is exhibited by alkylaziridines at room temperature.[11] The rate constant for the nitrogen inversion process of XXXVI is calculated to be about 30 sec.$^{-1}$ at −40°. In vinyl chloride as solvent, no new bands appeared in the spectrum of XXXVI in the temperature range −80° to −160°.[57] The symbol, T_c, after the formulas, stands for coalescence temperature.

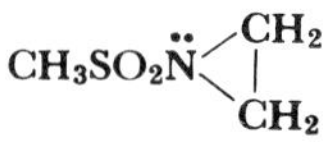

$CH_3SO_2\ddot{N}\langle CH_2-CH_2$ (aziridine ring)

XXXVI, $T_c = -40 \pm 2°$
$k \approx 30$ sec.$^{-1}$

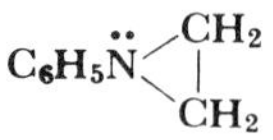

$C_6H_5\ddot{N}\langle CH_2-CH_2$ (aziridine ring)

XXXVII, $T_c = -40 \pm 2°$
$k \approx 40$ sec.$^{-1}$

c-$C_6H_{11}N\langle CH_2-CH_2$ (aziridine ring)

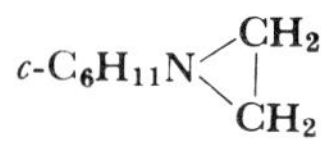

XXXVIII, $T_c = 95°$
$k \approx 51$ sec.$^{-1}$

The spectrum of XXXVII exhibited much the same changes with temperature as did XXXVI. The coalescence temperature proved to be close to −40°, and the rate constant for inversion was calculated to be about 40 sec.$^{-1}$ at −40°. By comparison, Bottini and Roberts[11] found the rate constant for inversion of *N*-cyclohexylaziridine (XXXVIII) to be 51 sec.$^{-1}$ at 95°.

Others[58] found that XXXVI still gave a single sharp line for its ring

[57] F. A. L. Anet and J. M. Osyany, private communication.
[58] T. G. Traylor, *Chem. & Ind. (London)*, p. 649 (1963).

protons at $-37°$. The spectra were taken in dichloromethane and in deuterated acetone at 40 Mc./sec. Furthermore, XXXVII gave a single line (apparently in the neat form) for the methylene protons down to $-77°$. In methanol solution a change to two bands occurred at $-60°$.[11]

Carbonyl compounds XXXIX, XL, and XLI were also examined in vinyl chloride (toluene was present in the solution of XLI).[57] As expected, $CH_3CO > CH_3OCO > (CH_3)_2NCO$ in ability to stabilize the transition state for inversion of the aziridine nitrogen. All of these carbonyl-containing groups exhibited inversion rates far in excess of that observed for the sulfonyl or phenyl groups.

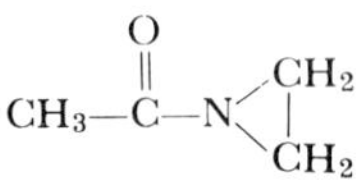

$CH_3-C(=O)-N(CH_2)_2$ — XXXIX, Line broadening below $-120°$

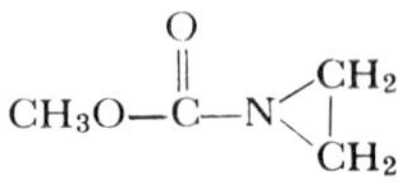

$CH_3O-C(=O)-N(CH_2)_2$ — XL, $T_c = -85 \pm 5°$, $k \approx 5$ sec.$^{-1}$

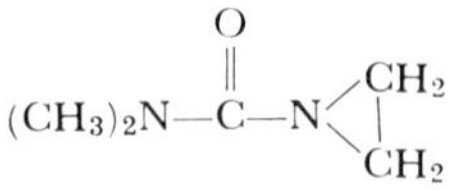

$(CH_3)_2N-C(=O)-N(CH_2)_2$ — XLI, $T_c = -75 \pm 5°$, $k \approx 20$ sec.$^{-1}$

In another study,[59] the approximate coalescence temperatures and rates of inversion for compounds XLII–XLVI were determined in deuterated chloroform, and that of XLVII was obtained in methylene dichloride.

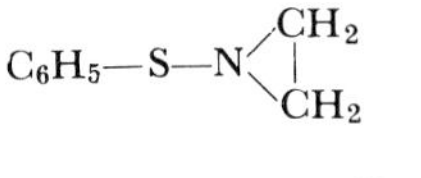

$C_6H_5-SO_2-N(CH_2)_2$ — XLII, $T_c = -30 \pm 1°$, $k \approx 36$ sec.$^{-1}$

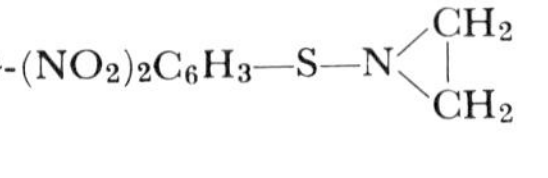

p-$CH_3C_6H_4-SO_2-N(CH_2)_2$ — XLIII, $T_c = -30 \pm 2°$, $k \approx 35.5$ sec.$^{-1}$

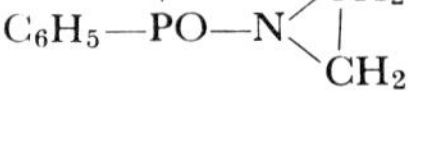

$C_6H_5-SO-N(CH_2)_2$ — XLIV, $T_c = 0 \pm 5°$, $k \approx 79$ sec.$^{-1}$

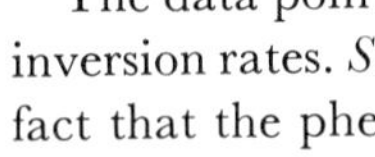

$C_6H_5-S-N(CH_2)_2$ — XLV, $T_c = -11 \pm 1°$, $k \approx 71$ sec.$^{-1}$

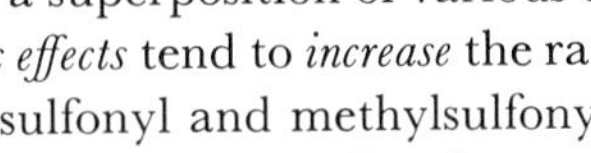

2,4-$(NO_2)_2C_6H_3-S-N(CH_2)_2$ — XLVI, $T_c = -18 \pm 1°$, $k \approx 63$ sec.$^{-1}$

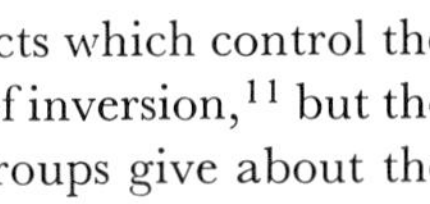

$(C_6H_5)_2PO-N(CH_2)_2$ — XLVII, $T_c = -108 \pm 2°$, $k \approx 48$ sec.$^{-1}$

The data point to a superposition of various effects which control the inversion rates. *Steric effects* tend to *increase* the rate of inversion,[11] but the fact that the phenylsulfonyl and methylsulfonyl groups give about the same rates indicates that such an effect is not important. The *inductive effect* should *decrease* the rate, since the greater the electron-withdrawing properties of the substituent attached to nitrogen, the more *p*-character nitrogen contributes to the bond, and the more *s*-character the orbital

[59] F. A. L. Anet, R. D. Trepka, and D. J. Cram, unpublished results.

has which contains the unshared electron pair of nitrogen. In the transition state for inversion, that electron pair has to go into a *p*-orbital. In those substituents that contain unshared electron pairs on the atoms attached directly to nitrogen, *electrostatic inhibition* of inversion effects should also *decrease* the rate. In the ground state, the electron pairs are *trans* to one another, whereas in the transition state, the electron pairs are brought closer together. *Overlap effects*, either *p-p* or *p-d*, should *increase* the rate of inversion, since overlap increases as the amount of *s*-character of the orbital containing the electron pair decreases.[49e]

The sulfur-containing groups exhibit the order $C_6H_5SO_2 > C_6H_5S > C_6H_5SO$ in their effect on the rate of inversion. However, the remarkable feature is that the rates of inversion of the aziridines with these substituents are so close together. This order does not match that of overall electron withdrawal (as measured by σ) in which $C_6H_5SO_2 > C_6H_5SO > C_6H_5S$. The last order matches that for both the inductive and overlap effects, each of which oppose one another in their effects on inversion rate. Furthermore, the unshared electron pairs on sulfur in C_6H_5S and C_6H_5SO provide electrostatic inhibition of inversion effects absent in $C_6H_5SO_2$. The observed order blends these opposing effects.

A more informative comparison involves the fact that although $(C_6H_5)_2PO$ and C_6H_5SO exhibit about the same overall acidifying effects on attached R_2C—H groups,[60] $(C_6H_5)_2PO$ enhances the rate of inversion of the aziridine by several powers of 10 compared to C_6H_5SO. The fact that sulfur of C_6H_5SO contains an unshared electron pair, and phosphorus of $(C_6H_5)_2PO$ does not, might be responsible. In this comparison, thc electrostatic inhibition of inversion should be less dominated by inductive and overlap effects than in the other comparisons.

An interesting set of comparisons of base-catalyzed hydrogen isotope exchange rates of cyclic and noncyclic carbon acids was made by Oae and co-workers.[61]

In *tert*-butyl alcohol–potassium *tert*-butoxide at 138°, bicyclic compound XLVIII was found to undergo exchange about 10^3 times as fast as its open-chain counterpart, XLIX. This experiment indicates that a pyramidal carbanion is at least as stabilized by three attached sulfur atoms as a similarly substituted carbanion which is free to assume its most stable configuration. This result contrasts with that of Doering and Levy,[50] who found the bicyclic sulfone XIX was a weaker acid than its open-chain analog, XVIII. The presence of large electrostatic effects

[60] (a) D. J. Cram and R. D. Partos, *J. Am. Chem. Soc.*, **85**, 1093 (1963); (b) D. J. Cram and S. H. Pine, *ibid.*, **85**, 1096 (1963).

[61] S. Oae, W. Tagaki, and A. Ohno, *J. Am. Chem. Soc.*, **83**, 5036 (1961).

in the sulfone carbanions and their absence in the sulfide carbanions accounts for the difference. The greater rate for the bicyclic sulfide system XLVIII as compared to XLIX is probably partially associated with the steric advantage for solvation of the transition state for proton removal. The authors[44a] invoked a special acidifying effect present only in the bicyclic system which involves 3*p*-3*d*-orbital sulfur–sulfur bonding, and which is favored by the enforced proximity of the three sulfurs to one another.

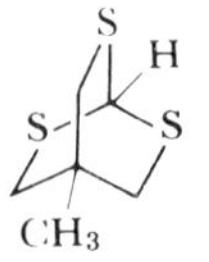

S H S S CH_3

XLVIII
Relative rate, 10^3

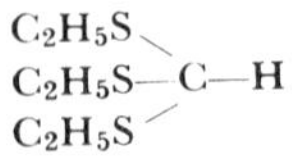

$(C_2H_5S)_3C{-}H$

XLIX
Relative rate, 1

A similar effect was used to explain why the five-membered cycle, L, gave a faster rate of exchange than either the open-chain model (LI) or the higher cyclic homologs, LII or LIII.[44a]

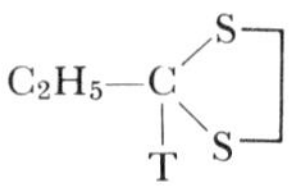

L
Relative rate, 19

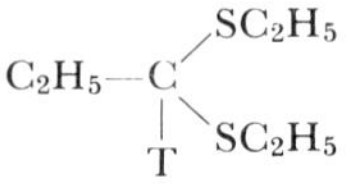

LI
Relative rate, 1

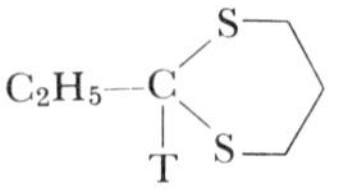

LII
Relative rate, 1.55

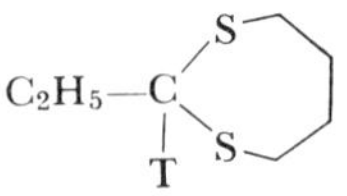

LIII
Relative rate, 5.37

CHAPTER III

Stereochemistry of Substitution of Carbon Acids and Organometallic Compounds

Much has been learned about the structure and immediate environment of reaction intermediates through use of stereochemical techniques. Particularly fruitful has been the study of substitution reactions at asymmetric carbon. The stereochemical course of substitution reactions that involve carbonium ions and radicals has been the subject of numerous investigations initiated before 1930. The first such studies of carbanions were made in 1936 and involved measurements of the relative rates of racemization and substitution of carbon acids. Wilson *et al.*[1a,c] observed equal rates for hydrogen–deuterium exchange and racemization of optically active 2-methyl-1-phenyl-1-butanone (I) in a solution of dioxane–deuterium oxide–sodium deuteroxide. A similar identity of rates was observed when the optically active, deuterated acid II was heated in aqueous sodium hydroxide.[1d,e] These results were interpreted as involving proton abstraction by base in the rate-controlling step, followed by deuteration[1a,c] (protonation[1d,e]) of a symmetrical ambident anion on either oxygen[1a] or carbon.[1d,e] In a parallel study,[1b] ketone I was found to undergo acetate ion-catalyzed bromination and racemization at the same rates. Again, a symmetrical carbanion which could be captured from either face with equal probability was invoked as a reaction intermediate.

The first report of stereospecificity that involved substitution of an organometallic reagent was made by Letsinger.[2] Optically active

[1] (a) C. L. Wilson, *J. Chem. Soc.*, p. 1550 (1936); (b) S. K. Hsu and C. L. Wilson, *ibid.*, p. 623 (1936); (c) S. K. Hsu, C. K. Ingold, and C. L. Wilson, *ibid.*, p. 78 (1938); (d) D. J. G. Ives and G. C. Wilks, *ibid.*, p. 1455 (1938); (e) D. J. G. Ives, *ibid.*, p. 81 (1938).

[2] R. L. Letsinger, *J. Am. Chem. Soc.*, **72**, 4842 (1950).

2-octyl iodide was treated with *s*-butyllithium at $-70°$, and the resulting 2-octyllithium was carbonated to give 2-methyloctanoic acid. The

$$C_6H_5—\underset{\underset{O}{\|}}{C}—\overset{*}{\underset{C_2H_5}{\overset{CH_3}{C}}}—H \qquad HO_2C—\overset{*}{\underset{C_6H_4CH_3\text{-}p}{\overset{C_6H_5}{C}}}—D$$

I II

reactions occurred with 20% overall retention of configuration and 80% racemization. The reactions were carried out in 94% hexane–6% ether, the small amount of ether being necessary for the metalation reaction. The organometallic intermediate when allowed to come to 0° before carbonation gave racemic acid. The question of whether carbanions intervene in reactions of this type will be discussed later in the chapter.

Several claims that carbanions were produced that retained their configurations [3] have been demonstrated to be without foundation.[4] The first evidence that carbanions could be generated and captured stereospecifically was reported in 1955,[5] and since that time a body of results and accompanying theory has appeared, which constitutes the subject of this chapter.

The first section is devoted to the stereochemistry of planar carbanions produced by base-catalyzed hydrogen–deuterium exchange of carbon acids. The second section treats the stereochemistry of systems that form unsymmetrical carbanions in the isotopic exchange reaction. The third section deals with the special case of carbanions (generated by proton abstraction) confined in three-membered rings. The fourth section is concerned with the unique symmetry properties of carbanions stabilized by homoconjugation. In the fifth section, the stereochemistry of substitution of organometallic compounds is treated.

CARBON ACIDS THAT FORM SYMMETRICAL OR NEAR SYMMETRICAL CARBANIONS

The possible configurations of trisubstituted carbanions were discussed in Chapter II, and a spectrum of configurations ranging from

[3] (a) R. Kuhn and H. Albrecht, *Ber.*, **60**, 1297 (1927); (b) R. L. Shriner and J. H. Young, *J. Am. Chem. Soc.*, **52**, 3332 (1930); (c) E. S. Wallis and F. H. Adams, *ibid.*, **55**, 3838 (1933).

[4] (a) N. Kornblum, J. T. Patton, and J. B. Woodman, *J. Am. Chem. Soc.*, **70**, 746 (1948); (b) G. Wittig, F. Vidal, and E. Bohnert, *Ber.*, **83**, 359 (1950).

[5] D. J. Cram, J. Allinger, and A. Langemann, *Chem. & Ind. (London)*, p. 919 (1955).

sp^2-p to sp^3 was considered possible, depending on the attached substituents. In this section, the stereochemical fate of carbanions thought to have planar or near planar configurations is discussed. These include carbanions stabilized by cyano-, nitro-, aryl-, and carbonyl-containing groups, as well as those that derive stabilization from formation of an aromatic π-system.

The simplest technique for studying carbanion stereochemistry involves a comparison of the relative rates of base-catalyzed racemization and hydrogen isotope exchange of optically active carbon acids. The maximum rotations of a number of deuterated and non-deuterated carbon acids have been found to be essentially the same. Therefore, no problem exists with regard to either maximum rotation or relative configurations of starting materials and products.

$$-\overset{*}{\underset{|}{\overset{|}{C}}}-H(D) + B-D(H) \xrightarrow{\bar{B}} -\underset{|}{\overset{|}{C}}-D(H) + B-H(D)$$

Four limiting ratios of k_e (rate constant for isotopic exchange) and k_α (rate constant for racemization) can be envisioned:

$k_e/k_\alpha \longrightarrow \infty$, stereochemistry $\longrightarrow$ 100% retention

$k_e/k_\alpha \longrightarrow 1$, stereochemistry $\longrightarrow$ 100% racemization

$k_e/k_\alpha \longrightarrow 0.5$, stereochemistry $\longrightarrow$ 100% inversion

$k_e/k_\alpha \longrightarrow 0$, stereochemistry $\longrightarrow$ 100% isoracemization

If exchange occurs with complete retention, clearly overall configuration is retained, and k_e/k_α approaches infinity. If exchange occurs with complete racemization, then each carbanion is captured front and back with equal probability, and k_e/k_α approaches unity. If exchange occurs with complete inversion, each carbanion is captured at the back side to give product of inverted configuration. Since this inverted-exchange product is optically active, but possesses a rotation opposite in sign but equal in magnitude to that of the starting material, exchange occurs half as fast as racemization, and k_e/k_α approaches 0.5. If racemization occurs without exchange, then k_e/k_α approaches 0. The latter condition might be fulfilled in either of two ways. (1) Hydrogen might be transferred by base from the front to the back side of an asymmetric carbon acid in the presence of external proton donors of acidity great enough to capture the free carbanion, should it be produced. Such a process is an *intramolecular* proton transfer, and is referred to as an *isoracemization.* (2) The reaction might be carried out in the absence of acids of sufficient strength to provide hydrogen to the carbanion even if it becomes completely free.

In such a case, the hydrogen originally associated with the carbon acid becomes reassociated with carbanions to form racemic material without isotopic exchange. Such a process involves an *intermolecular* proton transfer, and is more akin to an ordinary racemization process. Examples of all these processes are found in the following pages.

Retention Mechanisms

The retention mechanism is readily recognized experimentally, because any value of k_e/k_α in excess of unity points to its existence. Three distinct retention mechanisms have been identified. In the first of these, values of k_e/k_α that ranged from 1.0 to 150 were obtained when optically

D CH_3 ... $CON(CH_3)_2$ + HNR_2 ⟶ H CH_3 ... $CON(CH_3)_2$ + DNR_2

III ($pK_a \sim 21$)

TABLE I

Retention and Racemization in Base-Catalyzed Hydrogen Isotope Exchange Reactions

Run no.	Solvent	Base: Nature	Base: Conc. (M)	T (°C)	k_e/k_α
1	*tert*-BuOH	NH_3	0.8	200	>50
2	$(CH_2)_4O$	NH_3	0.3	145	148
3	$(CH_2)_4O$	NH_3	4	25	~2
4	$(CH_2)_4O$	NH_2Pr	0.5	145	>56
5	$(CH_2)_4O$	$NHPr_2$	0.6	145	>9
6	$(CH_3)_2SO$	NH_3	0.2	25	1.0
7	90% C_6H_6–10% C_6H_5OH	C_6H_5OK	0.1	75	18
8	90% C_6H_6–10% C_6H_5OH	$C_6H_5ON(CH_3)_4$	0.05	75	1.0
9	*tert*-BuOH	$N(Pr)_3$	0.5	200	>6
10	*tert*-BuOH	*tert*-BuOK	0.024	25	1.0

active system III was submitted to isotopic exchange with ammonia or primary or secondary amines.[6] Table I records the results (runs 1–6).

In solvents of low dielectric constant such as *tert*-butyl alcohol or tetrahydrofuran, and with proton-donating bases such as ammonia,

[6] (a) D. J. Cram and L. Gosser, *J. Am. Chem. Soc.*, **85**, 3890 (1963); (b) D. J. Cram and L. Gosser, *ibid.*, **86**, 5445 (1964).

propylamine, or dipropylamine, values of k_e/k_α in considerable excess of unity were observed (runs 1–5). In the presence of large concentrations of ammonia (run 3), or when a solvent good at dissociating ion pairs was used (dimethyl sulfoxide), values close to or equal to unity were obtained. The mechanism of Figure 1 explains these results.

The retention mechanism applies when k_2 and $k_{-1} > k_3$ (see Figure 1). In other words, rotation of the ammonium ion within an ion pair and collapse of the ion pair to the covalent state are faster than ion pair dissociation. In tetrahydrofuran or *tert*-butyl alcohol, which are poor

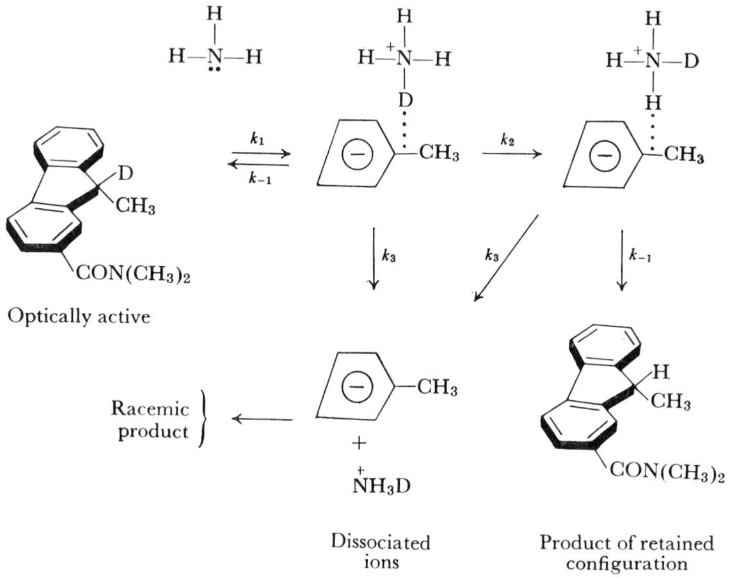

Optically active

Racemic product

$\overset{+}{N}H_3D$

Dissociated ions

Product of retained configuration

FIG. 1. Retention by ammonium ion rotation within ion pairs.

dissociating solvents, these conditions are fulfilled. Large concentrations of ammonia in tetrahydrofuran would understandably tend to increase the rate of ion-pair dissociation, increase the reaction rate, and decrease the value of k_e/k_α. The higher dielectric constant of dimethyl sulfoxide makes ion-pair dissociation faster than ammonium ion rotation within the ion pair ($k_3 > k_2$), the carbanion passes into a symmetrical environment, and k_e/k_α goes to unity (run 6).

Run 7, which involved potassium phenoxide–phenol in benzene as the base-solvent, provides an example of a second kind of retention mechanism. Here, $k_e/k_\alpha = 18$. This result of retention can be explained by a scheme similar to that of Figure 1, except that a potassium ion with its ligands of phenol molecules is put in place of ammonia (see Figure 2). Again, k_2 and $k_{-1} > k_3$. In other words, rotation of the potassium ion with its ligands and collapse of the ion pair to a covalent state are faster processes than dissociation. The fact that substitution of tetramethylammonium phenoxide for potassium phenoxide reduced k_e/k_α to unity (run 8) lends substance to this mechanism. This retention mechanism is dependent on the coordinating ability of the metal cation.

Run 9 provides the basis of the third mechanism. In this experiment, the solvent base was *tert*-butyl alcohol–tripropylamine, and $k_e/k_\alpha > 6$.

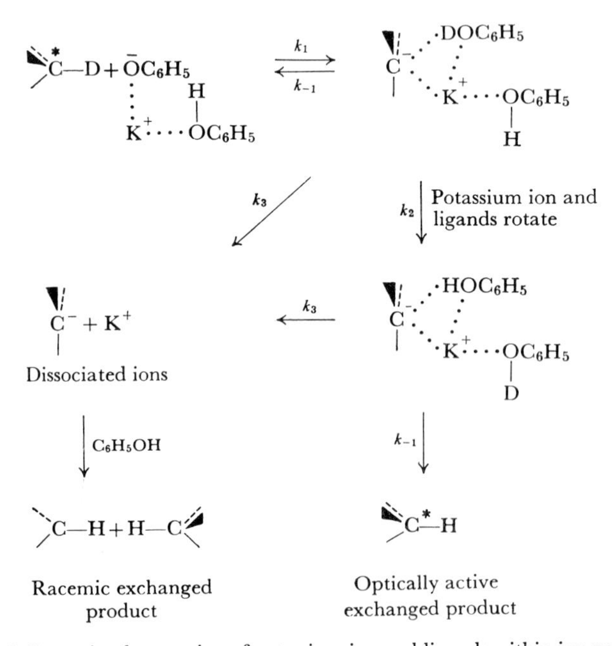

FIG. 2. Retention by rotation of potassium ion and ligands within ion pairs.

Although tripropylamine probably removed deuterium from substrate many times, the resulting hydrogen-bonded ion pair probably returned to unaltered starting material, and the reaction was invisible. A different and much more reactive form of base also undoubtedly was present, but in much lower concentration. The small equilibrium concentration of tripropylammonium *tert*-butoxide, when it removed deuterium from III, formed a new ion pair, tripropylammonium carbanide, which collapsed to exchanged product of retained configuration. In this mechanism (see Figure 3), k_2 and $k_4 > k_3$.

Credibility is given to this interpretation by the fact that substitution of potassium *tert*-butoxide for tripropylamine as base results in exchange occurring with racemization ($k_e/k_\alpha = 1$, run 10). In this experiment, the pK_a of the *tert*-butyl alcohol used as solvent (~ 19) and that of the substrate (~ 21) are too close together to provide the rate for proton

$$\geq C^*\text{—D} + \overset{+}{H}NPr_3 \cdots \bar{O}R \underset{k_{-1}}{\overset{k_1}{\rightleftharpoons}} \geq C^- \cdots DOR \cdots \overset{+}{H}NPr_3$$

$$\geq C^- \cdots DOR \xrightarrow{k_2} \geq C^- \cdots \overset{+}{H}NPr_3 \xrightarrow{k_4} \geq C^*\text{—H}$$

$$\geq C^- \cdots \overset{+}{H}NPr_3 \xrightarrow{k_3} C^- + H\text{—}\overset{+}{N}Pr_3 \; (k_3)$$

$$C^- + H\text{—}\overset{+}{N}Pr_3 \rightarrow \geq C\text{—H} + H\text{—}C\leq$$

Dissociated ions

Racemic exchanged product

Optically active exchanged product

FIG. 3. Retention with tripropylammonium alkoxide ion pair as base.

transfer needed for carbanion capture before it passes into a symmetrical environment, at least in a low dielectric constant medium. In the mechanisms formulated in Figures 1–3, ΔpK_a between proton donor

$$ROH \cdots C^- \cdots HOR \; (M^+) \longrightarrow R\bar{O} \quad H\text{—}C\leq \quad M^+ \cdots \overset{H}{O}R$$

Solvated ion pair

Product-separated ions

$$\longrightarrow \geq C\text{—H} + \overset{+\,-}{MOR} + ROH$$

Ion pair

(ammonium ion or phenol) and the conjugate acid of the proton acceptor (carbon acid) is about 9. This difference is probably great enough to make the proton transfer to the carbanion extremely high.

These retention mechanisms appear to operate only in solvents of low dielectric constant. In all cases, the stereospecific mechanisms which depend on ion pairing compete with nonstereospecific mechanisms which depend on ion-pair dissociation. The rates of the latter processes are lower in solvents of low than in those of high dielectric constant. The same effect also must inhibit proton capture at the rear face of the carbanion within the ion pair. Such a process produces *product-separated ions*, and in a solvent of low dielectric constant the rate of this reaction should be comparable to that of ordinary ion-pair dissociation, which is slower than ion-pair reorganization and collapse.

Racemization Mechanisms

Values of $k_e/k_\alpha = 1$ indicate that racemization mechanisms are operative. In principle, proper combinations of retention and inversion, or of retention and isoracemization, can provide k_e/k_α values of unity. Examples of the latter combination are exceptional, and will be cited in the next sections. In most cases, racemization mechanisms are recognized by their discrete character; values of unity produced under a variety of conditions require too great a degree of fortuity to involve delicate balancing of two competing stereospecific processes.

A variety of racemization mechanisms can be envisioned, most of which have been recognized experimentally. In those cases in which the carbanion is planar and symmetrical, racemization accompanies any process which leads to a symmetrical environment for the anion. Table I records three kinds of circumstances that produced this situation with fluorene III as substrate. (1) Either neutral or anionic bases in dissociating but non-hydroxylic solvent gave $k_e/k_\alpha = 1$. For example, system III with ammonia in dimethyl sulfoxide gave complete racemization (run 6). In this medium, either dissociated ions were the active bases, or ion pairs, if produced, dissociated faster than they gave optically active exchanged product. (2) Use of a quaternary ammonium base in a nondissociating solvent gave $k_e/k_\alpha = 1$. In run 8, tetramethylammonium phenoxide in benzene–phenol gave complete racemization. Again the life of the carbanion was extended to the point where it passed into a symmetrical solvent envelope. (3) In a nondissociating solvent, when the proton donors were not acidic enough to capture the carbanion before its surroundings became symmetrical, racemization was observed. Run 10 carried out in *tert*-butyl alcohol–potassium *tert*-butoxide provides an example.

A comparison of the behavior of fluorene systems III and IV focuses attention on the connection between the lifetime and stereochemical fate of carbanions. In tetrahydrofuran–propylamine, III gave $k_e/k_\alpha > 56$

(run 4). The estimated ΔpK_a between III and propylammonium ion is about 11. This difference in acidities is great enough so that proton capture by the carbanion occurs at a rate far in excess of any symmetrizing process. In the same medium and base, IV gave $k_e/k_\alpha = 1$.[7] Here the estimated ΔpK_a between IV and propylammonium ion is only 8. When aniline was substituted for *n*-propylamine in the same solvent, $k_e/k_\alpha = 6$. The estimated pK_a difference between IV and anilinium ion is about 14. Apparently in this medium of low dielectric constant, a ΔpK_a difference greater than 10 is required before the proton capture rate is fast enough to produce retention of configuration.

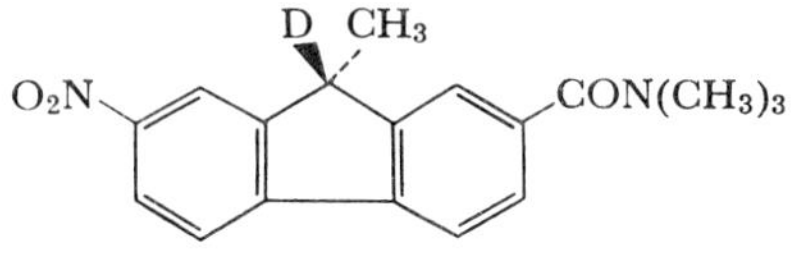

D CH$_3$

O_2N — CON(CH$_3$)$_3$

IV ($pK_a \sim 18$)

Another type of racemization mechanism is observed with ambient carbanions. With a variety of solvents and bases, compounds V to VIII

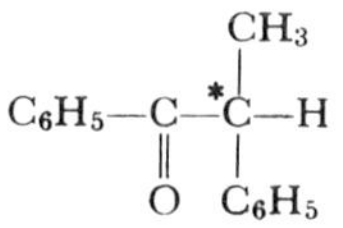

C_6H_5—C(=O)—$\overset{*}{C}$H(CH_3)(C_6H_5)

V

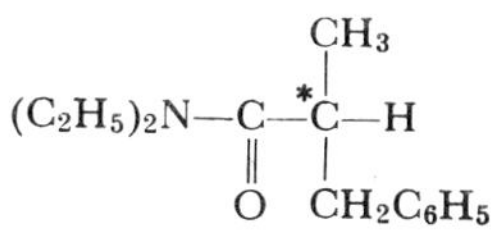

$(C_2H_5)_2N$—C(=O)—$\overset{*}{C}$H(CH_3)($CH_2C_6H_5$)

VI

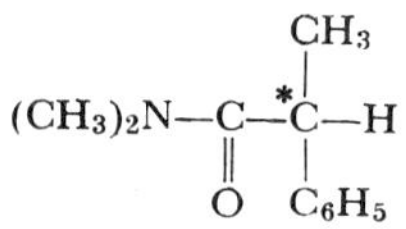

$(CH_3)_2N$—C(=O)—$\overset{*}{C}$H(CH_3)(C_6H_5)

VII

$(CH_3)_3CO$—C(=O)—$\overset{*}{C}$H(CH_3)(C_6H_5)

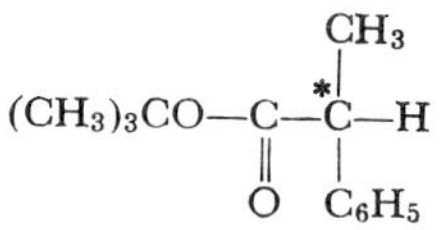

VIII

gave k_e/k_α values of unity in deuterated media.[7,8] In all of these systems, the charge of the carbanion is largely concentrated on the oxygen of the carbonyl group (the more electronegative element), and proton capture undoubtedly occurs there much faster than on carbon.[9] Subsequent base-catalyzed hydrogen exchange reactions on the enol forms of these compounds probably occur much faster than the enols revert to the thermodynamically more stable carbonyl forms. The enol tautomers are symmetrical, last long enough to pass into symmetrical environments,

[7] D. J. Cram and L. Gosser, *J. Am. Chem. Soc.*, **86**, 2950 (1964).
[8] D. J. Cram, C. A. Kingsbury, and P. Haberfield, *J. Am. Chem. Soc.*, **83**, 3678 (1961).
[9] M. Eigen, *Angew. Chem.* (*Intern. Ed. Engl.*), **3**, 1 (1964).

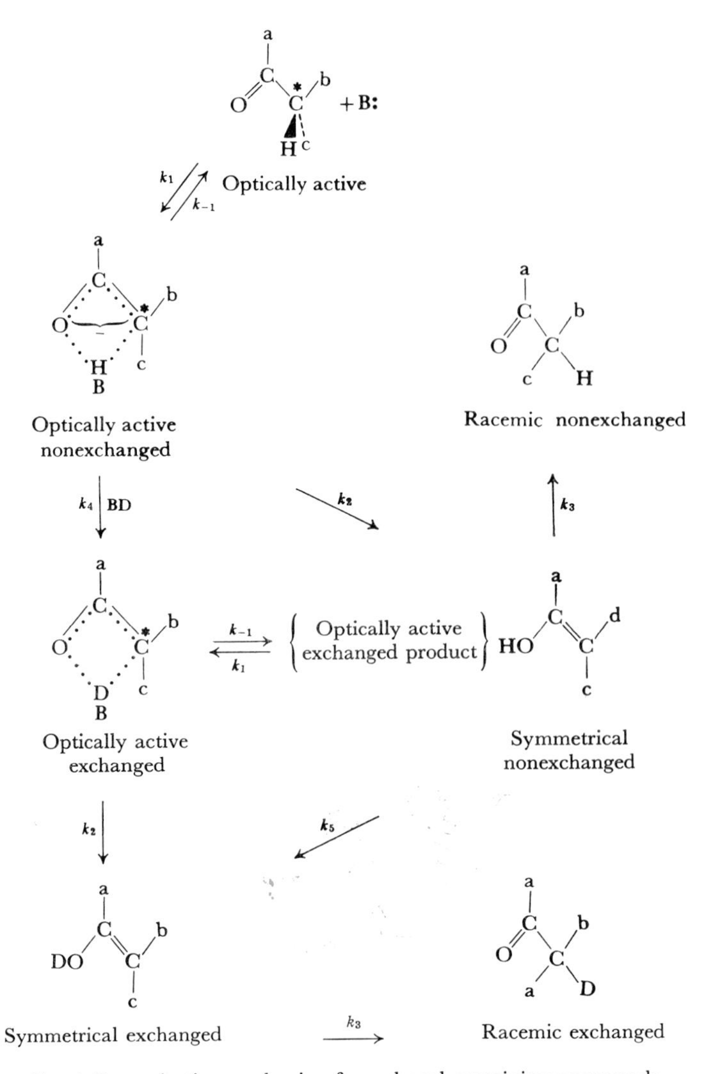

FIG. 4. Racemization mechanism for carbonyl-containing compounds.

and as a result k_e/k_α values of unity are observed. In terms of Figure 4, in which this racemization mechanism is formulated, $k_2 \gg k_{-1}$ and $k_5 \gg k_3$. Should $k_3 > k_5$, k_e/k_α would be less than unity, and isoracemization would result, which is not observed in any of these systems. Should k_4 and $k_{-1} \gg k_2$, k_e/k_α would be greater than unity, and exchange with retention would result.

The question arises as to whether benzyl anions have ambident character with respect to proton donors. In this connection, Russell quenched 2-phenyl-2-propylpotassium with various deuterium acids, and examined the cumene produced for deuterium in the benzene ring.[10] Deuteration at the benzyl position predominated over deuteration in the *p*-position of the ring by about the following rate factors: D_2O in ether, > 500; DOAc in ether, 65; DCl in ether, 4.5; DCl in dimethoxyethane, > 500. Deuteration in the *p*-position was greater than at the *o*-position of the ring. Although results of this sort are needed in many more systems, it seems probable that carbanion capture occurs predominantly at those sites that restore the aromatic character of the system.

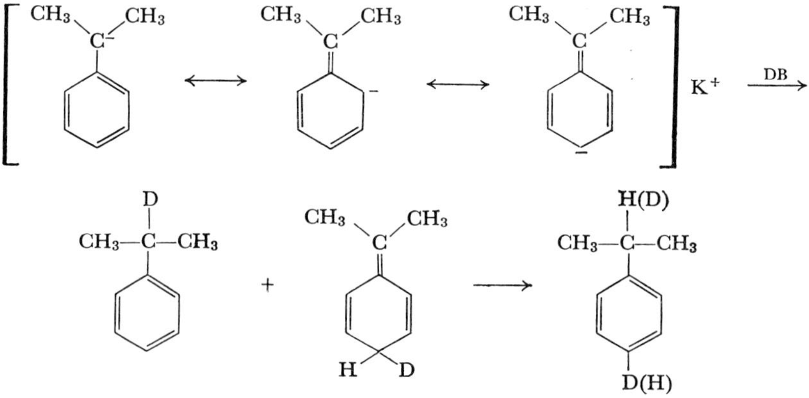

CH_3 CH_3 C^- ⟷ CH_3 CH_3 C ⟷ CH_3 CH_3 C K^+ $\xrightarrow{DB}$

D, CH_3—C—CH_3 + CH_3 CH_3 C, H D ⟶ H(D), CH_3—C—CH_3, D(H)

In systems whose carbanions are pyramidal and rapidly inverting, racemization mechanisms must be different. This subject is discussed in a future section.

Inversion Mechanism

Values of k_e/k_α which lie between 0.5 and 1 point to either of two general possibilities; first, appropriate blends of inversion and racemization or retention; second, proper blends of isoracemization and retention or racemization. In dissociating and proton-donating solvents, such as methanol or ethylene glycol, mixtures of inversion and racemization have been encountered in a number of systems. In non-dissociating solvents, such as *tert*-butyl alcohol or tetrahydrofuran, mixtures of isoracemization and retention or racemization have been found. The first type of process is discussed here, and the second is taken up in the next section.

[10] G. A. Russell, *J. Am. Chem. Soc.*, **81**, 2017 (1959).

TABLE II

Inversion and Racemization in Hydrogen Isotope Exchange

	Compound	Run no.	Solvent	Base	T (°C)	k_e/k_α
III	(fluorene, 9-D, 9-CH_3 *, $CON(CH_3)_2$)	1	CH_3OH	Pr_3N	75	0.65
		2	CH_3OH	CH_3OK	25	0.69
		3	CH_3OH–25% H_2O	Pr_3N	75	0.78
		4	$(CH_2OH)_2$	Et_3N	50	0.94
IX	$N{\equiv}C{-}\overset{*}{C}(C_2H_5)(C_6H_5){-}D(H)$	5	CH_3OH	Pr_3N	50	0.84
		6	$(CH_2OH)_2$	$KHCO_3$	25	0.87
X	$F_3C{-}\overset{*}{C}(C_2H_5)(C_6H_5){-}D(H)$	7	CH_3OH	CH_3OK	120	0.80
		8	CH_3OH	CH_3OLi	120	0.65

In Table II are reported results of base-catalyzed isotopic exchange studies carried out in methanol and ethylene glycol on compounds III, IX,[11] and X.[12] Values of k_e/k_α were obtained that ranged from 0.65 to 0.94. The same general results were obtained whether *tert*-amines, potassium carbonate, potassium methoxide, or lithium methoxide were used as bases. The results are interpreted in Figure 5.

In this scheme, deuterium is abstracted from carbon by either the amine or a free alkoxide anion at the same time that a solvent molecule becomes hydrogen bonded at the carbanion face remote from the leaving deuterium. The planar intermediate produced is hydrogen bonded at the back by a solvent molecule and at the front by either a trialkylammonium deuteride ion or a deuterated solvent molecule. Deuterium capture at the front leads to unaltered starting material, whereas proton capture at the back gives inverted-exchanged product. Replacement of the species hydrogen bonded at the front by a solvent molecule gives a

[11] D. J. Cram and L. Gosser, *J. Am. Chem. Soc.*, **86**, 5457 (1964).
[12] D. J. Cram and A. S. Wingrove, *J. Am. Chem. Soc.*, **86**, 5490 (1964).

symmetrically solvated anion, which can give only racemic-exchanged product.

Methanol and ethylene glycol dissociate their respective lithium and potassium alkoxides.[13] Therefore these solvents are expected to promote the ion pair dissociative processes when the asymmetrically solvated anion (amine as base) either collapses to inverted product or goes to

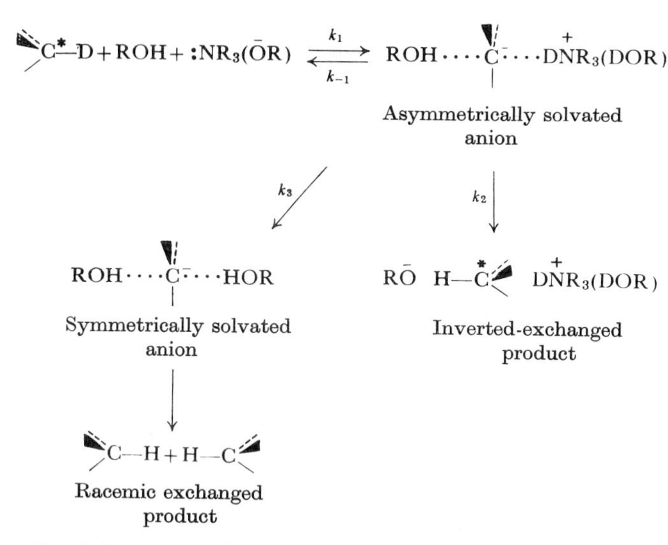

FIG. 5. Inversion and racemization in competition with each other.

symmetrically solvated carbanion. The fact that values for k_e/k_α are higher than 0.5 but less than unity indicates that $k_2 \sim k_3$. Since collapse to inverted product is a process similar to collapse to starting material, $k_2 \sim k_{-1}$ when alkoxide is base. When *tert*-amine is base, it is likely that $k_{-1} \gg k_2$, but, of course, this would be difficult to demonstrate.

It seems unlikely that retention and inversion mechanisms compete in dissociating solvents such as methanol or ethylene glycol, since net retention has never been observed in these kinds of media.

In these inversion solvents, the ΔpK_a ($\sim$ 5 units) for the medium and the carbon acid is less than that required to produce retention in solvents of low dielectric constant. This difference in ΔpK_a requirement for stereospecificity may reflect differences in rate for the competing racemization reactions rather than for the proton-capturing processes.

[13] D. J. Cram, B. Rickborn, C. A. Kingsbury, and P. Haberfield, *J. Am. Chem. Soc.*, **83**, 3678 (1961).

Isoracemization Mechanism

Values of k_e/k_α of less than 0.5 indicate the presence of a mechanism in which racemization occurs faster than exchange. Operation of such a mechanism is not surprising when an optically active, deuterated carbon acid is treated with a non-proton-donating base, such as tripropylamine in the absence of any proton pool. For example, when fluorene system III was heated to 145° with an 0.5 *M* solution of triethylamine in dry tetrahydrofuran, a k_e/k_α value of 0.23 was observed. The presence of a trace of adventitious moisture probably accounts for the fact that a small amount of exchange occurred. Aside from such moisture, the only acids strong enough to protonate the carbanion were other molecules of the carbon acid itself, or the deuterated triethylammonium ions. Probably triethylammonium carbanide ion pairs were generated, but usually collapsed back to starting material. Occasionally, the ion pair dissociated enough so that the carbanion could turn over and capture a proton from the triethylammonium ion on the back face of the carbanion. In other words, $k_{-1} \gg k_2$.

$$(a)(b)(c)C\text{—}D + :NEt_3 \underset{k_{-1}}{\overset{k_1}{\rightleftharpoons}} (a)(b)(c)\bar{C}\cdots D\overset{+}{N}Et_3 \underset{}{\overset{k_2}{\rightleftharpoons}} (a)(c)(b)\bar{C}\cdots D\overset{+}{N}Et_3 \xrightarrow{k_{-1}} (a)(c)(b)C\text{—}D + :NEt_3$$

This experiment, carried out in the absence of a proton pool, provides little information about the real character of the racemization process. It could occur either by a process of ion pair dissociation, or by an *ion pair reorganization* reaction. The term *isoracemization* is reserved for the latter process, and has been detected in experiments with the three systems, IV,[7] IX,[11] and XI.[14] Table III contains the relevant data.

Runs 1, with deuterated fluorene system IV, and 6, with deuterated nitrile IX, were carried out in tetrahydrofuran, 1.5 *M*, in *tert*-butyl alcohol in the presence of tripropylammonium iodide and tripropylamine as base. Values of k_e/k_α of 0.1 and 0.19 were obtained, respectively. In runs 2 and 5, tetrabutylammonium iodide replaced the deuterated tripropylammonium iodide, with little effect on k_e/k_α values. In runs 4 and 3, deuterated nitrile IX was racemized with tripropylamine in tetrahydrofuran, 1.5 *M*, in *tert*-butyl alcohol, and in *tert*-butyl alcohol. Values of k_e/k_α of 0.05 and 0.2 were obtained, respectively.

The racemization undoubtedly occurs through deuterated tripropylammonium carbanide ion pairs as intermediates. These racemize and collapse to covalent product faster than they either dissociate or react

[14] D. J. Cram and H. P. Fischer, unpublished results.

TABLE III

Isoracemization in Hydrogen Isotope Exchange

Compound	Run no.	Medium	Base	k_e/k_α
IV (O_2N, D, *, CH_3, $CON(CH_3)_2$)	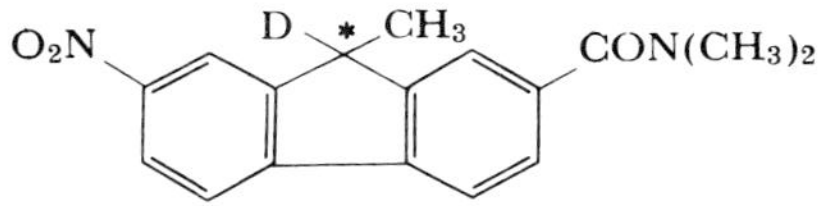 1	$(CH_2)_4O$–1.5*M* *tert*-BuOH–10^{-4}*M* Pr_3NHI	Pr_3N	0.1
	2	$(CH_2)_4O$–1.5*M* *tert*-BuOH–10^{-4}*M* *n*-Bu_4NI	Pr_3N	0.1
IX ($N{\equiv}C{-}C(C_2H_5)(C_6H_5){-}D$)	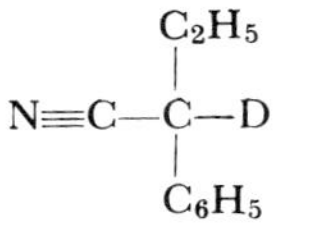 3	*tert*-BuOH	Pr_3N	0.2
	4	$(CH_2)_4O$–1.5*M* *tert*-BuOH	Pr_3N	0.05
	5	$(CH_2)_4O$–1.5*M* *tert*-BuOH–0.1*M* *n*-Bu_4NI	Pr_3N	0.09
	6	$(CH_2)_4O$–1.5*M* *tert*-BuOH–0.1*M* Pr_3NHI	Pr_3N	0.19
XI ($N{\equiv}C{-}C^{*}(CH_3)(C(C_6H_5)_3){-}H$)	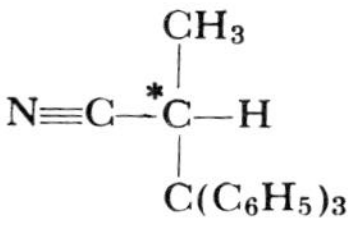 7	$(CH_2)_4O$–1.1*M* *tert*-BuOD	*tert*-BuOK	0.52
	8	C_6H_6–1.1*M* *tert*-BuOD	*tert*-BuOK	0.51
	9	$(CH_2)_6$–0.5*M* *tert*-BuOD	*tert*-BuOK	0.50

directly with their isotopic counterparts in the medium, tripropylammonium iodide or *tert*-butyl alcohol molecules. The rate of exchange of deuterated tripropylammonium iodide with *tert*-butyl alcohol in the presence of tripropylamine was undoubtedly fast enough to prevent accumulation of any deuterated salt. The rate constant for exchange of protons between ammonia and ammonium ions in water at 25° has been found to be 10^9 sec.$^{-1}$ M^{-1}.[15] The reaction at hand is of similar character.

In runs 7–9, nitrile XI was submitted to the action of potassium *tert*-butoxide in tetrahydrofuran, benzene, or cyclohexane containing a reservoir of *tert*-butyl alcohol-*O-d*. Values of $k_e/k_\alpha = 0.50$–0.52 were obtained. Again, an isoracemization of the same general mechanism as described above must have been operative.

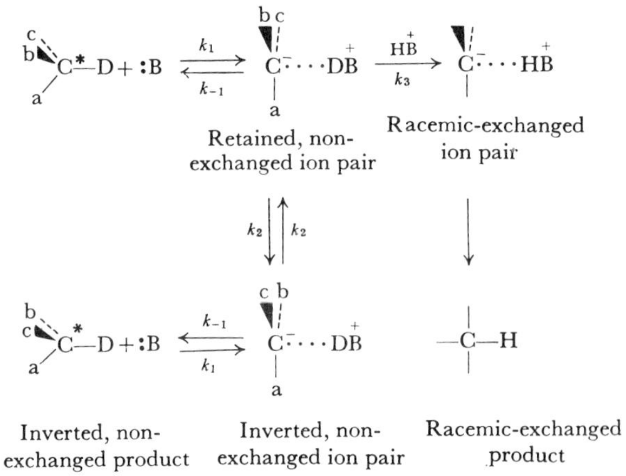

$$\text{abc}\overset{*}{\text{C}}\text{—D} + \text{:B} \underset{k_{-1}}{\overset{k_1}{\rightleftarrows}} \text{abc}\bar{\text{C}}\cdots\overset{+}{\text{D}}\text{B} \xrightarrow[k_3]{\overset{+}{\text{H}}\text{B}} \text{abc}\bar{\text{C}}\cdots\overset{+}{\text{H}}\text{B}$$

Retained, non-exchanged ion pair — Racemic-exchanged ion pair

k_2 ⇅ k_2 ↓

$$\text{abc}\overset{*}{\text{C}}\text{—D} + \text{:B} \underset{k_1}{\overset{k_{-1}}{\leftrightarrows}} \text{acb}\bar{\text{C}}\cdots\overset{+}{\text{D}}\text{B} \qquad \text{—C—H}$$

Inverted, non-exchanged product — Inverted, non-exchanged ion pair — Racemic-exchanged product

FIG. 6. General isoracemization mechanism.

The general mechanism for isoracemization is formulated in Figure 6. In this scheme, the data are accommodated only if k_{-1} and $k_2 \gtrsim k_3$. In other words, ion pair racemization and collapse to the covalent state occur faster than ion pair cation exchange with cations of the opposite isotopic type in the medium. What cationic exchange does occur could do so directly between two ion pairs, or by ion pair dissociative processes. When the conditions of run 2 (Table III) were duplicated, except that the concentration of tetrabutylammonium iodide was increased from

[15] (a) M. T. Emerson, E. Grunwald, and R. A. Kromhout, *J. Chem. Phys.*, **33**, 547 (1960); (b) E. Grunwald, P. J. Karabatsos, R. A. Kromhout, and E. L. Purlee, *ibid.*, **33**, 556 (1960).

10^{-4} to 0.03 *M*, the rate of exchange increased markedly, and $k_e/k_\alpha = 1$.[7] This suggests that exchange occurs largely by cationic exchange reactions between ion pairs. However, the fact that some exchange occurs in the absence of added salt suggests that a small amount of exchange could involve ion pair dissociation.

Two more intimate mechanisms for isoracemization can be envisioned. In the first, the carbanion simply rotates 180° with respect to its counterion around any axis that lies in the plane of the carbanion. The second, which fulfills the conditions of the first but is more specific, provides a means for rotation without the hydrogen bonds between the two ions being completely destroyed. In this scheme, which may be called a "conducted tour mechanism," the deuterium or hydrogen originally attached to carbon becomes hydrogen bonded, or even covalently bonded to oxygen or nitrogen to form tautomers as discrete intermediates. These intermediates are symmetrical, and revert to racemized starting material without having undergone isotopic exchange. In other words, the basic catalyst takes hydrogen or deuterium on a "conducted tour" of the substrate from one face of the molecule to the opposite face, from one hydrogen bonding site to the next without ever wandering out into the medium. Although external molecules of the opposite isotopic variety may also hydrogen-bond other electron-rich sites during the conducted tour, reaction does not occur at these sites because charge would be separated should this occur. Charge would not be separated, but discharged (in the case of a *tert*-amine as base) by proton transfer to any site of the carbanion. The negative charge, in a sense, must partially follow the positive charge around the carbanion during the "conducted tour." These ideas can be equally well applied to the nitrile systems, IX and XI, or to the nitrofluorene system, IV. Figure 7 shows how this mechanism might apply to these systems.

Support for the conducted tour mechanism as applied to fluorene system IV is found in the observation that conversion of compound XII to its aromatic tautomer occurs 97% intramolecularly with tripropylamine in deuterated triethylcarbinol-*O-d*, 0.1 *M*, in tripropylammonium iodide.[16] Support for the conducted tour mechanism as applied to nitriles IX and XI derives from the fact that in dimethyl sulfoxide–methanol–triethylenediammonium iodide, acetylene XIII undergoes triethylenediamine-catalyzed isomerization to its allene tautomer with 88% intramolecularity. This and other intramolecular base-catalyzed allylic proton transfers are treated in Chapter V.

[16] D. J. Cram, F. Willey, H. P. Fischer and D. A. Scott, *J. Am. Chem. Soc.*, **86**, 5510 (1964).

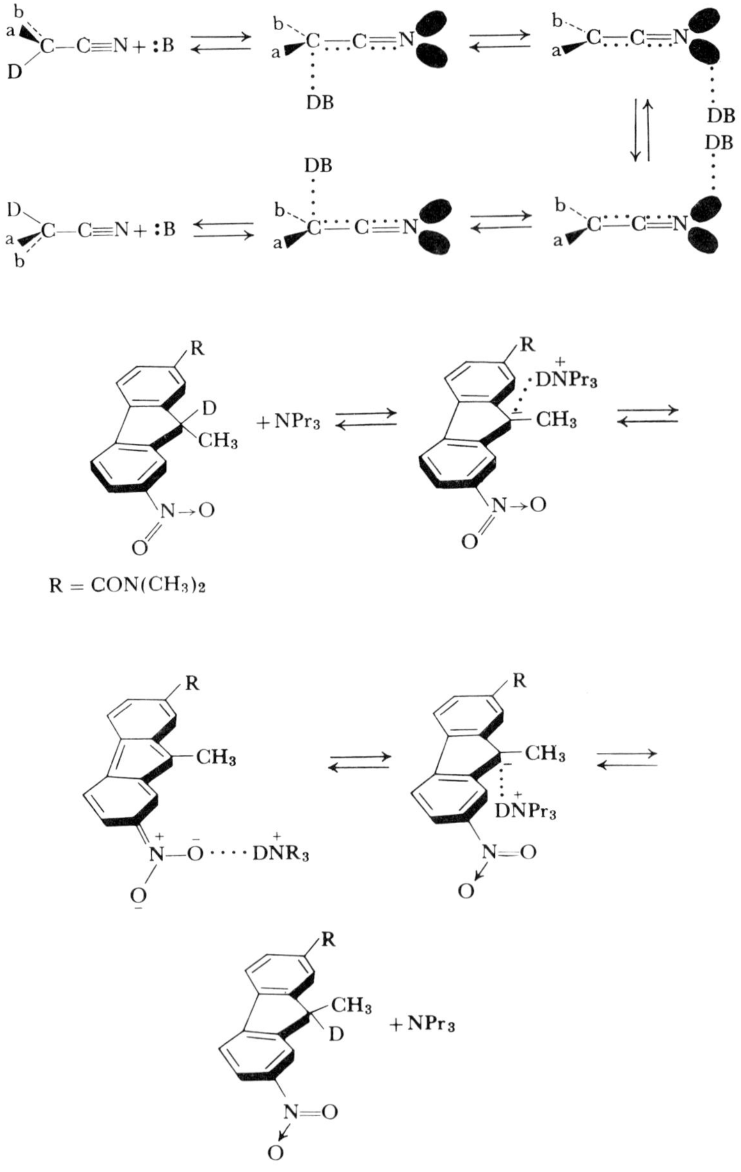

R = CON(CH$_3$)$_2$

FIG. 7. Conducted tour mechanisms for isoracemization.

$$\text{XII} \xrightarrow[\substack{(C_2H_5)_3COD \\ Pr_3NDI\ 65^\circ}]{Pr_3N} p\text{-}(C_6H_5)_2C{-}H\text{-}C_6H_4\text{-}C(CH_3)_2CO_2C_2H_5$$

$$\underset{\text{XIII}}{(C_6H_5)_2C(D){-}C{\equiv}C\,C_6H_5} \xrightarrow[\substack{(CH_3)_2SO{-}CH_3OH \\ N(CH_2CH_2)_3NHI}]{N(CH_2CH_2)_3N} (C_6H_5)_2C{=}C{=}CDC_6H_5$$

Effect of Structure on Mechanism

Comparison of the behavior of the two fluorene systems, III and IV, and nitrile IX in nondissociating solvents is instructive (see Table IV).[6,7,11] In tetrahydrofuran with propylamine as base, fluorene-amide (III) in run 1 gave high retention ($k_e/k_\alpha > 56$), whereas nitro-fluorene-amide (IV) and nitrile IX in runs 4 and 6 gave total racemization ($k_e/k_\alpha = 1$). In 90% benzene–10% phenol–potassium phenoxide, fluorene-amide (III) in run 2 again gave high retention ($k_e/k_\alpha = 18$), whereas nitrile IX in run 7 gave $k_e/k_\alpha = 0.76$. The latter value probably represents a combination of isoracemization and ordinary racemization (or retention). In *tert*-butyl alcohol–tripropylamine, fluorene-amide (III) in run 3 gave retention ($k_e/k_\alpha > 6$), whereas in tetrahydrofuran, 1.5 *M*, in *tert*-butyl alcohol–tripropylamine, nitro-fluorene-amide (IV) in run 5 gave isoracemization ($k_e/k_\alpha = 0.1$) and nitrile IX (run 8) gave isoracemization ($k_e/k_\alpha = 0.05$).

Clearly, fluorene-amide is predisposed toward retention and nitro-fluorene-amide and nitrile toward isoracemization mechanisms. In terms of the mechanistic schemes set forth above, it appears that in the ion pairs formed by the fluorene-amide system, whose cations contain proton donors, the cation is inclined to rotate faster than the anion (runs 1 and 2). In the corresponding ion pairs formed by the nitro-fluorene-amide or the nitrile systems, both cations and anions rotate at comparable rates to give net racemization or near racemization (runs 4, 6, and 7). In the ion pairs whose cations do not contain proton donors, the anions of the nitro-fluorene-amide and nitrile systems rotate faster than

TABLE IV

Competition Between Retention and Isoracemization Mechanisms

Compound	Run no.	Medium	Base	k_e/k_α
III (D, *, CH_3; $CON(CH_3)_2$)	1	$(CH_2)_4O$	$PrNH_2$	>56
	2	90% C_6H_6–10% C_6H_5OH	C_6H_5OK	18
	3	*tert*-BuOH	Pr_3N	>6
IV (O_2N; D, *, CH_3; $CON(CH_3)_2$)	4	$(CH_2)_4O$	$PrNH_2$	1.0
	5	$(CH_2)_4O$–1.5*M* *tert*-BuOH	Pr_3N	0.1
IX ($N{\equiv}C{-}\overset{*}{C}(C_2H_5)(C_6H_5){-}D$)	6	$(CH_2)_4O$	$PrNH_2$	1.0
	7	90% C_6H_6–10% C_6H_5OH	C_6H_5OK	0.76
	8	$(CH_2)_4O$–1.5*M* *tert*-BuOH	Pr_3N	0.05

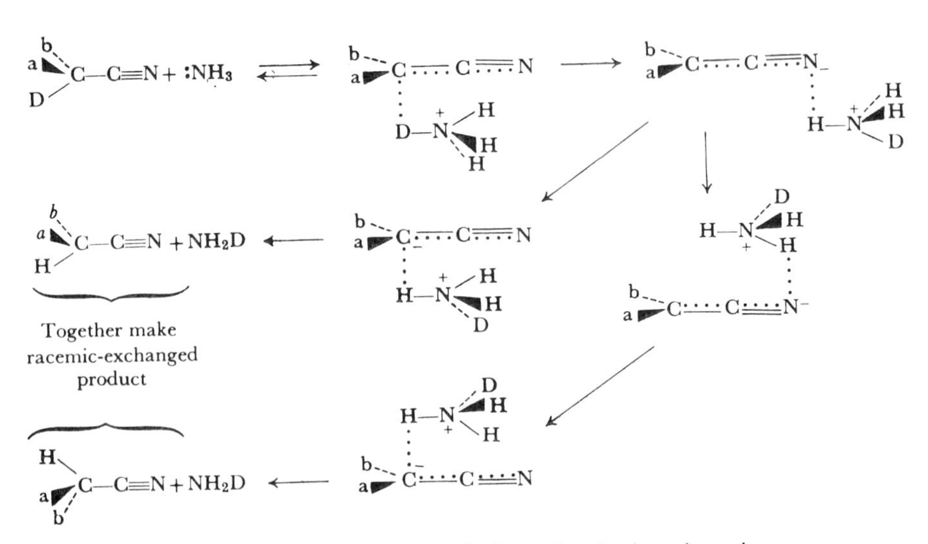

FIG. 8. Simultaneous rotation of anion and cation in an ion pair.

other competing processes occur, whereas with the anions of the fluorene-amide, no anion rotation is evident. Simultaneous rotation of anion and cation in an ammonium-cyanocarbanide ion pair is formulated in Figure 8 (nitrile IX with ammonia in tetrahydrofuran gave $k_e/k_\alpha = 1.2$).[11]

The data and interpretations indicate that in nondissociating solvents, the cyano and nitro as carbanion-stabilizing groups promote the iso-racemization "conducted tour mechanism," whereas the amide group promotes a retention mechanism, at least when attached to the fluorene ring. The reason for this difference is obscure.

SYSTEMS THAT FORM UNSYMMETRICAL CARBANIONS

In the previous section of this chapter, the stereochemical capabilities of intrinsically planar, symmetrical carbanions have been discussed. Asymmetry was induced either by ion pairing or by asymmetric solvation. In all cases, the carbanions were stabilized by groups capable of dispersing charge, in some cases over large unsaturated systems.

In this section, the behavior of carbanions stabilized by functional groups centered around second-row elements is treated. Such functional groups stabilize charge both by the inductive effect and by *d*-orbital involvement (see Chapter II). The fact that these carbanions exhibit properties different from those that are stabilized largely by *p*-orbital delocalization effects justifies their treatment in a separate section.

The first indication that carbanions stabilized by groups centered about second-row elements were different from those previously studied was the observation that optically active 2-octyl phenyl sulfone underwent base-catalyzed hydrogen-deuterium exchange with high retention, even in solvents of high dielectric constant previously classified as "racemization" or "inversion solvents."[17] In subsequent work, k_e/k_α values were carefully measured in methanol, dimethyl sulfoxide–methanol, ethylene glycol, *tert*-butyl alcohol,[18] and 67% ethanol–33% water.[19] A third set of investigators also reported examination of k_e/k_α values in methanol.[20] The values of this ratio of rates varied from a low of 10 in dimethyl sulfoxide–methanol and in methanol, to a high of 1980 in *tert*-butyl alcohol under the right conditions.[18]

A survey was made of seven other systems that contained the 2-octyl

[17] D. J. Cram, W. D. Nielsen, and B. Rickborn, *J. Am. Chem. Soc.*, **82**, 6415 (1960).
[18] D. J. Cram, D. A. Scott, and W. D. Nielsen, *J. Am. Chem. Soc.*, **83**, 3696 (1961).
[19] E. J. Corey and E. T. Kaiser, *J. Am. Chem. Soc.*, **83**, 490 (1961).
[20] H. L. Goering, D. L. Towns, and B. Dittmer, *J. Org. Chem.*, **27**, 736 (1962).

group attached to various carbanion-stabilizing groups centered around sulfur[21] and phosphorus.[21a,c,22] These results, as well as those obtained with the 2-octyl phenyl sulfone system, are recorded in Table V. In runs 1–7 with the sulfone system, k_e/k_α values vary by a factor of almost 200 as solvent and base are changed. Comparison of runs 2 and 3 reveals that substitution of tetramethylammonium hydroxide for potassium *tert*-butoxide in *tert*-butyl alcohol as solvent reduces k_e/k_α by a factor of 7–30. Comparison of runs 2, 6, and 7 indicates that substitution of methanol or dimethyl sulfoxide–methanol for *tert*-butyl alcohol reduces k_e/k_α values by factors of 14 to 100. In the systems that give planar carbanions (discussed earlier), these types of solvent and base changes reduced k_e/k_α values to unity or less. With the sulfone, the minimum value of the ratio is 10. Clearly the solvent and base effects visible with the planar carbanions are present in the sulfonyl carbanions, but they are superimposed on those of carbanions that are, themselves, asymmetric.

The eight systems represented in Table V fall into either of two categories: (1) those like the sulfone that give k_e/k_α values of 10 or higher;

TABLE V

Ratios of k_e (Isotopic Exchange Rate Constant) to k_α (Racemization Rate Constant) for 2-Octyl Systems Attached to Functional Groups Centered Around Sulfur and Phosphorus

$$\overset{*}{HR} = H-\overset{*}{C}(CH_3)-C_6H_{13}\text{-}n \qquad \overset{*}{DR} = D-\overset{*}{C}(CH_3)-C_6H_{13}\text{-}n$$

Run no.	Compound	Solvent	Base	T (°C)	k_e/k_α
1	$\overset{*}{HR}-SO_2-C_6H_5$	*tert*-BuOD	*tert*-BuOK	25	73–1200
2	$\overset{*}{DR}-SO_2-C_6H_5$	*tert*-BuOH	*tert*-BuOK	25	139–1980
3	$\overset{*}{DR}-SO_2-C_6H_5$	*tert*-BuOH	$(CH_3)_4NOH$	25	22–64
4	$\overset{*}{HR}-SO_2-C_6H_5$	$(CH_2OD)_2$	$DOCH_2CH_2OK$	100	15
5	$\overset{*}{DR}-SO_2-C_6H_5$	$(CH_2OH)_2$	$HOCH_2CH_2OK$	100	32
6	$\overset{*}{DR}-SO_2-C_6H_5$	CH_3OH	CH_3OK	100	10

[21] (a) D. J. Cram, R. D. Partos, S. H. Pine, and H. Jäger, *J. Am. Chem. Soc.*, **84**, 1742 (1962); (b) D. J. Cram and S. H. Pine, *ibid.*, **85**, 1096 (1963); (c) D. J. Cram, R. D. Trepka, and P. St. Janiak, *ibid.*, **86**, 2731 (1964).

[22] D. J. Cram and R. D. Partos, *J. Am. Chem. Soc.*, **85**, 1093 (1963).

TABLE V (*cont.*)

Run no.	Compound	Solvent	Base	T (°C)	k_e/k_α
7	$D\overset{*}{R}—SO_2—C_6H_5$	$(CH_3)_2SO—CH_3OH$	CH_3OK	25	10
8	$H\overset{*}{R}—SO_2—N(CH_3)C_6H_5$	*tert*-BuOD	*tert*-BuOK	25	~19
9	$D\overset{*}{R}—SO_2—N(CH_3)C_6H_5$	*tert*-BuOH	*tert*-BuOK	25	~37
10	$D\overset{*}{R}—SO_2—N(CH_3)C_6H_5$	$(CH_2OH)_2$	$HOCH_2CH_2OK$	100	~17
11	$D\overset{*}{R}—SO_2—N(CH_3)C_6H_5$	$(CH_3)_2SO—CH_3OH$	CH_3OK	25	~34
12	$H\overset{*}{R}—SO_2—OCH_3$	*tert*-BuOD	*tert*-BuOK	25	$\gtrsim 28$
13	$H\overset{*}{R}—PO_2KC_6H_5$	*tert*-BuOD	*tert*-BuOK	225	~17
14	$D\overset{*}{R}—SO—C_6H_5$	*tert*-BuOH	*tert*-BuOK	60	~2.4
15	$D\overset{*}{R}—SO—C_6H_5$	$(CH_3)_2SO—CH_3OH$	CH_3OK	60	~1.0
16	$D\overset{*}{R}—SO_3K$	*tert*-BuOH	*tert*-BuOK	146	1.6
17	$D\overset{*}{R}—SO_3K$	*tert*-BuOD	*tert*-BuOK	146	1.7
18	$H\overset{*}{R}—PO(C_6H_5)_2$	*tert*-BuOD	*tert*-BuOK	100	3.3
19	$D\overset{*}{R}—PO(C_6H_5)_2$	*tert*-BuOH	*tert*-BuOK	100	3.4
20	$H\overset{*}{R}—PO(C_6H_5)_2$	$(CH_2OD)_2$	$DOCH_2CH_2OK$	175	1.2
21	$D\overset{*}{R}—PO(C_6H_5)_2$	$(CH_2OH)_2$	$HOCH_2CH_2OK$	175	1.4
22	$D\overset{*}{R}—PO(C_6H_5)_2$	CH_3OH	CH_3OK	75	1.1
23	$D\overset{*}{R}—PO(C_6H_5)_2$	$(CH_3)_2SO—CH_3OH$	CH_3OK	75	1.0
24	$H\overset{*}{R}—PO(OC_2H_5)_2$	*tert*-BuOD	*tert*-BuOK	100	1.2

(2) those that provide values not far from unity. What little stereospecificty is observed for the latter group seems to be associated with ion pair or solvent effects, and these seem to be less important than those discussed earlier which involved symmetrical carbanions. In fact, no examples of exchange occurring with net inversion of configuration were found. One can conclude that the symmetry properties of all these carbanions are somewhat different from those of the symmetrical

carbanions discussed earlier, but that the difference is more pronounced for one class than for the other.

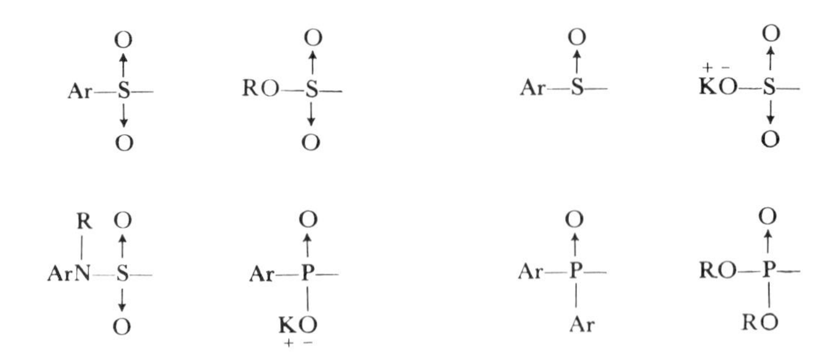

Functional groups that induce carbanion asymmetry

Functional groups that visibly do not induce carbanion asymmetry

Comparison of the structural features of these two classes of groups indicates that those that induce asymmetry have two unsubstituted oxygens bound to the second-row element, and those that do not induce observable carbanion asymmetry have either one or three unsubstituted oxygens attached to the second-row element. The presence or absence of a formal negative charge on one of these oxygens seems to make no qualitative difference.

Production of the sulfonyl carbanion by use of carbon as leaving group gives results similar to those obtained with hydrogen or deuterium as leaving group.[23] The base-catalyzed decarboxylation reaction as well as the alkoxide cleavage gave 2-octyl phenyl sulfone of 98–100% optical purity, irrespective of what solvent-base system was used. This result was observed in *tert*-butyl alcohol, ethylene glycol, methanol, and water with the corresponding potassium bases, as well as with dimethyl sulfoxide–methanol with either potassium methoxide or tetramethylammonium hydroxide as base. Although the relative configurations of the starting materials and products in these reactions have not been determined, it is safe to say the reaction occurred with retention of configuration by analogy with hydrogen as the leaving group.

Two other experiments have a bearing on the configuration of the anion of the sulfone.[24] Comparison of the values of k_e/k_α for sulfones XIV

[23] (a) D. J. Cram and A. S. Wingrove, *J. Am. Chem. Soc.*, **84**, 1496 (1962); (b) D. J. Cram and A. S. Wingrove, *ibid.*, **85**, 1100 (1963). See also reference 24.

[24] E. J. Corey, H. Konig, and T. H. Lowry, *Tetrahedron Letters*, **12**, 515 (1962).

and XV in 2-to-1 ethanol–water provided values not far from each other. Thus steric effects, at least at these levels of structural ramification, have only a minor effect on the stereospecificity of the exchange reaction.

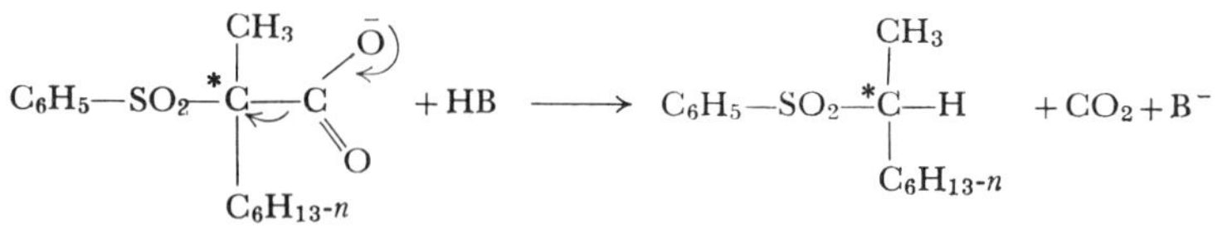

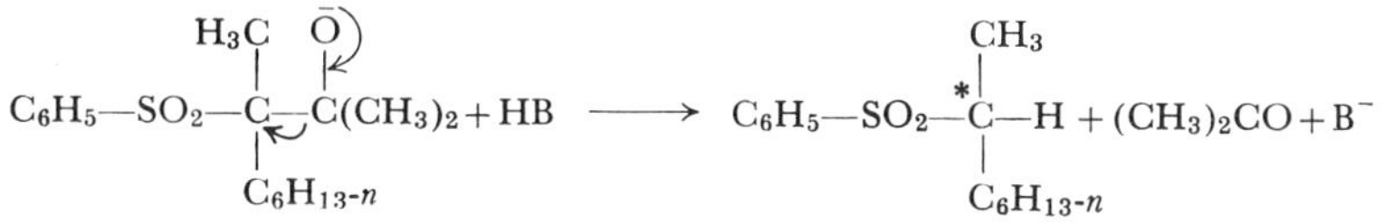

$$C_6H_5-SO_2-\overset{*}{C}(CH_3)(C_6H_{13}\text{-}n)-CO_2^- + HB \longrightarrow C_6H_5-SO_2-\overset{*}{C}H(CH_3)(C_6H_{13}\text{-}n) + CO_2 + B^-$$

$$C_6H_5-SO_2-C(H_3C)(C_6H_{13}\text{-}n)-C(O^-)(CH_3)_2 + HB \longrightarrow C_6H_5-SO_2-\overset{*}{C}H(CH_3)(C_6H_{13}\text{-}n) + (CH_3)_2CO + B^-$$

In the second experiment, optically active acid XVI was found to decarboxylate to give racemic XVII at 150–180° in ethanol–water.

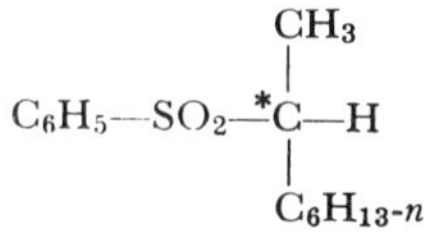

$$C_6H_5-SO_2-\overset{*}{C}H(CH_3)(C_6H_{13}\text{-}n)$$

XIV

$k_e/k_a = 42$

$$C_6H_5-SO_2-CH(CH_3)(C(CH_3)_3)$$

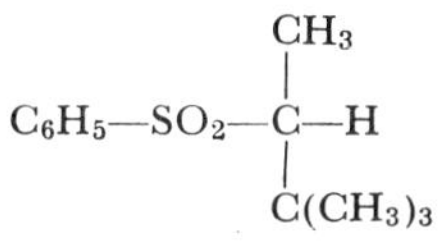

XV

$k_e/k_a = 59$

Unfortunately, neither the acid XVI nor the product XVII was demonstrated to be optically stable under the conditions of this experiment. Thus conclusions derived from the experiment should be accepted with reservations. It is conceivable that at 180°, the salt of XVI racemizes via a biradical faster than decarboxylation occurs. Attention is called to the racemization of system XVIII through radical cleavage and recombination, a process that competes with ionic cleavage.[25]

These data taken together suggest that carbanions stabilized by functional groups centered around sulfur and phosphorus are themselves asymmetric, at least if not confined within a ring system. When the functional group contains two unsubstituted oxygens, the rate of

[25] D. J. Cram, A. Langemann, W. Lwowski, and K. R. Kopecky, *J. Am. Chem. Soc.*, **81**, 5760 (1959).

racemization of the anion is low compared with the rate of proton capture, and high net retention is observed in all solvent-base systems. When this condition is not fulfilled, the rate of racemization of the anion is high compared to the rate of proton capture.

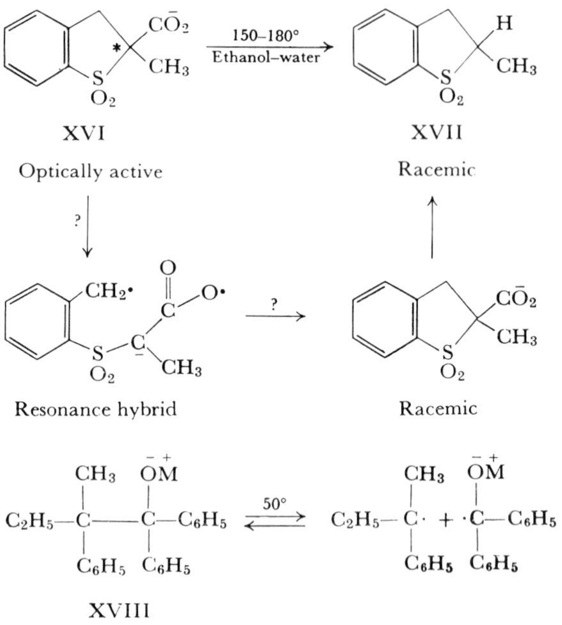

It was concluded in Chapter II that overlap between the *d*-orbitals of second-row elements and the orbital occupied by the two electrons of the carbanion did not have configurational or conformational requirements that outweighed electrostatic or solvation requirements. Two general electrostatic explanations are available to account for the asymmetry of these carbanions.[23] In the first, the carbanions are presumed to be pyramidal, and the presence on the adjacent atom of two negative oxygens provides an electrostatic barrier to inversion great enough to depress carbanion racemization rates and make proton capture the faster process. The presence of either one or three negative oxygens on the functional group provides less electrostatic driving force for a pyramidal configuration to start with, and less of an electrostatic barrier to inversion of a pyramidal carbanion (see Figure 9).

In the second explanation, the carbanion is presumed to be trigonal, and asymmetric in those systems that contain two equivalent negative oxygens. The asymmetry is due to conformation, and that conformation which minimizes electrostatic repulsion is selected by the system for both formation and consumption of the carbanion. Both electrostatic and steric effects inhibit rotation of the carbanion to give a symmetrical

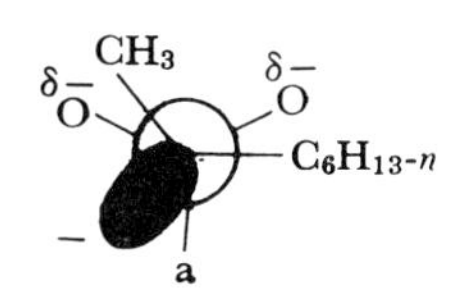

Conformation and hybridization that minimize electrostatic repulsion

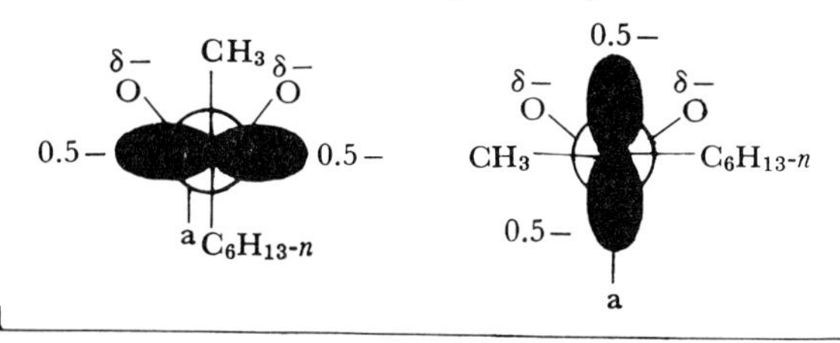

Possible transition states for pyramidal carbanion inversion; charges concentrated compared to ground state

δ−
O
CH3
C6H13-n
a
a
−

Conformation and hybridization that minimize electrostatic repulsion

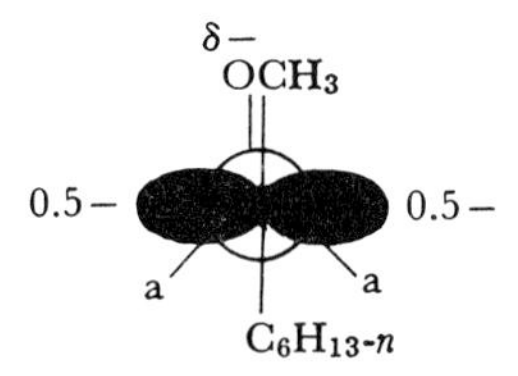

Transition state that minimizes electrostatic repulsion; not much different electrostatically from the ground state

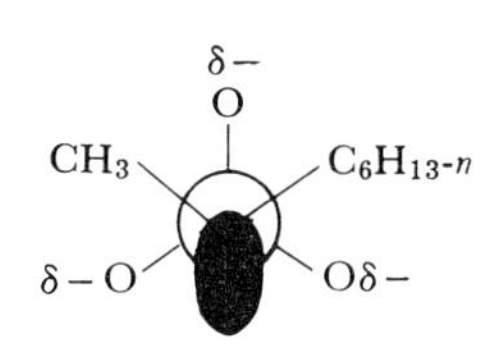

One of 3 conformations and the hybridization that minimizes electrostatic repulsion

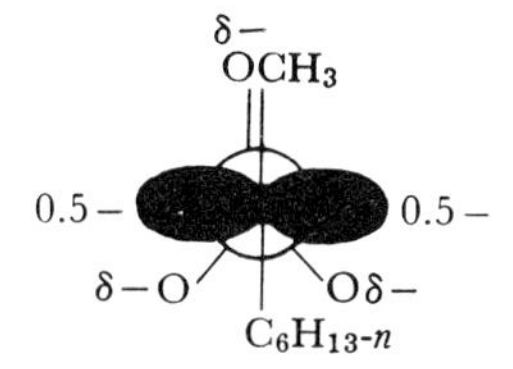

One of 3 transition states that minimizes electrostatic repulsion; not much different electrostatically from ground state

FIG. 9. Pyramidal carbanions stabilized by functional groups centered around second-row elements.

conformation, and proton capture becomes faster than racemization. In those systems that contain either one or three unsubstituted oxygens, the electrostatic driving force for forming the carbanion in an asymmetric conformation disappears. When only one oxygen is present, even if an asymmetric carbanion were first formed, the electrostatic barrier to rotation to a symmetrical conformation is expected to be small, and rotation to be faster than proton capture (see Figure 10).

Data are not yet available which allow these two mechanisms to be differentiated. The fact that exchanges occur with about the same values of k_e/k_α for compounds XIV and XV, even though a *tert*-butyl is substituted for a *n*-hexyl group on the carbanion, can be interpreted in

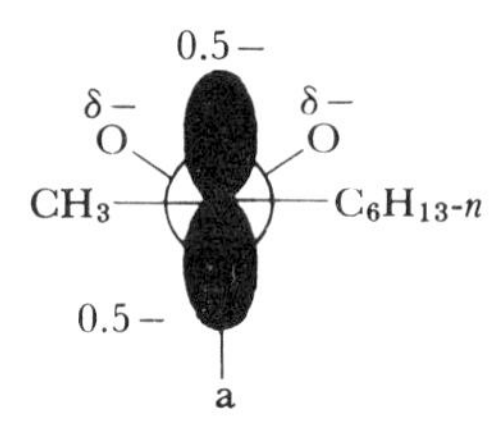

Asymmetric conformation that minimizes electrostatic repulsion in transition state for carbanion formation and consumption

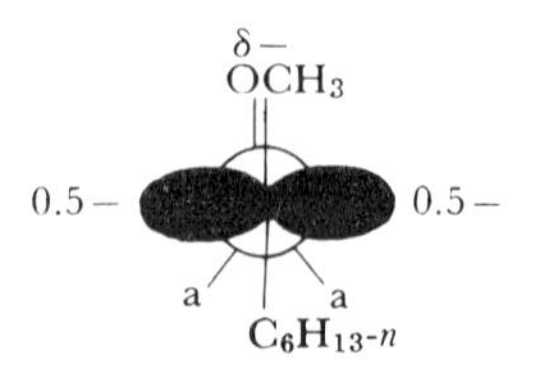

Symmetric conformation that minimizes electrostatic repulsion in transition state for carbanion formation and consumption

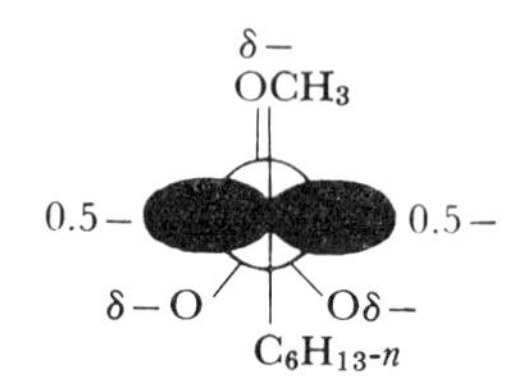

Symmetric conformation that minimizes electrostatic repulsion in transition state for carbanion formation and consumption

FIG. 10. Trigonal carbanions stabilized by functional groups centered around second-row elements.

terms of either mechanism. If the carbanion is pyramidal, then inversion rates are expected to be speeded up by B-strain effects, but slowed down by back-side steric inhibition of solvation effects, and these two effects might balance each other. If the carbanion is trigonal, steric inhibition of rotation should be larger for the *tert*-butyl than for the *n*-hexyl group, but the effect might be small.

If controls prove that XVI does decarboxylate with complete racemization, then each of the above mechanisms provides an equally good explanation. The five-membered ring imposes a conformation for the transition state that is electrostatically unfavorable. If the carbanion is pyramidal, electrostatic inhibition of inversion of the pyramid should be low in the enforced conformation, and the carbanion racemization rate higher than that for proton capture. If the carbanion is planar, a symmetrical conformation is enforced, and racemic product results.

SPECIAL CASE OF CARBANIONS CONFINED IN THREE-MEMBERED RINGS

Nuclear magnetic resonance studies on *N*-ethylethyleneimine systems demonstrated that the rate of inversion of the unshared pair of electrons on nitrogen was vastly reduced by confinement of the nitrogen in the three-membered ring.[26] Unsaturated groups connected to the nitrogen caused the inversion rates to increase, presumably because the planar transition state is stabilized relative to the nonplanar ground state by electron delocalization of the electron pair on nitrogen into the unsaturated system. Furthermore, hydrogen bonding solvents such as water or alcohols substantially reduced the inversion frequencies of the amine, which indicates that hydrogen bonds tend to anchor the unshared electron pair on nitrogen.

The reason that ethyleneimines invert more slowly than do their open-chain counterparts is associated with the angle strain inherent in a three-membered ring. The transition state for inversion of ammonia involves sp^2 hybridization of nitrogen with a H—N—H bond angle of 120°. Confinement of the nitrogen in a three-membered ring reduces the normal bond angles of 109° to 60°. The transition state for inversion of an ethyleneimine involves reducing a normal bond angle of 120° to 60°. Hence, angle strain should be considerably greater in the transition state for inversion than in the ground state. The resulting increased activation energy for inversion of a nitrogen in a three-membered ring makes such inversion slower than in an open-chain system where this effect is absent.[27]

These effects have been demonstrated to carry over to carbanions[28a] which are isoelectronic with amines. Although the k_e/k_α values of open-chain nitrile XIX in methanol-*O-d*–potassium methoxide is equal to unity,[22] the value for this ratio of rate constants is 8,080 for the cyclic

[26] A. T. Bottini and J. D. Roberts, *J. Am. Chem. Soc.*, **78**, 5126 (1956); *ibid.*, **80**, 5203 (1958).

[27] J. F. Kincaid and F. C. Henriques, Jr., *J. Am. Chem. Soc.*, **62**, 1474 (1940).

[28] (a) H. M. Walborsky, A. A. Youssef, and J. M. Motes, *J. Am. Chem. Soc.*, **84**, 2465 (1962); (b) H. M. Walborsky, private communication.

nitrile XX.[28] These experiments suggest that the carbanion derived from the open-chain nitrile is made planar and symmetrical by delocalization of the electron pair onto nitrogen. As indicated in earlier sections, protonation on nitrogen may occur more rapidly than on carbon. The

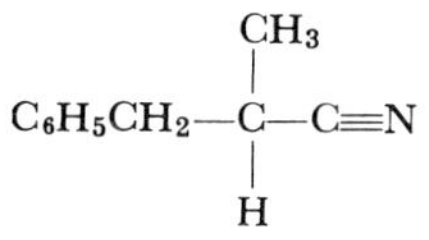

$C_6H_5CH_2—C(CH_3)(H)—C{\equiv}N$

XIX

$(C_6H_5)_2C$ / $CH_2—C(H)—C{\equiv}N$

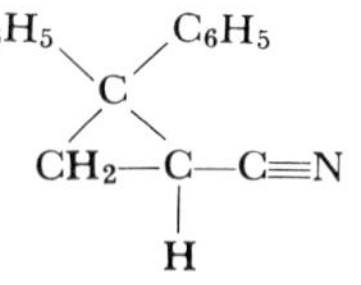

XX

carbanion derived from cyclic nitrile XX appears to retain its pyramidal character, with its inversion rate being slower than its rate of proton capture. In dimethyl sulfoxide–potassium methoxide, $k_e/k_\alpha \approx 60$.[28b]

HOMOCONJUGATED CARBANIONS

Optically active camphenilone XXI was observed to undergo base-catalyzed hydrogen–deuterium exchange (rate constant k_e) at about the same rate as the system racemized (rate constant k_α).[29] The existence of these two processes indicated the generation of carbanion intermediates, and the fact that $k_e/k_\alpha = 1$ indicated that the two processes were linked

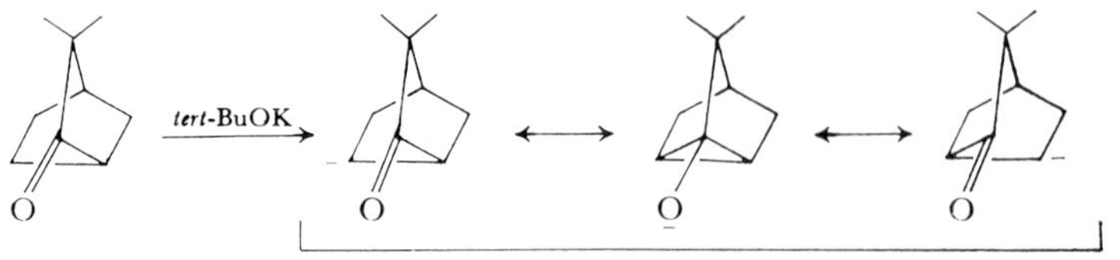

tert-BuOK

XXI
Optically active

XXII
Symmetrical

tert-BuOD

Racemic-deuterated product

to a common, symmetrical intermediate, the homoenolate anion XXII, which was visualized as being a bridged anion with contributing open ion resonance structures.

Confirmation of this interpretation was found in a stereochemical study in which 1-acetoxynortricyclene (XXIII) was converted to 6-deutero-2-norbornanone (XXIV) by hydrolysis in basic deuterated

[29] A. Nickon and J. L. Lambert, *J. Am. Chem. Soc.*, **84**, 4604 (1962).

alcohol.[30] The deuterium was demonstrated to enter the molecule from the *exo*-side with at least 95% stereospecificity in *tert*-butyl alcohol-*O-d*–potassium *tert*-butoxide, in *tert*-butyl alcohol-*O-d*–tetramethylammonium hydroxide-*O-d*, in methanol-*O-d*–potassium methoxide, or in one-to-one dimethyl sulfoxide–methanol-*O-d*–potassium methoxide. Thus, in this polycyclic system, electrophilic substitution occurred with very high inversion, and the steric course did not vary with solvent and base. The carbonyl group is enough bonded to the carbanion at its front face to force proton capture to occur at the back face of the anion. This

OAc KOR/DOR → D H O

XXIII XXIV

result implies that proton abstraction from XXI by base should occur largely from the *exo*-side, and that the carbonyl group should aid in the proton removal.

β α OD D_2SO_4 / AcOD—D_2O → H D O

XXV XXVI

In contrast to this stereochemical result, treatment of alcohol XXV with deuterated sulfuric acid in deuterated acetic acid–deuterium oxide gave XXVI,[30] in which deuterium had entered the molecule from the *endo*-side with at least 90% stereospecificity.[30] In this reaction, electrophilic substitution occurred with high retention of configuration. Quite possibly the conjugate acid of XXV serves as an intermediate in this reaction, and provides an internal electrophile for attack on the α,β-bonding electrons.

In a previous study, *cis*-1-methyl-2-phenylcyclopropanol (XXVII) was found to isomerize to 4-phenyl-2-butanone (XXVIII) by different stereospecific paths in acidic and basic solutions.[31] The authors indicated that the base-catalyzed reaction probably proceeded with inversion, and the acid-catalyzed reaction with retention. The subsequent work of

[30] A. Nickon, J. H. Hammons, J. L. Lambert, and R. O. Williams, *J. Am. Chem. Soc.*, **85**, 3713 (1963).

[31] C. H. DePuy and F. W. Breitbeil, *J. Am. Chem. Soc.*, **85**, 2176 (1963).

Nickon *et al.*[30] in system XXV supports this conclusion. Thus, these two stereospecific cleavage reactions are not unique properties of the polycyclic system XXV, but appear to be associated with the three-membered cyclanols.

$$\underset{\text{XXVII}}{C_6H_5(H)\overset{*}{C}{-}CH_2{-}C(OH)(CH_3)\ (\text{cyclopropane})} \xrightarrow[\text{or DCl, } D_2O\text{, dioxane}]{\text{NaOD, } D_2O\text{, dioxane}} \underset{\text{XXVIII}}{C_6H_5{-}\overset{*}{C}(H)(D){-}CH_2{-}C(=O){-}CH_3}$$

STEREOCHEMISTRY OF SUBSTITUTION OF ORGANOMETALLIC COMPOUNDS

Study of the stereochemistry of substitution reactions of organometallic compounds was long plagued by the fact that attempts to form optically active organometallics in which the metal was bonded to a single asymmetric carbon resulted in racemic products.[32] Many of the organometallic compounds of interest are highly reactive, and cannot be isolated and their stereochemical structure examined. Therefore, they must be prepared and consumed immediately. As a consequence, although the overall stereochemical course of preparation and reaction of the substances could be studied, the stereochemical course of each individual reaction was long delayed.

Letsinger's[2] preparation of optically active 2-octyllithium by metalation of 2-octyl iodide was the first reported preparation of an optically active organometallic compound that contained a single asymmetric center bound to the metal. This compound racemized completely when warmed to 0°, a fact that suggests a carbanion intermediate. At −70°, the combined metalation and carbonation steps produced 20% overall retention of configuration. In principle, this result could reflect either two retention or two inversion stages, but recent work indicates that each stage occurred with net retention.

Organomercury compounds can be isolated and handled, and therefore are more ideal substrates for stereochemical studies. However, the very feature that makes organomercury compounds convenient to

[32] (a) R. H. Prichard and J. Kenyon, *J. Chem. Soc.*, **99**, 65 (1911); (b) A. M. Schwartz and J. R. Johnson, *J. Am. Chem. Soc.*, **53**, 1063 (1931); (c) C. W. Porter, *ibid.*, **57**, 1436 (1935); (d) D. S. Tarbell and M. Weiss, *ibid.*, **61**, 1203 (1939); (e) K. Ziegler and A. Wenz, *Ber.*, **83**, 354 (1950); (f) G. Wittig, F. Vidal, and E. Bohnert, *ibid.*, **83**, 359 (1950); see also (g) G. Roberts and C. W. Shoppee, *J. Chem. Soc.*, p. 3418 (1954); (h) H. L. Goering and F. H. McCarron, *J. Am. Chem. Soc.*, **80**, 2287 (1958).

handle (covalent carbon–metal bond character) also makes it problematic as to whether they can form carbanions.

$$C_6H_{13}\text{—}\overset{*}{\underset{}{C}}H(CH_3)\text{—}I + C_2H_5\text{—}CH(CH_3)\text{—}Li \xrightarrow[\text{ether, } 70^\circ]{\text{Hexane-}} C_6H_{13}\text{—}\overset{*}{C}H(CH_3)\text{—}Li \xrightarrow[-70^\circ]{CO_2} C_6H_{13}\text{—}\overset{*}{C}H(CH_3)\text{—}CO_2Li$$

In the following sections, the stereochemistry of saturated organomercury compounds is discussed first, followed by the stereochemistry of saturated alkyllithium and magnesium compounds. The third section is concerned with vinyl organometallic compounds and related systems.

Organomercury Compounds

The earliest studies of the stereochemistry of organomercury compounds involved systems in which the carbon attached to mercury was asymmetric, but the molecule contained other asymmetric centers as well.[33–35] Although neighboring group and asymmetric induction effects potentially could have complicated their investigations of the stereochemical course of substitution of such compounds, it is noteworthy that all three groups of investigators concluded that electrophilic substitution of their organomercurials occurred with *retention of configuration.* The same conclusion was reached by later investigators employing simpler systems.

The only portion of this early work that will be reviewed here deals with electrophilic substitution at a bridgehead carbon attached to mercury. Winstein and Traylor[35c] observed that di-4-camphylmercury underwent electrophilic substitution at bridgehead carbon with mercuric chloride or acetic acid–perchloric acid as electrophile, and that 4-camphylmercuric iodide was similarly substituted with triiodide ion. These substitutions necessarily proceed with retention of configuration, occurring at carbon in a ring-enforced pyramidal configuration. These bridgehead derivatives were not especially unreactive, and occupied a

[33] G. F. Wright, *Can. J. Chem.*, **30**, 268 (1952).

[34] (a) A. N. Nesmeyanov, O. A. Reutov, and S. S. Poddubraya, *Dokl. Akad. Nauk SSSR*, **88**, 479 (1953); (b) O. A. Reutov, I. P. Belelskaya, and E. L. Mardaleishvili, *Dokl. Akad. Nauk SSSR*, **116**, 617 (1957).

[35] (a) S. Winstein, T. G. Traylor, and C. S. Garner, *J. Am. Chem. Soc.*, **77**, 3741 (1955); (b) S. Winstein and T. G. Traylor, *ibid.*, **77**, 3747 (1955); (c) S. Winstein and T. G. Traylor, *ibid.*, **78**, 2597 (1956).

position intermediate in reactivity between the neophyl and butyl analogs. All three reactions were first order in organometallic and in the electrophile. Radical substitution reactions were also recognized, but were differentiated from those of electrophilic substitution.[35c]

$$\text{(C}_6\text{H}_{11})\text{—Hg—(C}_6\text{H}_{11}) \xrightarrow[\text{Ether}]{HgCl_2} \text{(C}_6\text{H}_{11})\text{—Hg—Cl}$$

$$\downarrow \text{(1) } H_2\overset{+}{O}Ac \quad \text{(2) LiCl}$$

$$\text{(C}_6\text{H}_{11})\text{—Hg—Cl} + \text{(C}_6\text{H}_{11})\text{—H}$$

$$\text{(C}_6\text{H}_{11})\text{—Hg—I} + I_3^- \xrightarrow[\text{(Wet)}]{\text{Dioxane}} \text{(C}_6\text{H}_{11})\text{—I}$$

The authors formulated non-carbanionic mechanisms for these electrophilic substitution reactions of organomercury compounds as they had in their earlier studies.[35a, b] In the reactions that involved mercuric chloride or triiodide as electrophiles, cyclic transition states for the substitution reaction were envisioned (one-stage SE_i or substitution, electrophilic, internal mechanism), whereas when protonated acetic acid served as electrophile, an open transition state was formulated (SE_2 or substitution, electrophilic, bimolecular mechanism).

In more recent work, *sec*-butylmercuric hydroxide was resolved through its mandelate derivative,[36a, b, 37a, c] and converted by anionic

[36] (a) H. B. Charman, E. D. Hughes, and C. K. Ingold, *Chem. & Ind. (London)*, p. 1517 (1958); (b) H. B. Charman, E. D. Hughes, and C. K. Ingold, *J. Chem. Soc.*, pp. 2523, 2530 (1959); (c) H. B. Charman, E. D. Hughes, C. K. Ingold, and F. G. Thorpe, *ibid.*, p. 1121 (1960); (d) E. D. Hughes, C. K. Ingold, F. G. Thorpe, and H. C. Volger, *ibid.*, p. 1133 (1961); (e) H. B. Charman, E. D. Hughes, C. K. Ingold, and H. C. Volger, *ibid.*, p. 1142 (1961); (f) E. D. Hughes and H. C. Volger, *ibid.*, p. 2359 (1961).

[37] (a) F. R. Jensen, L. D. Whipple, D. K. Wedegaertner, and J. A. Landgrebe, *J. Am. Chem. Soc.*, **81**, 1262 (1959); (b) F. R. Jensen and L. H. Gale, *ibid.*, **81**, 1261 (1959); (c) F. R. Jensen, L. D. Whipple, D. K. Wedegaertner, and J. A. Landgrebe, *ibid.*, **82**, 2466 (1960); (d) F. R. Jensen, *ibid.*, **82**, 2469 (1960); (e) F. R. Jensen and L. H. Gale, *ibid.*, **82**, 145, 148 (1960); (f) F. R. Jensen and J. A. Landgrebe, *ibid.*, **82**, 1004 (1960).

exchange reactions on mercury to various optically active *sec*-butylmercury salts. Both groups of investigators, through one device or

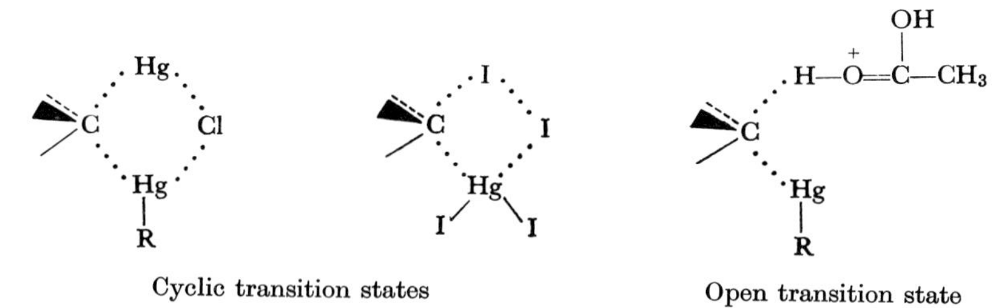

Cyclic transition states Open transition state

another, determined the relative configurations of starting materials and products in a series of stereochemical studies in which the asymmetric carbon of the *sec*-butyl group was substituted with a number of different electrophiles. Within experimental error, the electrophilic reactions proceeded with *complete retention of configuration* at carbon, and all available evidence supported mechanisms which did not involve carbanions as intermediates. The reactions in which stereochemistry was studied are listed as (1–5).

$$C_2H_5\text{—}\overset{*}{C}H(CH_3)\text{—}Hg\text{—}Br + Br_2 \xrightarrow{\text{Pyridine}} C_2H_5\text{—}\overset{*}{C}H(CH_3)\text{—}Br \qquad (1)$$

(Jensen *et al.*[37a,c])

$$C_2H_5\text{—}\overset{*}{C}H(CH_3)\text{—}Hg\text{—}\overset{*}{C}H(CH_3)\text{—}C_2H_5 + HgBr_2 \xrightarrow[\text{or acetone}]{\text{Ethanol}} C_2H_5\text{—}\overset{*}{C}H(CH_3)\text{—}HgBr \qquad (2)$$

Optically active diastereomer (Jensen *et al.*[37a, d]; Charman *et al.*[36b])

$$C_2H_5\text{—}\overset{*}{C}H(CH_3)\text{—}Hg\text{—}Br + Mg \longrightarrow C_2H_5\text{—}\overset{*}{C}H(CH_3)\text{—}Hg\text{—}\overset{*}{C}H(CH_3)\text{—}C_2H_5 + MgBr_2 + Hg \qquad (3)$$

(Jensen and Landgrebe[37f])

$$C_2H_5\text{—}\overset{*}{C}H(CH_3)\text{—}\overset{*}{Hg}\text{—}X + C_2H_5\text{—}CH(CH_3)\text{—}Hg\text{—}CH(CH_3)\text{—}C_2H_5 \xrightarrow{C_2H_5OH} \qquad (4)$$

$\overset{*}{Hg}$ is radiolabeled mercury Not optically active

$$C_2H_5\text{—}CH(CH_3)\text{—}Hg\text{—}X + C_2H_5\text{—}\overset{*}{C}H(CH_3)\text{—}\overset{*}{Hg}\text{—}\overset{*}{C}H(CH_3)\text{—}C_2H_5$$

(Charman *et al.*[36c])

Rate of loss of optical activity in *sec*-butylmercuric bromide (acetate, nitrate) equalled rate of loss of radiolabeled mercury. "Spent" *sec*-butylmercuric bromide contained one third of initial activity.

$$(5)\quad C_2H_5\text{—}\overset{*}{C}H(CH_3)\text{—}Hg\text{—}X + \overset{*}{Hg}X_2 \xrightarrow{C_2H_5OH} C_2H_5\text{—}CH(CH_3)\text{—}\overset{*}{Hg}\text{—}X + HgX_2$$

(Hughes *et al.*[36d])

In a parallel series of studies Reutov and co-workers[38] resolved 5-methyl-2-hexylmercuric and *sec*-butylmercuric salts as their ethyl tartrates[38a] and observed complete retention of configuration in reactions (7) and (8).[38b]

$$(6)\quad (CH_3)_2CHCH_2CH_2\text{—}CH(CH_3)\text{—}Hg\text{—}CH(CH_3)\text{—}CH_2CH_2CH(CH_3)_2 + \overset{*}{Hg}Br_2 \xrightarrow{C_2H_5OH} 2(CH_3)_2CHCH_2CH_2\text{—}CH(CH_3)\text{—}\overset{*}{Hg}\text{—}Br$$

$$(7)\quad (CH_3)_2CHCH_2CH_2\text{—}\overset{*}{C}H(CH_3)\text{—}\overset{*}{Hg}\text{—}Br + (CH_3)_2CHCH_2CH_2\text{—}CH(CH_3)\text{—}Hg\text{—}CH(CH_3)\text{—}CH_2CH_2CH(CH_3)_2 \underset{}{\overset{C_2H_5OH}{\rightleftarrows}} (CH_3)_2CHCH_2CH_2\text{—}CH(CH_3)\text{—}Hg\text{—}Br + (CH_3)_2CHCH_2CH_2\text{—}\overset{*}{C}H(CH_3)\text{—}\overset{*}{Hg}\text{—}\overset{*}{C}H(CH_3)\text{—}CH_2CH_2CH(CH_3)_2$$

$$(8)\quad (CH_3)_2CHCH_2CH_2\text{—}\overset{*}{C}H(CH_3)\text{—}Hg\text{—}X + HgX_2 \longrightarrow (CH_3)_2CHCH_2CH_2\text{—}\overset{*}{C}H(CH_3)\text{—}Hg\text{—}\overset{*}{X} + HgX_2$$

[38] (a) O. A. Reutov and E. U. Uglova, *Izv. Akad. Nauk SSSR, Otdel. Khim. Nauk.* p. 757 (1959); (b) O. A. Reutov and E. U. Uglova, *ibid.*, p. 1691 (1960); (c) O. A. Reutov, T. P. Karpov, E. U. Uglova, and V. A. Malyanov, *Tetrahedron Letters*, **19**, 6 (1960); (d) O. A. Reutov, *Angew. Chem.*, **72**, 198 (1960); (e) O. A. Reutov, *Record Chem. Progr.* (*Kresge-Hooker Sci. Lib.*), **22**, 1 (1961).

As did Winstein and Traylor in their earlier work,[35b,c] the Hughes-Ingold group recognized through kinetic techniques both SE_2 and non-carbanionic SE_i mechanisms.[36] For example, reactions such as (2) were found to be bimolecular, and their rates increased markedly in the order of the ionic character of the attacking electrophile, namely $Hg(NO_3)_2 > Hg(OAc)_2 > Hg(Br)_2 > LiHgBr_3$. This behavior was interpreted as favoring the SE_2 mechanism, since this order is opposite to that expected if the attacking reagent had to provide a collaborating nucleophilic attack on the expelled mercury by some potentially anionic component in the reagent. The SE_i mechanism would require binding of the attacking reagent to the leaving group in the transition state for breaking the carbon–mercury bond. Studies of reactions such as (4) and (5) led to the conclusion that SE_2 mechanisms also applied to these transformations. The transition states of the two latter reactions appeared to be much more polar than the starting states.[36c,d]

SE_2 Mechanism:

$$\overset{+}{E} + -\overset{|}{\underset{|}{C}}-Hg-X \longrightarrow \left[-\overset{|}{\underset{|}{C}} \begin{matrix} \cdot\cdot\cdot HgX \\ \\ \cdot\cdot\cdot E \end{matrix} \right]^{+} \longrightarrow -\overset{|}{\underset{|}{C}}-E + \overset{+}{HgX}$$

Open transition state

SE_i Mechanism (non-carbanionic variety):

$$E-Nu + -\overset{|}{\underset{|}{C}}-Hg-X \longrightarrow \left[-\overset{|}{\underset{|}{C}} \begin{matrix} \cdot\cdot\cdot HgX \cdot\cdot\cdot \\ \\ \cdot\cdot\cdot E \cdot\cdot\cdot \end{matrix} Nu \right] \longrightarrow -\overset{|}{\underset{|}{C}}-E + HgXNu$$

Closed transition state

The mercury exchange between alkylmercuric and radioactive mercuric salts (reaction (5)) was found to be catalyzed by anions that coordinate strongly with mercury. Thus $I^- > Br^- > Cl^- > AcO^-$ in effectiveness. Two catalytic processes were kinetically identified, one of which involved one extra anion, and the other two. Both of the basic processes were shown to be bimolecular apart from catalysis, and to occur with retention of configuration at carbon. These results were interpreted to be examples of operation of the SE_i mechanism[36e] (non-carbanionic variety). The transition states for the two substitution reactions are formulated.

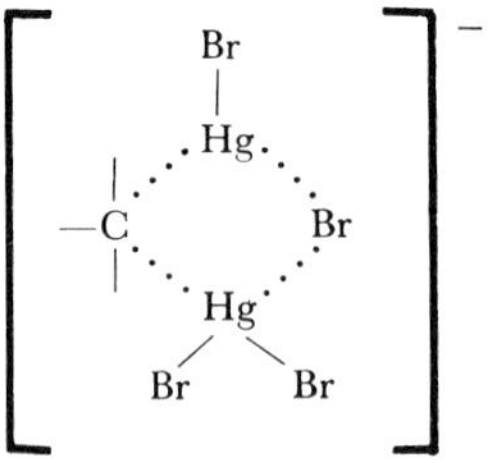

Transition state for one-anion catalysis

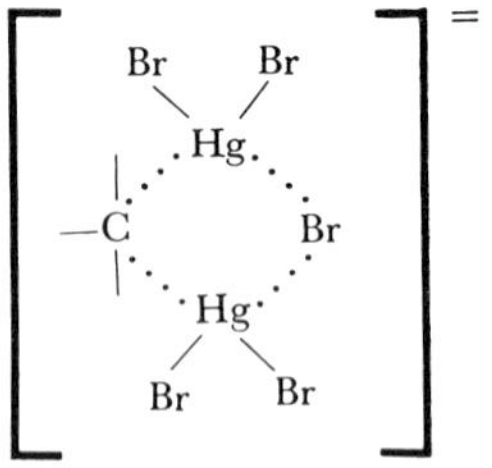

Transition state for two-anion catalysis

In another study, Hughes and Volger[36f] studied the effect on rate of varying R in the mercury exchange reaction between RHgX and $\overset{*}{Hg}X_2$, where X is bromide or acetate in ethanol as solvent. Under conditions of operation of the SE_2 mechanism, the relative rates for various alkyl groups were as follows: *sec*-butyl, 1; neopentyl, 5.5; ethyl, 7; methyl, 17. Comparison of the rates of methyl and neopentyl compounds under conditions of the SE_2, SE_i one-anion and SE_i two-anion mechanisms provided the following rate factors: SE_2 mechanism, $k_{CH_3}/k_{CH_2C(CH_3)_3} = 3$; SE_i one-anion mechanism, $k_{CH_3}/k_{CH_2C(CH_3)_3} = 33$; SE_i two-anion mechanism, $k_{CH_3}/k_{CH_2C(CH_3)_3} = 108$. The authors attribute these rate factors to the changes in steric compression of the transition states involved. They conclude that polar effects play little or no role because the carbon atom undergoing substitution carries little or no charge in the transition state.

Allylic organomercurials react with hydrochloric acid in ether or ethyl acetate by what seems to be and SE_i' (substitution, electrophilic, internal, with rearrangement) mechanism, and with perchloric acid in acetic acid by an SE_2' (substitution, electrophilic, bimolecular with rearrangement) mechanism.[39] Crotylmercuric bromide reacts with hydrochloric acid at a rate roughly 10^7 times that of its *n*-butyl analog, and the product is $> 99\%$ 1-butene. The high rate, coupled with the almost exclusive production of rearranged material, points to the concerted, cyclic SE_i' mechanism. The reaction of crotylmercuric bromide with acetic acid–perchloric acid also leads almost exclusively to 1-butene) $> 98.9\%$).

[39] P. D. Sleezer, S. Winstein, and W. G. Young, *J. Am. Chem. Soc.*, **85**, 1890 (1963).

$$CH_3CH{=}CHCH_2{-}HgBr \xrightarrow{HCl} [\text{cyclic transition state: } CH_3CH\cdots C(H)\cdots CH_2;\ CH\cdots H\cdots Cl\cdots HgBr\cdots CH_2] \longrightarrow CH_3CH_2CH{=}CH_2$$

Crotylmercuric bromide — Transition state for $SE_{i'}$ mechanism — 1-Butene

$\downarrow AcOH_2^+$

$$[CH_3CH\cdots C(H)\cdots CH_2;\ CH\cdots H{-}\overset{+}{O}{=}C(OH){-}CH_3;\ CH_2\cdots HgBr] \longrightarrow CH_3CH_2CH{=}CH_2$$

Transition state for SE_2, mechanism

Organolithium and Organomagnesium Compounds

The availability of simple optically active organomercury compounds provides a potential route to optically active organomagnesium or lithium compounds. Curtin and Koehl[40] treated optically active di-*sec*-butylmercury[36a, b, 37a, c] with racemic 2-octyllithium in pentane at temperatures below 0° and carbonated the product to give 2-methylbutanoic acid. Depending on the length of time before carbonation and the temperature of the reaction, the overall stereochemical course of the two reactions (metal exchange and carbonation) varied between 13 and 83% net retention. The shorter times and lower temperatures favored the more stereospecific result. When 94% pentane–6% ether was used as solvent, the acid obtained was completely racemic. Substitution of *tert*-butyllithium for 2-octyllithium led to an inconveniently slow metal exchange reaction, presumably for steric reasons.

The *s*-butyllithium once formed in an optically active state appeared to racemize at −8° at an appreciable rate. This fact, coupled with the observation that ether promotes the production of racemic material,

[40] (a) D. Y. Curtin and W. J. Koehl, Jr., *Chem. & Ind.* (*London*), p. 262 (1960); (b) D. Y. Curtin and W. J. Koehl, Jr., *J. Am. Chem. Soc.*, **84**, 1967 (1962).

points to the racemization occurring through ionization of the organometallic to lithium carbanide ion pairs which lose their configurational identity.

$$C_2H_5-\overset{\overset{CH_3}{|}}{\overset{*}{C}H}-Hg-\overset{\overset{CH_3}{|}}{\overset{*}{C}H}-C_2H_5 + C_6H_{13}\overset{\overset{CH_3}{|}}{C}H-Li \xrightarrow[-3\text{ to }40^\circ]{\text{Pentane}}$$

$$C_2H_5-\overset{\overset{CH_3}{|}}{\overset{*}{C}H}-Li \xrightarrow[(2)\ H_3O^+]{(1)\ CO_2} C_2H_5-\overset{\overset{CH_3}{|}}{\overset{*}{C}H}-CO_2H$$

In a different study, Winstein and Traylor[35c] prepared 4-camphyllithium by metalation of 4-chlorocamphane, and observed that this bridgehead organolithium compound underwent normal reactions with carbon dioxide, iodine, or mercuric chloride to give the expected products. The enforced pyramidal character of carbanionic intermediates in these reactions appeared to have little effect on their tendency to occur.

$$\equiv R-X \qquad R-Cl \xrightarrow{Li} R-Li \xrightarrow[(2)\ H_3O^+]{(1)\ CO_2} R-CO_2H$$

$$\xrightarrow{I_2} R-I$$

$$\xrightarrow{HgCl_2} R-Hg-Cl$$

The tendency of the cyclopropyl anion to preserve its configuration when it is an intermediate in hydrogen–deuterium exchange reactions[28] has been discussed earlier in the chapter. In anticipation that cyclopropyllithium compounds might also preserve their configurations, Applequist and Peterson[41] examined the behavior of *cis*- and *trans*-2-methylcyclopropyllithium. The organolithium compounds were prepared by exchange of *cis*- and *trans*-2-methylcyclopropyl bromide with isopropyllithium in media varying from pure pentane to 66% ether–34% pentane at temperatures varying from −70° to 32°. These organometallic compounds were carbonated and brominated to give the corresponding carboxylic acids and bromides.

The facts that *cis*-bromide gave only *cis*-acid and *trans*-bromide only *trans*-acid (94% pentane–6% ether) indicate that both the exchange and carbonation reactions were highly stereospecific. However, brominolysis

[41] D. E. Applequist and A. H. Peterson, *J. Am. Chem. Soc.*, **83**, 862 (1961).

of each organometallic reagent gave mixtures of *cis*- and *trans*-2-methyl-cyclopropyl bromide, the relative amounts of each varying with whichever isomer was used, with the exchange solvent, and the exchange temperature. However, no variation was observed with the time elapsed after the exchange reaction, despite the fact that the time varied by about a factor of 100. Clearly, the *cis*- and *trans*-cyclopropyllithium compounds are formed stereospecifically with retention, they do not interconvert under the conditions employed, and the carbonation reactions go stereospecifically with retention. On the other hand, the bromination occurs by a mixture of retention (electrophilic) and non-stereospecific (possibly radical) mechanisms.

CH_3 Br / H H $\xrightarrow{i\text{-}C_3H_7Li}$ CH_3 Li / H H $\xrightarrow[(2)\ H_3O^+]{(1)\ CO_2}$ CH_3 CO_2H / H H

$\xrightarrow{Br_2}$ CH_3 Br / H H + CH_3 H / H Br

Predominant isomer

CH_3 H / H Br $\xrightarrow{i\text{-}C_3H_7Li}$ CH_3 H / H Li $\xrightarrow[(2)\ H_3O^+]{(1)\ CO_2}$ CH_3 H / H CO_2H

$\xrightarrow{Br_2}$ CH_3 H / H Br + CH_3 Br / H H

Predominant isomer

The first study of the optical stability of optically active cyclopropyl-organometallics was made by Walborsky and Impastato.[42a] Optically

[42] (a) H. M. Walborsky and F. J. Impastato, *J. Am. Chem. Soc.*, **81**, 5835 (1959); (b) H. M. Walborsky and A. E. Young, *ibid.*, **83**, 2595 (1961); (c) H. M. Walborsky, *Record Chem. Progr.* (*Kreskge-Hooker Sci. Lib.*), **23**, 75 (1962); (d) H. M. Walborsky, L. Barash, A. E. Young, and F. J. Impastato, *J. Am. Chem. Soc.*, **83**, 2517 (1961); (e) H. M. Walborsky and C. G. Pitt, *ibid.*, **84**, 4831 (1962); (f) F. J. Impastato and H. M. Walborsky, *ibid.*, **84**, 4838 (1962); (g) H. M. Walborsky, F. J. Impastato, and A. E. Young, *ibid.*, **86**, 3283 (1964); (h) H. M. Walborsky and A. E. Young, *ibid.*, **86**, 3288 (1964).

active bromide XXIX was prepared,[42d] and its configuration established relative to those of hydrocarbon XXXI, iodide XXXII, and carboxylic acid XXXIII.[42d,e,f] Treatment of optically active bromide XXIX with butyllithium in a variety of solvents gave the optically active cyclopropyllithium compound, XXX. This compound was protonated with methanol, carbonated, brominated, and iodinated. The steric course and stereospecificity of each two-step reaction (metal exchange and production of stable product) was examined.[42g] The overall stereochemical course for each reaction pair is as follows: metalation and protonation, 79–100% retention (minimum values); metalation and carbonation, 100% retention; metalation and iodination 100% retention; metalation and bromination, 95% retention. In the metalation-protonation experiments, the solvent varied from hexane–dimethoxyethane to ether to ether–benzene–hexane; time varied from 10 to 30 min.; and temperature from 0° to 35° without any discernible effect on steric course. In the carbonation experiments, the solvent was changed from ether to tetrahydrofuran, time from 10 to 30 min. and temperature from $-8°$ to $28°$ without change in steric result. Thus the cyclopropylorganometallic can be prepared and made to undergo electrophilic substitution with essentially complete stereospecificity, and with overall retention. Clearly, each individual reaction must also occur with retention. The cyclopropylorganometallic showed no tendency to racemize in several solvents at room temperature.[42g]

C_6H_5 C_6H_5 / CH_3 / Br — XXIX $\underset{Br_2}{\overset{BuLi}{\rightleftarrows}}$ C_6H_5 C_6H_5 / CH_3 / Li — XXX $\xrightarrow{CH_3OH}$ C_6H_5 C_6H_5 / CH_3 / H — XXXI

XXX $\underset{BuLi}{\overset{I_2}{\rightleftarrows}}$ C_6H_5 C_6H_5 / CH_3 / I — XXXII

XXX $\xrightarrow{CO_2}$ C_6H_5 C_6H_5 / CH_3 / CO_2H — XXXIII

The authors[42g] point to the high stereospecificity of these four reactions as being consistent with the SE_2 or SE_i (non-carbanion variety) mechanisms. They are equally consistent with lithium carbanide ion pair mechanisms, the cyclopropyl anion being optically stable because of the high steric barrier to its inversion. Superimposed on the intrinsic tendency of this cyclic carbanion to be pyramidal is the fact that its inversion in an ion pair would involve charge separation, which in solvents of low dielectric constant would involve considerable activation energy. Even in the form of small polymers,[43] cyclopropyllithium upon ionization would be expected to collapse back to the covalent state faster than inversion and polymer reorganization occurred.

The optically active Grignard reagent of the same cyclic system was also prepared (XXXIV).[42c, h] Treatment of optically active bromide XXIX with magnesium in tetrahydrofuran and carbonation of the resulting Grignard reagent gave a mixture of hydrocarbon XXXI and acid XXXIII, each of the two overall reactions having occurred with 56% retention of configuration. The nonstereospecific component of the reaction must have involved formation of the Grignard reagent. When prepared by treatment of optically active lithium compound XXX with anhydrous magnesium bromide, the Grignard reagent XXXIV, when carbonated, produced acid with complete overall retention of configuration.

C_6H_5 C_6H_5 / CH_3 / Br — XXIX $\xrightarrow[(CH_2)_4O]{Mg}$ C_6H_5 C_6H_5 / CH_3 / MgBr — XXXIV $\xrightarrow[(2)\ H_3O^+]{(1)\ CO_2}$ C_6H_5 C_6H_5 / CH_3 / H — XXXI, 56% Retention + C_6H_5 C_6H_5 / CH_3 / CO_2H — XXXIII, 56% Retention

XXIX $\xrightarrow{\text{BuLi | Ether}}$

C_6H_5 C_6H_5 / CH_3 / Li — XXX $\xrightarrow[(2)\ (CH_3OCH_2)_2]{(1)\ MgBr_2}$ C_6H_5 C_6H_5 / CH_3 / MgBr — XXXIV $\xrightarrow[(2)\ H_3O^+]{(1)\ CO_2}$ C_6H_5 C_6H_5 / CH_3 / CO_2H — XXXIII, 100% Retention

[43] (a) T. L. Brown, D. W. Dickerhoof, and D. A. Bafus, *J. Am. Chem. Soc.*, **84**, 1371 (1962); (b) M. A. Weiner and R. West, *ibid.*, **85**, 485 (1963); (c) J. F. Eastham and J. W. Gibson, *ibid.*, **85**, 2171 (1963).

Roberts and co-workers,[44] using nuclear magnetic resonance (NMR) techniques, have concluded that cyclopropyl and 2,2-dimethylcyclopropylmagnesium bromide (XXXV) are configurationally stable *on the NMR time scale* in diethyl ether solution at 175°, and possibly higher. Likewise, 3,3-dimethylcyclobutylmagnesium bromide (XXXVI) is configurationally stable *on the NMR time scale* at 125° and possibly higher in diethyl ether. In contrast to this behavior of these cyclic secondary Grignard reagents, 3,3-dimethylbutylmagnesium bromide and chloride (XXXVII and XXXVIII) undergo rapid inversion in ether at room temperature, inversion being more rapid with the chloride, which becomes configurationally stable *on the NMR time scale* below −70°. The reaction is higher than first order, there being a marked decrease in rate with dilution. The dialkylmagnesium analog (XXXIX) proved configurationally stable *on the NMR time scale* below 30°. The activation energy for inversion of dialkylmagnesium compound XXXIX proved to be 20 Kcal./mole, and for chloride XXXVIII, 11 Kcal./mole. The inversion rate of chloride XXXVIII decreased as the solvent was changed in the order diethyl ether > tetrahydrofuran > dimethoxyethane.

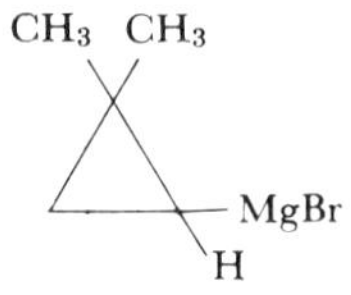

CH_3 CH_3 / MgBr / H

XXXV

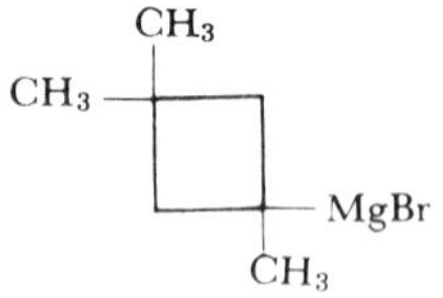

CH_3 / CH_3 / MgBr / CH_3

XXXVI

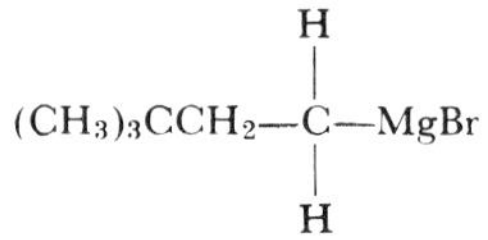

$$(CH_3)_3CCH_2—CH_2—MgBr$$

XXXVII

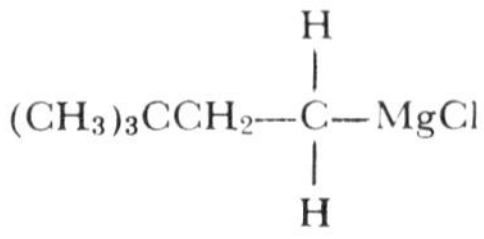

$$(CH_3)_3CCH_2—CH_2—MgCl$$

XXXVIII

$$[(CH_3)_3CCH_2—CH_2—]_2Mg$$

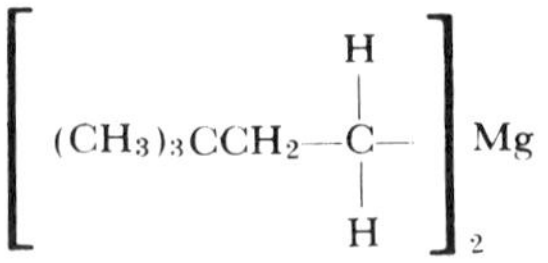

XXXIX

[44] (a) G. M. Whitesides, F. Kaplan, and J. D. Roberts, *J. Am. Chem. Soc.*, **85**, 2167 (1963); (b) J. D. Roberts, G. M. Whitesides, and M. Witanowski, private communication; (c) M. E. Howden, Ph.D. Thesis, California Institute of Technology, 1962.

By *configurationally stable on the NMR time scale* is meant that inversion must occur slower than approximately 2 times per second. By comparison, the rate of inversion would have to be slower by about 3 powers of 10 to be observable with conventional techniques.

With the same technique, Roberts and co-workers[44] compared the configurational stabilities of alkyllithium, magnesium, aluminum, and mercury compounds, XL, XXXIX, XLI, and XLII in diethyl ether.

$$(CH_3)_3CCH_2-CH_2-Li$$

XL

$$[(CH_3)_3CCH_2-CH_2-]_3Al$$

XLI

$$[(CH_3)_3CCH_2-CH_2-]_2Hg$$

XLII

The lithium compound was found to undergo rapid inversion above 0°, and the magnesium compound above 30°. No evidence for inversion was observed for aluminum compound XLI even at 147°, or for mercury compound XLII at 163°.

To the extent that data are available, the rates of inversion of the organometallic compounds decrease with decreasing percent ionic character of the carbon metal bonds[45]: C—Li > C—Mg > C—Al > C—Hg in ionic character. The rates of inversion were faster for primary than for secondary, and for cyclobutyl than for cyclopropyl. These generalizations correlate with what is expected on the basis of ease of carbanion formation from the organometallic compound, and with the ease of inversion of the derived carbanion (in an ion pair) once formed.

Reutov[46] reported that optically active mercury compound XLIII could be converted to optically active magnesium compound XLIV, which, in turn, gave optically active XLV. Apparently the secondary magnesium compound was formed by at least a partially stereospecific

[45] (a) E. G. Rochow, D. T. Hurd, and R. N. Lewis, "The Chemistry of Organometallic Compounds," Wiley, New York, 1957, p. 18; (b) J. Hinze and H. H. Jaffé, *J. Am. Chem. Soc.*, **84**, 540 (1962).

[46] O. A. Reutov, *Bull. Soc. Chim. France*, p. 1383 (1963).

exchange reaction, maintained at least some of its configuration during its lifetime, and was at least partially stereospecifically converted to acid XLV. The relative configurations of the starting materials and products were not determined, nor was the degree of stereospecificity of each of the two electrophilic substitution reactions.

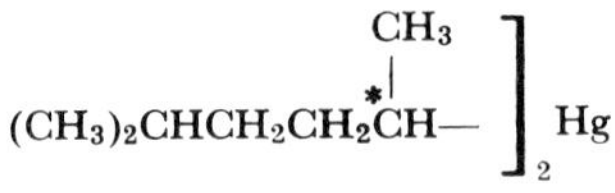

$$\left[(CH_3)_2CHCH_2CH_2\overset{*}{C}H(CH_3)-\right]_2Hg$$
XLIII

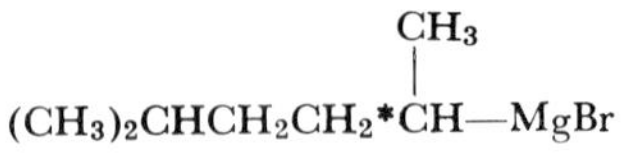

$$(CH_3)_2CHCH_2CH_2{*}CH(CH_3)-MgBr$$
XLIV

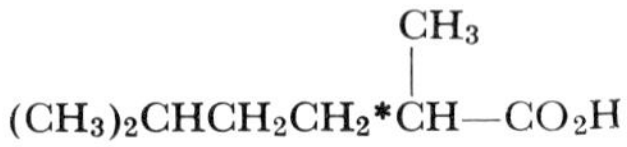

$$(CH_3)_2CHCH_2CH_2{*}CH(CH_3)-CO_2H$$
XLV

Stereochemistry of Vinyl Organometallic Compounds

The stereochemistry of the unsaturated organometallic compounds centers about the problems of the configurational stability, steric course of preparation, and reactions of *cis*- and *trans*-vinyllithium compounds.

In parallel studies, Curtin and Harris,[47a] and Braude and Coles[48] first attacked the problem of the stereochemistry of vinyllithium compounds. The former authors observed that *cis*- and *trans*-XLVI and *cis*- and *trans*-XLVII could be converted to their lithium derivatives (bromine–lithium exchange). These were submitted to electrophilic substitution to give products with a high degree of overall retention of configuration at the double bond. The latter authors[48] reported that *cis*-propenyl

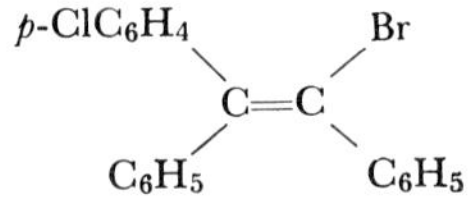

$$(p\text{-}ClC_6H_4)(C_6H_5)C{=}C(Br)(C_6H_5)$$
cis-XLVI

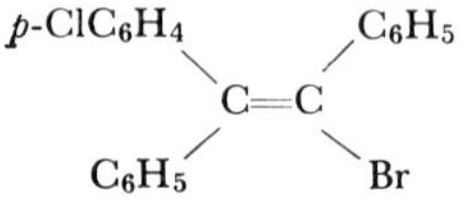

$$(p\text{-}ClC_6H_4)(C_6H_5)C{=}C(C_6H_5)(Br)$$
trans-XLVI

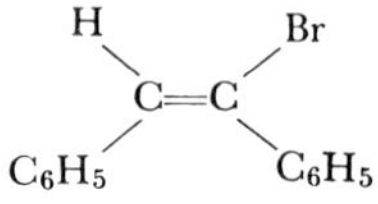

$$(H)(C_6H_5)C{=}C(Br)(C_6H_5)$$
cis-XLVII

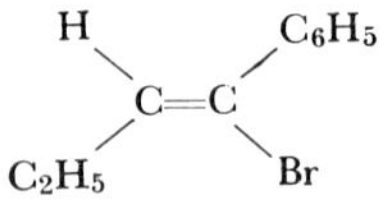

$$(H)(C_2H_5)C{=}C(C_6H_5)(Br)$$
trans-XLVII

[47] (a) D. Y. Curtin and E. E. Harris, *J. Am. Chem. Soc.*, **73**, 2716 (1951); (b) D. Y. Curtin and E. E. Harris, *ibid.*, **73**, 4519 (1951); (c) D. Y. Curtin, H. W. Johnson, Jr., and E. C. Steiner, *ibid.*, **77**, 4566 (1955).

[48] E. A. Braude and J. A. Coles, *J. Chem. Soc.*, pp. 2078, 2085 (1951).

bromide reacted with lithium in ether to give an organometallic compound, which with carbonyl compounds gave condensation products of *cis*-configuration about the double bond.

Somewhat later, Dreiding and Pratt[49] treated *cis*- and *trans*-2-bromo-2-butene with lithium in ether and then with carbon dioxide to give carboxylic acids. About 70–90% overall preservation of configuration at vinyl carbon was observed for the two reactions (formation and consumption of the organometallic). Bordwell and Landis[50] treated *cis*- and *trans*-2-bromo-2-butene both with lithium in ether and with butyllithium and the derived organometallics were submitted to electrophilic substitution with *p*-tolyl disulfide. The stereochemistry of the products indicated that the two steps (formation and consumption of the organometallic) occurred with 75–85% preservation of configuration at vinyl carbon irrespective of the method used for preparing the organometallic. When the organometallic prepared from *cis*-bromide was allowed to stand in ether at $-40°$ for 2 hr., no change was observed in the overall stereochemical result.

Nesmeyanov and co-workers[51] reported that *cis*-1,2-diphenylvinyllithium readily isomerizes to the *trans*-isomer at temperatures above $-40°$ in benzene–ether, and Curtin and co-workers[52] found that *cis*- and *trans*-2-*p*-chlorophenyl-1,2-diphenylvinyllithium underwent interconversions at a moderate rate in ether at 0° or above. The observations of Nesmeyanov *et al.* removed any possible doubt as to the configuration of the organolithium compound formed from *cis*-bromostilbene. Had *trans*-1,2-diphenylvinyllithium been obtained directly, it certainly would not tend to isomerize to the *cis*-isomer, since the *trans*-isomer should be the more thermodynamically stable by a large factor. Although little doubt existed from the start that the individual steps of formation and reactions of these vinyl organometallics had each occurred with retention, the work of the Russian workers provided a new kind of persuasive evidence on this point.

Curtin and co-workers[40,53] studied the geometric stability of *cis*- and *trans*-vinyllithium compounds as structure and solvent were varied. They observed that *cis*- and *trans*-propenyllithium prepared from the

[49] A. S. Dreiding and R. J. Pratt, *J. Am. Chem. Soc.*, **76**, 1902 (1954).

[50] F. G. Bordwell and P. S. Landis, *J. Am. Chem. Soc.*, **79**, 1593 (1957).

[51] (a) A. N. Nesmeyanov, A. E. Borisov, and N. A. Volkenau, *Izv. Akad. Nauk SSSR, Otdel. Khim. Nauk*, p. 992 (1954); (b) A. N. Nesmeyanov and A. E. Borisov, *Tetrahedron*, **1**, 158 (1957).

[52] D. Y. Curtin, H. W. Johnson, Jr., and E. C. Steiner, *J. Am. Chem. Soc.*, **77**, 4566 (1955).

[53] D. Y. Curtin and J. W. Crump, *J. Am. Chem. Soc.*, **80**, 1922 (1958).

corresponding bromides in ether were configurationally stable for periods of one hour in refluxing ether.[53] This stability contrasts with the

$$\begin{matrix} H & & H \\ & C{=}C & \\ CH_3 & & Li \end{matrix} \qquad\qquad \begin{matrix} H & & Li \\ & C{=}C & \\ CH_3 & & H \end{matrix}$$

cis-Propenyllithium   *trans*-Propenyllithium

instability of the *cis*-stilbenyllithium.[51] In a study of solvent effects, *cis*-stilbenyllithium was prepared from the corresponding mercury compound[51] by mercury–lithium exchange reactions in a variety of solvents. After 30 min., the organometallic was treated with either carbon dioxide or benzophenone, and the structures of the products were determined. In 3 to 1 ether–benzene at $-54°$ or in 1 to 1 benzene–pentane at 2°, only products of *cis*-configuration were produced. Thus the vinyllithium compound exhibited configurational stability. In tetrahydrofuran at $-45°$ or 1 to 1 ether–benzene at 3°, only products of *trans*-configuration were obtained, and hence the *cis*-vinyllithium isomer must have isomerized almost completely to the *trans*-isomer. In 1 to 1 benzene–pentane which was 0.5 to 1.1% in ether at 3°, products that were 71 to 89% *cis* in configuration were obtained. In benzene at 27°, products that were 29% *cis* were produced. Similar work with *cis*- and *trans*-2-*p*-chlorophenyl-1,2-diphenylvinyllithium gave similar results.[40]

Curtin *et al.*[40,53] drew the following conclusions. (1) Arylvinyllithium compounds isomerize much more readily that alkylvinyllithium compounds. (2) The rates of isomerization of arylvinyllithium compounds vary markedly with the solvent polarity, the rate decreasing in the order tetrahydrofuran > 3 to 1 ether–benzene > hydrocarbon solvents. (3) Varying amounts of sodium (up to 2.3%) in the lithium used to prepare the organometallic make no difference in the steric result. The authors[40] suggest that the isomerization of the arylvinyllithium compounds occurs by ionization of the partially covalent carbon–lithium bond; that the vinyl anion isomerizes through a linear transition state or intermediate in which charge is highly delocalized into the benzene ring; and that recapture of the lithium ion occurs to give isomerized organometallic.

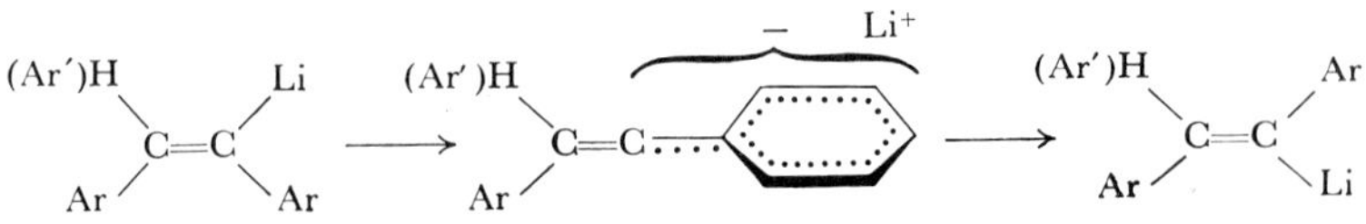

$$\begin{matrix} (Ar')H & & Li \\ & C{=}C & \\ Ar & & Ar \end{matrix} \longrightarrow \begin{matrix} (Ar')H & & \\ & C{=}C\cdots C_6H_5^{-}\ Li^{+} \\ Ar & & \end{matrix} \longrightarrow \begin{matrix} (Ar')H & & Ar \\ & C{=}C & \\ Ar & & Li \end{matrix}$$

Anionic transition state
or intermediate

Assignments of configuration have been made by measurement of spectral properties of vinyllithium compounds.[54] Allinger and Hermann,[54a] on the basis of infrared spectra, assigned structures to the *cis*- and *trans*-propenyllithium isomers, although the basis of this assignment has been criticized.[54b] Seyferth and Vaughan[54b] have used nuclear magnetic resonance spectroscopy to confirm configurational assignments of long standing, and observed that *cis*- and *trans*-propenyltrimethyltin undergo metal exchange with butyllithium in ether without loss of configuration.

The configurational stability of the vinyllithium compounds appear to be associated with that of vinyl anions. Information bearing on the latter problem is available from base-catalyzed hydrogen–deuterium exchange studies at vinyl carbon. Miller and Lee[55] observed that *cis*-dibromoethylene underwent methoxide ion-catalyzed hydrogen–deuterium exchange with methanol-*O-d* at a rate about 25 times as fast as that for elimination to bromoacetylene. Furthermore, *cis*- and *trans*-1,2-dichloroethylene underwent deuteroxide-catalyzed exchange with deuterium oxide at moderate temperatures (3–70°) without isomerization. On the basis of kinetic measurements, the authors calculated that the lower limits to the activation energy for isomerization of 1,2-dihalovinyl anions lie in the range of 25–35 Kcal./mole in the solvent employed. Clearly, these vinyl carbanions possess considerable configurational stability.

In a study of the configurational stability of the vinyl anions of the stilbene system, *cis*- and *trans*-stilbene were submitted to potassium *tert*-butoxide-catalyzed isotopic exchange with *tert*-butyl alcohol-*O-d* at 146°.[56] Vinyl isotopic exchange for the *cis*-isomer was found to occur at a rate about 3 powers of 10 faster than isomerization, about 1 power of 10 faster than vinyl exchange of the *trans*-isomer, and about 2 powers of 10 faster than either isomer underwent aryl exchange (per exchangeable hydrogen). These experiments demonstrate that different anions or ion pairs exist in the exchange of the two isomeric stilbenes. Two possibilities were visualized. In the first, *cis*- and *trans*-vinyl anions were the two intermediates. In the second, two intimate ion pairs were intermediates, each of which has the same anion and cation, but they differ in regard to whether the potassium ion is close to the face of the allenic anion occupied by the phenyl group, or to the face occupied by the hydrogen. The latter

[54] (a) N. L. Allinger and R. B. Hermann, *J. Org. Chem.*, **26**, 1040 (1961); (b) D. Seyferth and L. G. Vaughan, *J. Am. Chem. Soc.*, **86**, 883 (1964).

[55] S. I. Miller and W. G. Lee, *J. Am. Chem. Soc.*, **81**, 6316 (1959).

[56] D. H. Hunter and D. J. Cram, *J. Am. Chem. Soc.*, **86**, 5478 (1964).

is a distinct possibility in view of the fact that enantiomeric ion pairs have been demonstrated as intermediates in base-catalyzed hydrogen–deuterium exchange reactions of compounds (e.g., 2-phenylbutane) under comparable conditions (see first part of chapter).

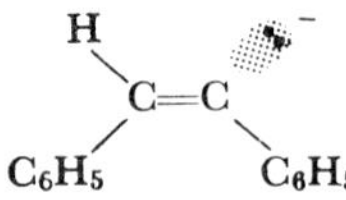

cis-Stilbenyl anion

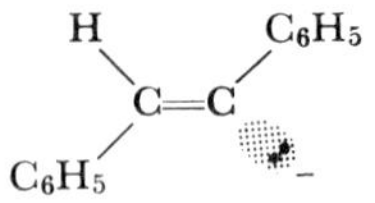

trans-Stilbenyl anion

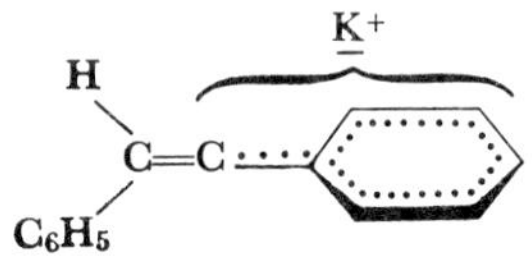

K^+

K^+

Ion pair from *cis*-stilbene

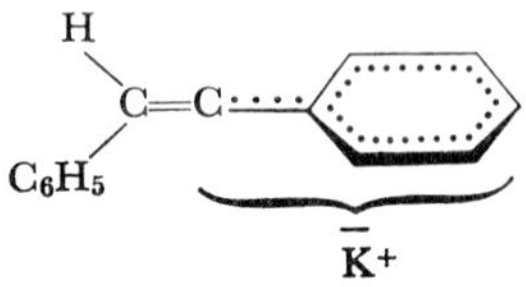

Ion pair from *trans*-stilbene

Curtin and co-workers[57] with NMR methods measured the change in rates at which XLVIII underwent inversion at nitrogen as X was changed from methyl to hydrogen to carboethoxy. A total spread in rate of about 30 was observed, the rate being slowest when X was methyl and fastest when X was carboethoxy. The reaction was estimated to have a ρ of 1.5 ± 0.5, which indicates that the reaction is moderately sensitive to substituents, and that the transition state for the reaction is stabilized by electron-withdrawing groups.

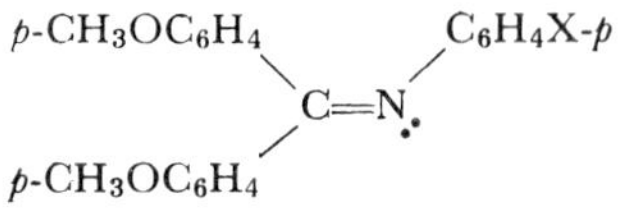

XLVIII

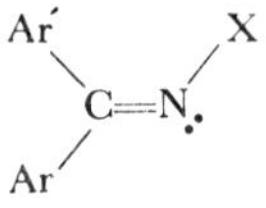

XLIX

These investigators[57] reported that the rates of inversion for compounds of the general structure XLIX decreased by 4 powers of 10 as X was changed from phenyl to methyl, and by an additional 4–5 powers of 10 (minimum) as X was changed from methyl to chloro or methoxy groups. Clearly the phenyl group stabilizes the transition state for inversion by electron delocalization. The inductive effect of the chloro and methoxy groups must destabilize the transition state for inversion

[57] (a) D. Y. Curtin and J. W. Hausser, *J. Am. Chem. Soc.*, **83**, 3474 (1961); (b) D. Y. Curtin and C. G. McCarty, *Tetrahedron Letters*, **26**, 1269 (1962).

through operation of the inductive effect. The stronger the electron-withdrawing inductive effect of X, the greater the *p*-character which nitrogen contributes to the N—X bond, and the richer in *s*-character becomes the orbital occupied by the electron pair on nitrogen. Since the electron pair must pass into a *p*-orbital in the transition state for inversion, the inductive effect increases the energy of activation and decreases the rate.

Electrostatic inhibition of inversion effects also probably tend to decrease the rate of inversion with the chloro and methoxy groups. Each has unshared pairs of electrons, and the transition would be destabilized by bringing the electrons on nitrogen and the substituent closer together than in the ground state.

CHAPTER IV

Stereochemistry of Carbanions Generated with Different Leaving Groups

The base-catalyzed hydrogen–deuterium exchange reactions of the previous chapter can be regarded as acid-base reactions, or as electrophilic substitution reactions at saturated carbon (SE reactions; S = substitution, E = electrophilic). The latter designation becomes useful when the results of the hydrogen-deuterium exchange reactions are correlated with those obtained when carbanions are generated and disposed of by other means. The general family of SE reactions is defined in Equation (1), in which L is a generalized *leaving group*, and E a generalized *electrophile*. Equation (1) specifies only the overall reaction, and the fact that the electrons that bond C to L are the same that finally bond C to E.

$$\text{(1)} \qquad -\overset{|}{\underset{|}{C}}-L + E \longrightarrow -\overset{|}{\underset{|}{C}}-E + L \qquad \text{SE Reaction}$$

The general reaction of Equation (1) might involve either a carbanion as an intermediate, or only a single transition state for the

$$-\overset{|}{\underset{|}{C}}-L \longrightarrow -\overset{|}{\underset{|}{C}}^{-} + L$$

$$\text{(2)} \qquad E + -\overset{|}{\underset{|}{C}}^{-} \longrightarrow -\overset{|}{\underset{|}{C}}-E$$

Monomolecular electrophilic substitution at saturated carbon, or SE_1 mechanism bond-breaking and bond-making processes. The former mechanism is sometimes referred to as SE_1 (substitution, electrophilic, monomolecular), and the latter as SE_2 (substitution, electrophilic, bimolecular). These mechanisms are outlined in Equations (2) and (3).

$$\text{E} + -\overset{|}{\underset{|}{\text{C}}}-\text{L} \longrightarrow -\overset{|}{\underset{|}{\text{C}}}-\text{E} + \text{L} \tag{3}$$

Bimolecular electrophilic substitution at saturated carbon, or SE_2 mechanism

In the acid-base reactions discussed in Chapter III, hydrogen or deuterium served as leaving groups, and deuteron or proton donors as electrophiles. The reactions were SE_1 since carbanions intervened as intermediates. In the first section, the stereochemical course of the SE_1 reaction is discussed in which carbon serves as leaving group, and proton or deuteron donors as electrophiles. In the second section, the stereochemical results obtained with other leaving groups are treated. The third section deals with a comparison of the stereochemical capabilities of carbanions and carbonium ions.

CARBON AS LEAVING GROUP IN ELECTROPHILIC SUBSTITUTION AT SATURATED CARBON

Because of the great structural variation possible with carbon as a leaving group in electrophilic substitution, a large number of different reactions can be used to generate carbanions. Likewise, the many special effects associated with the exact structure of the leaving group make the stereochemistry of the reaction both complex and interesting. Fortunately, many of the patterns of carbanion behavior detected in the hydrogen–deuterium exchange reactions of Chapter III are visible in the carbanions generated with carbon as leaving group.

Survey of Reactions

A variety of base-catalyzed reactions is known that involve cleavage of carbon–carbon bonds to generate carbanions. Reverse condensations such as the reverse aldol, Claisen, and Michael reactions as well as decarboxylation reactions fall into this category. The cleavage of ketones with sodamide has been reviewed.[1]

[1] F. W. Bergstrom and W. C. Fernelius, *Chem. Rev.*, **20**, 450 (1937).

The following reactions were surveyed[2] for their suitability for study of the stereochemical course of the SE_1 reaction. In all cases, 2-phenylbutane was the product. Particularly desirable were those reactions which went under conditions of base and temperature that preserved the configurational integrity of the product. The reactions (4) to (14) are listed in the order of increasing severity of the conditions (temperature) required to drive them to completion. Since the amide bases are much stronger than the alkoxides, the alkoxides (probably) are formed almost quantitatively from the alcohols. In general, the compounds fall in an order of decreasing size of the leaving group. This fact suggests that release of steric compression provides some of the driving force for this type of anionic cleavage.

$$\overset{*}{R} = C_2H_5\text{—}\overset{*}{C}(CH_3)(C_6H_5)\text{—} \qquad C_2H_5\text{—}\overset{*}{C}H(CH_3)(C_6H_5)$$

2-Phenylbutane

$$\text{(4)}\quad \overset{*}{R}\text{—}C(OH)(C_6H_5)\text{—}C_6H_5 + C_6H_5\bar{N}(\overset{+}{K})CH_3 \longrightarrow \overset{*}{R}\text{—}C(\bar{O}\overset{+}{K})(C_6H_5)\text{—}C_6H_5 \xrightarrow{BH} \overset{*}{R}\text{—}H + C_6H_5\text{—}C(=O)\text{—}C_6H_5$$

$$\text{(5)}\quad \overset{*}{R}\text{—}C(OH)(C_6H_5)\text{—}CH_3 + C_6H_5\bar{N}(\overset{+}{K})H \longrightarrow \overset{*}{R}\text{—}C(\bar{O}\overset{+}{K})(C_6H_5)\text{—}CH_3 \xrightarrow{BH} \overset{*}{R}\text{—}H + C_6H_5\text{—}C(=O)\text{—}CH_3$$

$$\text{(6)}\quad \overset{*}{R}\text{—}C(OH)(CH_3)\text{—}C_2H_5 + C_6H_5\bar{N}(\overset{+}{K})CH_3 \longrightarrow \overset{*}{R}\text{—}C(\bar{O}\overset{+}{K})(CH_3)\text{—}C_2H_5 \xrightarrow{BH} \overset{*}{R}\text{—}H + CH_3\text{—}C(=O)\text{—}C_2H_5$$

$$\text{(7)}\quad \overset{*}{R}\text{—}C(OH)(CH_3)\text{—}CH_3 + C_6H_5\bar{N}(\overset{+}{K})H \longrightarrow \overset{*}{R}\text{—}C(\bar{O}\overset{+}{K})(CH_3)\text{—}CH_3 \xrightarrow{BH} \overset{*}{R}\text{—}H + CH_3\text{—}C(=O)\text{—}CH_3$$

$$\text{(8)}\quad \overset{*}{R}\text{—}C{\equiv}N + C_6H_5\bar{N}(\overset{+}{K})CH_3 \longrightarrow \overset{*}{R}\text{—}C(N(CH_3)C_6H_5){=}\bar{N}\overset{+}{K} \xrightarrow{BH} \overset{*}{R}\text{—}H + C_6H_5N(CH_3)\text{—}C{\equiv}N$$

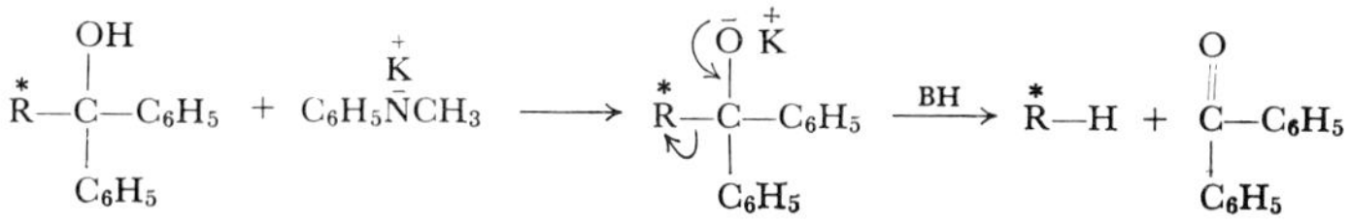

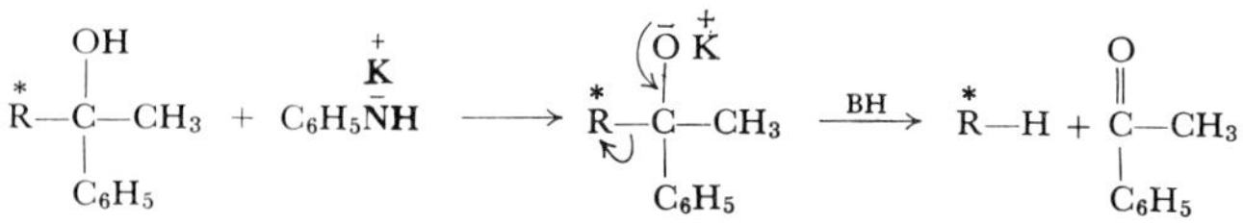

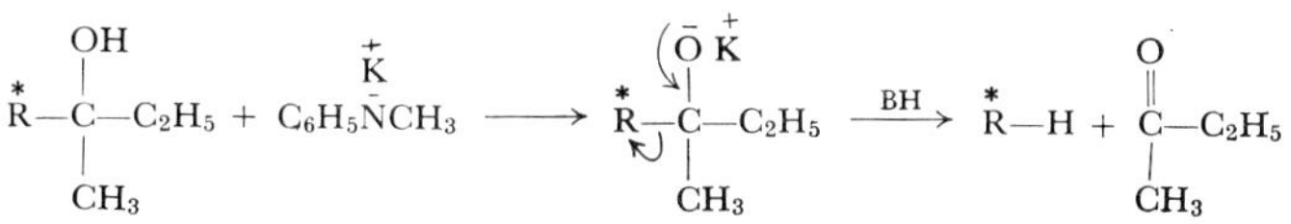

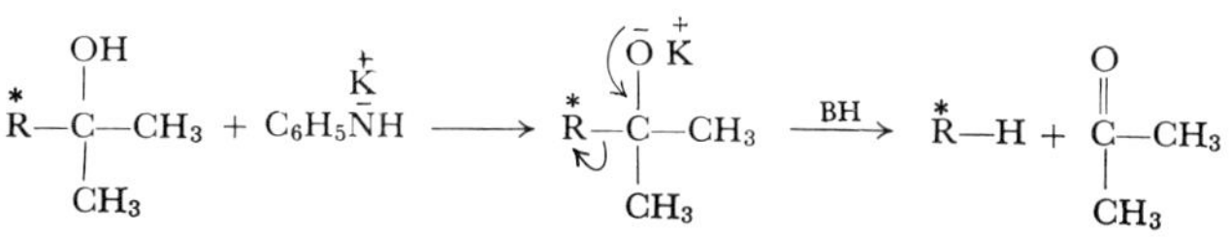

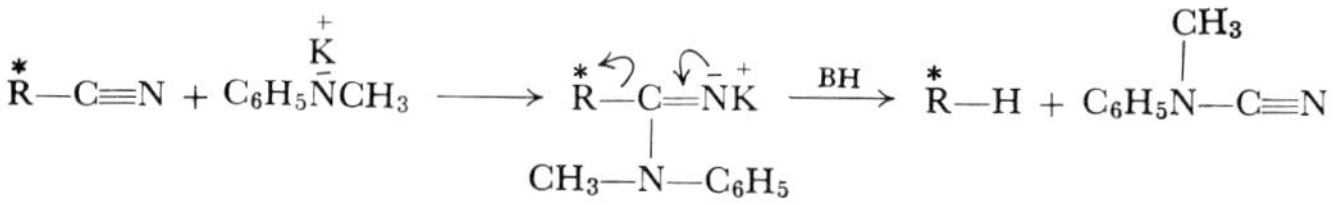

[2] D. J. Cram, A. Langemann, J. Allinger, and K. R. Kopecky, *J. Am. Chem. Soc.*, **81**, 5740 (1959).

$$(9)\quad \overset{*}{R}-\overset{\overset{O}{\|}}{C}-NH_2 + C_6H_5\overset{\overset{+}{K}}{\bar{N}}CH_3 \longrightarrow \overset{*}{R}-\overset{\overset{-\ +}{OK}}{\underset{CH_3-N-C_6H_5}{C}}-NH_2 \xrightarrow{BH} \overset{*}{R}-H + C_6H_5\overset{CH_3}{N}-\underset{\underset{O}{\|}}{C}-NH_2$$

$$(10)\quad \overset{*}{R}-\overset{\overset{O}{\|}}{C}-C_6H_5 + C_6H_5\overset{\overset{+}{K}}{\bar{N}}CH_3 \longrightarrow \overset{*}{R}-\overset{\overset{-\ +}{OK}}{\underset{CH_3-N-C_6H_5}{C}}-C_6H_5 \xrightarrow{BH} \overset{*}{R}-H + C_6H_5\overset{CH_3}{N}-\underset{\underset{O}{\|}}{C}-C_6H_5$$

$$(11)\quad \overset{*}{R}-\overset{OH}{CH}-C_6H_5 + C_6H_5\overset{\overset{+}{K}}{\bar{N}}CH_3 \longrightarrow \overset{*}{R}-\overset{\overset{-\ +}{OK}}{CH}-C_6H_5 \xrightarrow{BH} \overset{*}{R}-H + C_6H_5CHO$$

$$(12)\quad \overset{*}{R}-\overset{OH}{CH}-CH_3 + C_6H_5\overset{\overset{+}{K}}{\bar{N}}CH_3 \longrightarrow \overset{*}{R}-\overset{\overset{-\ +}{OK}}{CH}-CH_3 \xrightarrow{BH} \overset{*}{R}-H + CH_3CHO$$

$$(13)\quad \overset{*}{R}-CH_2-\overset{H}{CH}-\underset{\underset{O}{\|}}{C}-C_6H_5 + C_6H_5\overset{\overset{+}{K}}{\bar{N}}CH_3 \longrightarrow \overset{*}{R}-CH_2-\overset{\overset{+}{K}}{\bar{C}H}-\underset{\underset{O}{\|}}{C}-C_6H_5$$

$$\xrightarrow{BH} \overset{*}{R}-H + CH_2{=}CH-\underset{\underset{O}{\|}}{C}-C_6H_5$$

$$(14)\quad \overset{*}{R}-CH_2-\overset{H}{\underset{C_6H_5}{C}}-C_6H_5 + C_6H_4(CH_2)_3\overset{\overset{+}{K}}{\bar{N}} \longrightarrow \overset{*}{R}-CH_2-\overset{\overset{+}{K}}{\underset{C_6H_5}{\bar{C}}}-C_6H_5 \longrightarrow$$

$$\overset{*}{R}-H + CH_2{=}\underset{C_6H_5}{C}-C_6H_5$$

All of the reactions listed proceeded with some net retention of configuration, but in some of the reactions the conditions required partially racemized the product after it was formed. Only reactions (4–7) and (10–12) were found to occur in alcohols–potassium alkoxides under conditions which did not racemize the product.

Similarly, 1-phenylmethoxyethane was produced under conditions that did not racemize it once formed[3] (reactions (15) and (16)). The

[3] D. J. Cram, K. R. Kopecky, F. Hauck, and A. Langemann, *J. Am. Chem. Soc.*, **81**, 5754 (1959).

same was true of 1-deuterophenylethane[4] and 1-dimethylaminophenylethane,[5] the products of reactions (17) and (18):

$$\overset{*}{R} = CH_3O-\overset{*}{C}(CH_3)(C_6H_5) \qquad CH_3O-\overset{*}{C}(CH_3)(C_6H_5)-H$$

1-Phenylmethoxyethane

$$(15)\quad \overset{*}{R}-C(OH)(C_6H_5)-CH_3 \underset{}{\overset{\overset{-\ +}{ROK}}{\rightleftharpoons}} \overset{*}{R}-C(\overset{-}{O}\overset{+}{K})(C_6H_5)-CH_3 \xrightarrow{ROH} \overset{*}{R}-H + C(=O)(C_6H_5)-CH_3$$

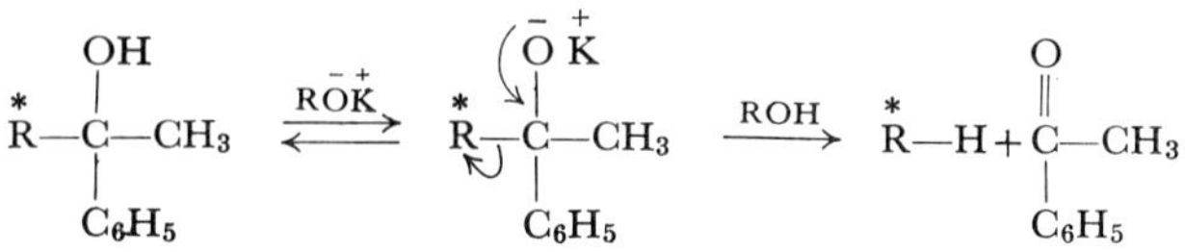

$$(16)\quad \overset{*}{R}-C(=O)-C_6H_5 \overset{\overset{-\ +}{ROK}}{\rightleftharpoons} \overset{*}{R}-C(\overset{-}{O}\overset{+}{K})(OR)-C_6H_5 \xrightarrow{ROH} \overset{*}{R}-H + ROC(=O)-C_6H_5$$

$$(17)\quad CH_3-\overset{*}{C}(H)(H_5C_6)-C(OD)(CH_3)-CH_3 \overset{\overset{-\ +}{ROK}}{\rightleftharpoons} CH_3-\overset{*}{C}(H)(H_5C_6)-C(\overset{-}{O}\overset{+}{K})(CH_3)-CH_3 \xrightarrow{ROD} CH_3-\overset{*}{C}(D)(H_5C_6)-H + (CH_3)_2CO$$

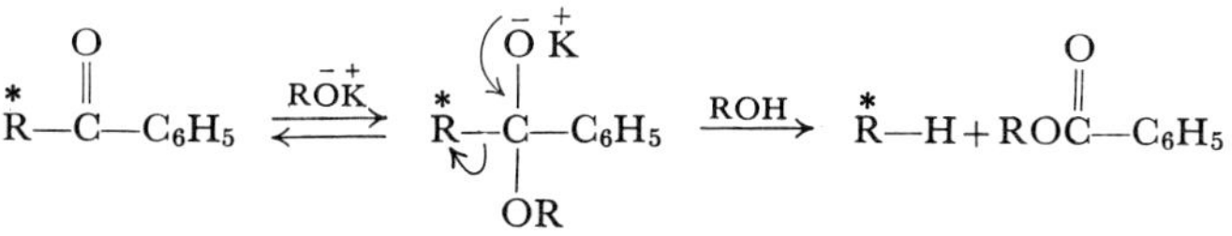

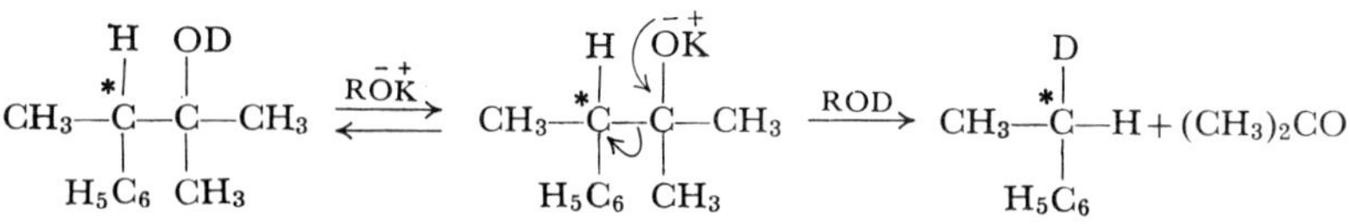

1-Deuterophenylethane

$$(18)\quad (CH_3)_2N-\overset{*}{C}(H_3C)(H_5C_6)-C(OH)(CH_3)-CH_3 \overset{\overset{-\ +}{ROK}}{\rightleftharpoons} (CH_3)_2N-\overset{*}{C}(H_3C)(H_5C_6)-C(\overset{-}{O}\overset{+}{K})(CH_3)-CH_3 \xrightarrow{ROH}$$

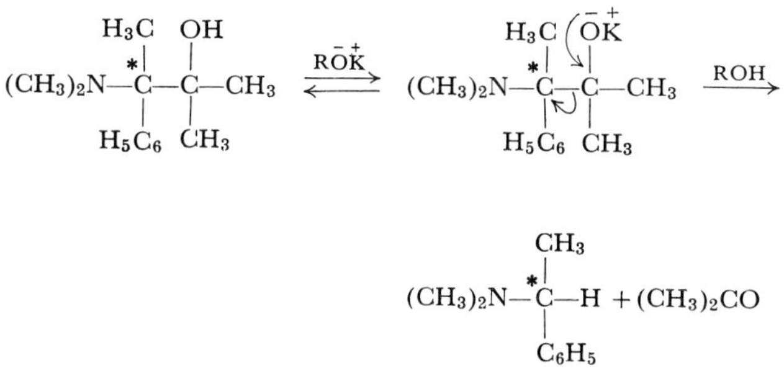

$$(CH_3)_2N-\overset{*}{C}(CH_3)(C_6H_5)-H + (CH_3)_2CO$$

1-Dimethylaminophenylethane

[4] D. J. Cram and B. Rickborn, *J. Am. Chem. Soc.*, **83**, 2178 (1961).
[5] D. J. Cram, L. K. Gaston, and H. Jäger, *J. Am. Chem. Soc.*, **83**, 2183 (1961).

Generation of 2-phenylbutane by a base-catalyzed decarboxylation reaction under conditions that preserved the optical activity of the product failed.[1] However, 2-cyano-2-phenylbutanoic acid underwent reaction (19) to give 2-phenylbutyronitrile.[6] The reaction did not occur

$$(19)\quad C_2H_5\text{—}\overset{*}{C}(CN)(C_6H_5)\text{—}CO_2H \underset{}{\overset{\overset{-}{R}O\overset{+}{K}}{\rightleftarrows}} C_2H_5\text{—}\overset{*}{C}(CN)(C_6H_5)\text{—}CO_2^{-}\overset{+}{K} \xrightarrow{ROH} C_2H_5\text{—}\overset{*}{C}(CN)(C_6H_5)\text{—}H + CO_2$$

2-Phenylbutyronitrile

$$(20)\quad C_6H_5CH_2\text{—}\overset{*}{C}(CN)(CH_3)\text{—}CO_2H \overset{\overset{-}{R}O\overset{+}{K}}{\rightleftarrows} C_6H_5CH_2\text{—}\overset{*}{C}(CN)(CH_3)\text{—}\overset{-}{C}O_2\overset{+}{K} \xrightarrow{ROH} C_6H_5CH_2\text{—}\overset{*}{C}(CN)(CH_3)\text{—}H + CO_2$$

2-Methyl-3-phenylpropionitrile

$$(21)\quad C_6H_5CH_2\text{—}\overset{*}{C}(NC)(H_3C)\text{—}C(OH)(CH_3)\text{—}CH_3 \overset{\overset{-}{R}O\overset{+}{K}}{\rightleftarrows} C_6H_5CH_2\text{—}\overset{*}{C}(NC)(H_3C)\text{—}C(\overset{-}{O}\overset{+}{K})(CH_3)\text{—}CH_3 \longrightarrow C_6H_5CH_2\text{—}\overset{*}{C}(CN)(CH_3)\text{—}H + (CH_3)_2CO$$

without base, but if the reaction mixture became basic, the product once formed racemized. The reaction was best run in the presence of 0.2 equivalent of added base and was stopped before the mixture became basic (about 70% completion). Reaction (20) was conducted under similar conditions with preservation of the configuration of the product as was (21).[7]

[6] D. J. Cram and P. Haberfield, *J. Am. Chem. Soc.*, **83**, 2354 (1961).
[7] D. J. Cram and P. Haberfield, *J. Am. Chem. Soc.*, **83**, 2364 (1961).

The usefulness of these transformations for stereochemical studies of the SE_1 reaction depends on a knowledge of the maximum rotations and relative configurations of starting materials and products. Since all of the starting materials were prepared from the corresponding acids without configurational modification, the problem was reduced to the maximum rotations and the relative configurations of the acids and the products of the cleavage reactions.

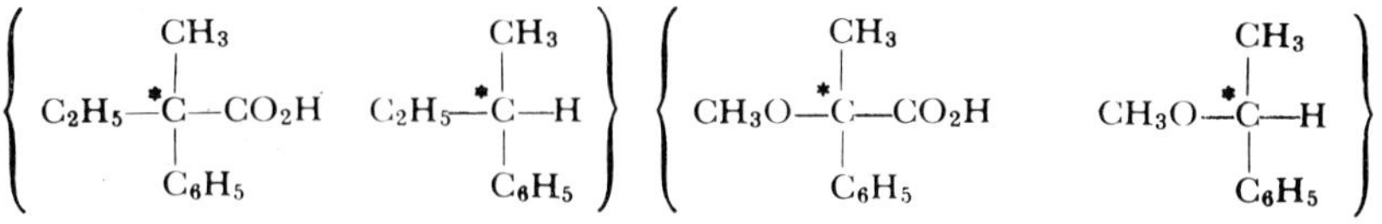

$$\left\{ C_2H_5-\overset{CH_3}{\underset{C_6H_5}{\overset{|}{\underset{|}{C^*}}}}-CO_2H \quad C_2H_5-\overset{CH_3}{\underset{C_6H_5}{\overset{|}{\underset{|}{C^*}}}}-H \right\} \quad \left\{ CH_3O-\overset{CH_3}{\underset{C_6H_5}{\overset{|}{\underset{|}{C^*}}}}-CO_2H \quad CH_3O-\overset{CH_3}{\underset{C_6H_5}{\overset{|}{\underset{|}{C^*}}}}-H \right\}$$

I II III IV

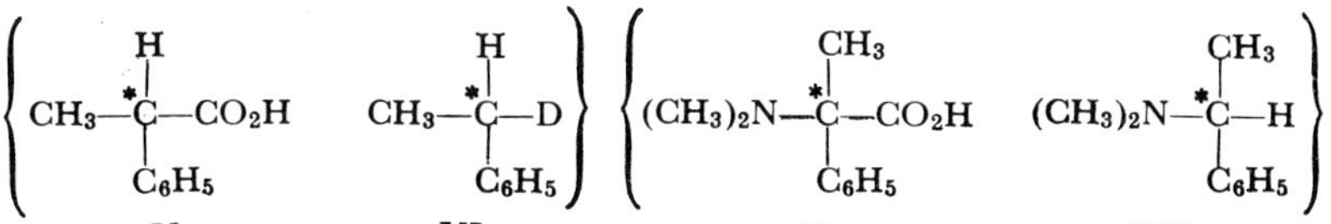

$$\left\{ CH_3-\overset{H}{\underset{C_6H_5}{\overset{|}{\underset{|}{C^*}}}}-CO_2H \quad CH_3-\overset{H}{\underset{C_6H_5}{\overset{|}{\underset{|}{C^*}}}}-D \right\} \quad \left\{ (CH_3)_2N-\overset{CH_3}{\underset{C_6H_5}{\overset{|}{\underset{|}{C^*}}}}-CO_2H \quad (CH_3)_2N-\overset{CH_3}{\underset{C_6H_5}{\overset{|}{\underset{|}{C^*}}}}-H \right\}$$

V VI VII VIII

$$\left\{ C_2H_5-\overset{CN}{\underset{C_6H_5}{\overset{|}{\underset{|}{C^*}}}}-CO_2H \quad C_2H_5-\overset{CN}{\underset{C_6H_5}{\overset{|}{\underset{|}{C^*}}}}-H \right\} \quad \left\{ C_6H_5CH_2-\overset{CN}{\underset{CH_3}{\overset{|}{\underset{|}{C^*}}}}-CO_2H \quad C_6H_5CH_2-\overset{CN}{\underset{CH_3}{\overset{|}{\underset{|}{C^*}}}}-H \right\}$$

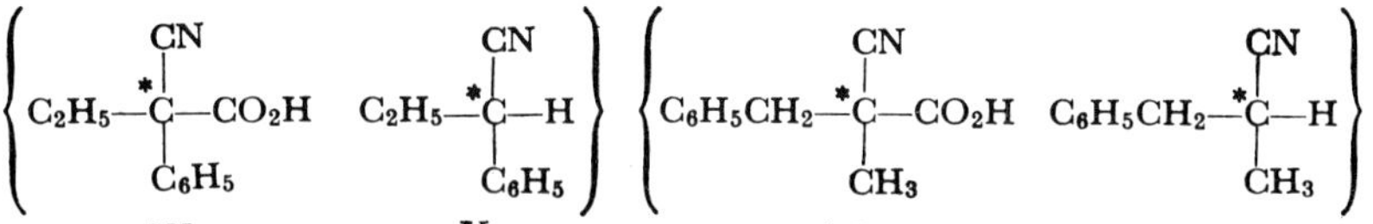

IX X XI XII

The maximum rotations of the acids listed were established by repeated exhaustive resolution,[2–7] usually of both enantiomers. Those of the products of cleavage were established by synthesis of the compounds from other substances of known optical purity by reactions that did not racemize the asymmetric center.[2–7] The relative configurations of three of the pairs of compounds listed were determined by interconversions through reactions of known stereochemical course (I and II, III and IV, and V and VI).[2–4,8] The configurations of VII and VIII were related by correlations between signs and magnitudes of rotation of compounds of known configuration with the signs and rotations of VII and VIII.[5] No direct evidence was obtained for the relative configurations of the pairs of compounds, IX and X,[6] and XI and XII.[7] However, the patterns of stereochemical results obtained in the cleavage reactions (see next sections) leave no reasonable doubt as to the relative configurations of these pairs of compounds.

[8] D. J. Cram and J. Allinger, *J. Am. Chem. Soc.*, **76**, 4516 (1954).

Stereochemical Course of Cleavage Reactions

In the first report of SE_1 reactions which occurred stereospecifically,[9] a series of alcohols and one ketone were cleaved in basic solutions to give 2-phenylbutane. The steric course varied from 93% retention (7% racemization) to 51% inversion (49% racemization). Later studies of these and other cleavage reactions established the following limiting values for the steric course of the reactions: 99% retention[10]; 100% racemization[11]; and 65% net inversion (35% racemization).[4]

TABLE I

Correlation of Stereochemical Course of Cleavage Reaction with Dielectric Constant of Medium

$$C_2H_5-\overset{*}{C}(CH_3)(C_6H_5)-C(OH)(CH_3)(C_2H_5) \xrightarrow[HB]{KB} C_2H_5-\overset{*}{C}(CH_3)(C_6H_5)-H$$

Solvent	$\epsilon^{°C}$	Molarity of H donors	$k_{ret.}/k_{inv.}$
C_6H_6	2^{20}	1	28
$CH_3NHC_6H_5$	6^{20}	7	17
$(CH_3)_3COH$	11^{19}	10	17
$CH_3CH_2CHOHCH_3$	19^{20}	10	12
$CH_3(CH_2)_3OH$	18^{25}	10	6
C_2H_5OH	27^{20}	19	4
$H_2NCH_2CH_2NH_2$	16^{18}	63	1.5
CH_3OH	34^{18}	28	1.2
$O(CH_2CH_2OH)_2$	35^{20}	16	0.56
$(CH_2OH)_2$	35^{20}	29	0.35
$HOCH_2CHOHCH_2OH$	46^{20}	30	0.33

The stereochemical courses of the cleavage reactions of tertiary and secondary alcohols and of ketones were most sensitive to the character of the solvent. A sample of the data is found in Table I.[12] The temperatures of the runs listed varied from 72° to 237°, the concentrations of starting

[9] D. J. Cram, J. Allinger, and A. Langemann, *Chem. & Ind.* (*London*), p. 919 (1955).

[10] D. J. Cram, F. Hauck, K. R. Kopecky, and W. D. Nielsen, *J. Am. Chem. Soc.*, **81**, 5767 (1959).

[11] D. J. Cram, J. L. Mateos, F. Hauck, A. Langemann, K. R. Kopecky, W. D. Nielsen, and J. Allinger, *J. Am. Chem. Soc.*, **81**, 5774 (1959).

[12] D. J. Cram, A. Langemann, and F. Hauck, *J. Am. Chem. Soc.*, **81**, 5750 (1959).

materials from 0.1 to 2 M, the concentrations of basic catalysts from 0.1 to 2 M, and the pK_a's of the electrophile from 16 to about 27. In spite of this wide range of variation in reaction conditions, there exists a good correlation between the steric course of the reaction and the dielectric constants of the solvents. Solvents with low dielectric constants poor at dissociating ion pairs give retention, and solvents with high dielectric constants good at dissociating ion pairs give inversion. The steric course also correlates roughly with the molarities of proton donors in the media. The lower the concentration of proton donors, the higher the value of $k_{ret.}/k_{inv.}$ (ratio of rate constants for producing product of retained and inverted configuration).

The concentration of proton donors in solvents of low dielectric constant does not seem to be important to the retention mechanism. For example, cleavage of 2,3-diphenyl-3-methyl-2-pentanol at 150° in *tert*-butyl alcohol–potassium *tert*-butoxide gave 2-phenylbutane with 95% net retention, whereas cleavage at 150° in dioxane, 0.4 M in diethylene-glycol–potassium glycoxide gave 91% net retention. The molarity of proton donors was varied in the two experiments by a factor of about 20 with only a small change in steric result.[3]

In media that gave high retention, variation of the cation of the base from potassium to sodium to lithium had only small effects on the steric course of the cleavage reactions. Changes to quaternary ammonium ion reduced the steric result from high retention to almost complete racemization (see Table II). Thus the retention mechanism seems dependent

TABLE II

Dependence of Steric Course of Cleavage on Cation of Basic Catalyst

$$C_2H_5-\overset{CH_3}{\underset{C_6H_5}{\overset{*|}{\underset{|}{C}}}}-\overset{OH}{\underset{C_6H_5}{\overset{|}{\underset{|}{C}}}}-R \xrightarrow[BH]{Base} C_2H_5-\overset{CH_3}{\underset{C_6H_5}{\overset{*|}{\underset{|}{C}}}}-H$$

R	Solvent	Base	Steric course (net)
CH_3	$O(CH_2CH_2)_2O$, *tert*-BuOH	*tert*-BuOK	96% Retention
CH_3	$O(CH_2CH_2)_2O$, *tert*-BuOH	*tert*-BuOLi	95% Retention
CH_3	*tert*-BuOH	*tert*-BuOK	84% Retention
CH_3	*tert*-BuOH	$C_6H_5CH_2N(CH_3)_3OH$	100% Racemization
C_6H_5	*tert*-BuOH	*tert*-BuOK	>90% Retention
C_6H_5	*tert*-BuOH	$(CH_3)_4NOH$	100% Racemization
C_6H_5	$HO(CH_2CH_2O)_2H$	$HO(CH_2CH_2O)_2K$	16% Inversion
C_6H_5	$HO(CH_2CH_2O)_2H$	$(CH_3)_4NOH$	20% Inversion

on the presence of metal cations. In contrast, the inversion mechanism appears not to depend on the nature of the cation, since both potassium and quaternary ammonium bases gave about the same steric result in diethylene glycol (see Table II).[10,11]

Use of dimethyl sulfoxide as solvent with a small amount of alcohol present to act as electrophile and potassium *tert*-butoxide as base gave complete racemization with the four systems listed.[5,11] The first system

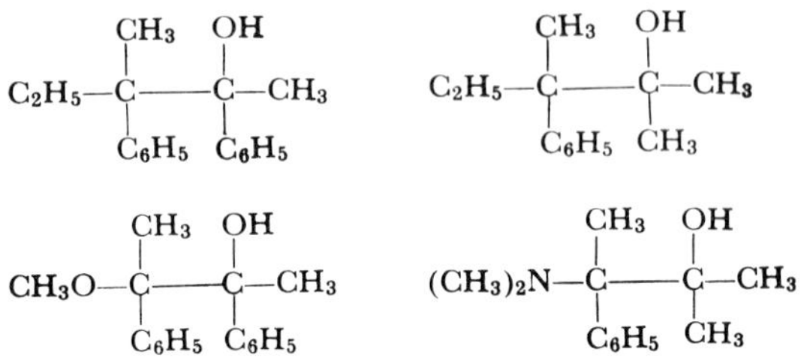

$$
\begin{array}{ccccc}
 & CH_3 & & OH & \\
 & | & & | & \\
C_2H_5- & C & - & C & -CH_3 \\
 & | & & | & \\
 & C_6H_5 & & C_6H_5 &
\end{array}
\qquad
\begin{array}{ccccc}
 & CH_3 & & OH & \\
 & | & & | & \\
C_2H_5- & C & - & C & -CH_3 \\
 & | & & | & \\
 & C_6H_5 & & CH_3 &
\end{array}
$$

$$
\begin{array}{ccccc}
 & CH_3 & & OH & \\
 & | & & | & \\
CH_3O- & C & - & C & -CH_3 \\
 & | & & | & \\
 & C_6H_5 & & C_6H_5 &
\end{array}
\qquad
\begin{array}{ccccc}
 & CH_3 & & OH & \\
 & | & & | & \\
(CH_3)_2N- & C & - & C & -CH_3 \\
 & | & & | & \\
 & C_6H_5 & & CH_3 &
\end{array}
$$

with potassium, sodium, or lithium *tert*-butoxide as base in dimethyl sulfoxide was observed to give 100% racemization, but the rates differed by orders of magnitude in the order *tert*-BuOK $\gg$ *tert*-BuONa $\gg$ *tert*-BuOLi.[11] Dimethyl sulfoxide possesses a high dielectric constant ($49^{20°}$),[13] but is a poor proton donor, having a pK_a of about 36 on the Streitwieser scale,[14] and 31 on the Steiner scale. Thus it possesses neither the properties exhibited by the retention solvents (low dielectric constant) nor those of inversion solvents (high dielectric constant, high proton donor concentration). The marked dependence of the rate of cleavage on the kind of metal cation probably reflects the difference in dissociation constants of the three *tert*-butoxides in dimethyl sulfoxide (see Chapter I).

Appropriate mixtures of proton-donating retention solvents and dimethyl sulfoxide produce media which give low net inversion[15a] (Table III). These experiments illustrate the fact that high concentrations of proton donors and high dielectric constant media are needed for operation of the inversion mechanism.

[13] (a) H. L. Schlafer and W. Schaffernicht, *Angew. Chem.*, **72**, 618 (1960); (b) A. J. Parker, *Quart. Rev.* (*London*), **16**, 163 (1962).

[14] (a) A. Streitwieser, Jr., J. I. Brauman, J. H. Hammons, and A. H. Pudjaatmaka, *J. Am. Chem. Soc.*, **87**, 384 (1965); (b) E. C. Steiner, J. M. Gilbert, *ibid.*, **87**, 382 (1965).

[15] (a) D. J. Cram and W. D. Nielsen, *J. Am. Chem. Soc.*, **83**, 2174 (1961); (b) P. G. Gassman and F. V. Zalar, *Tetrahedron Letters*, **44**, 3251 (1964); (c) D. J. Cram and H. P. Fischer, unpublished work.

TABLE III

Mixtures of Retention and Racemization Solvents Give Low Net Inversion

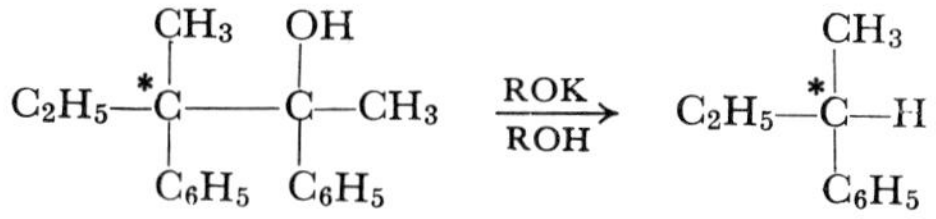

$$C_2H_5-\overset{*}{C}(CH_3)(C_6H_5)-C(OH)(CH_3)(C_6H_5) \xrightarrow[ROH]{ROK} C_2H_5-\overset{*}{C}(CH_3)(C_6H_5)-H$$

Solvent	Base	Steric result (net)
tert-BuOH	*tert*-BuOLi	85% Retention
70 Mole % *tert*-BuOH, 30 Mole % $(CH_3)_2SO$	*tert*-BuOLi	2% Inversion
$(CH_3)_2SO$	*tert*-BuOLi	100% Racemization
$CH_3CH_2CH_2OH$	$CH_3CH_2CH_2OLi$	13% Retention
80 Mole % $CH_3CH_2CH_2OH$, 20 Mole % $(CH_3)_2SO$	$CH_3CH_2CH_2OLi$	14% Inversion
$(CH_3)_2SO$	$CH_3CH_2CH_2OLi$	100% Racemization

Gassman and Zalar[15b] observed that nortricyclanone in dimethyl sulfoxide–potassium *tert*-butoxide gave [3.1.0] bicyclohexane-3-carboxylic acid with almost complete retention of configuration. These authors point to operation of either a mechanism with cyclopropyl anions as intermediates, or one in which the carbon–carbon bond cleavage occurs in the same transition state that involves carbon–hydrogen bond formation. In each, a proton donor is generated at the front side of the incipient carbanion by attack on the ketone by a dimethyl sulfoxide anion.

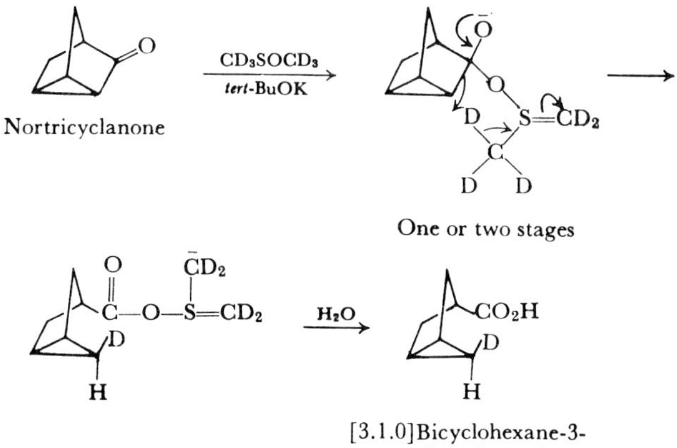

Nortricyclanone $\xrightarrow[\textit{tert}\text{-BuOK}]{CD_3SOCD_3}$ [intermediate: O⁻, O, D, S=CD₂, C, D D] $\longrightarrow$

One or two stages

[product: C(=O)–O–S(=CD₂)–$\bar{C}D_2$, D, H] $\xrightarrow{H_2O}$ [CO_2H, D, H]

[3.1.0]Bicyclohexane-3-carboxylic acid

Similar treatment at 25° of optically active 1,2-diphenyl-2-methyl-1-butanone (0.16 *M*) in dimethyl sulfoxide with potassium *tert*-butoxide (0.21 *M*) gave 2-phenylbutane in 54% yield.[15c] The reaction proceeded with 48% net retention of configuration. When it was carried out with dimsylsodium as base in place of potassium *tert*-butoxide, only a 2% yield of 2-phenylbutane was obtained, and no starting material could be recovered. When the reaction was carried out at 25° in a solution of dimethyl sulfoxide, 0.42 *M* in *tert*-butyl alcohol and 0.51 *M* in potassium *tert*-butoxide, the same optically active ketone (0.25 *M*) gave 2-phenylbutane in 74% yield. The reaction proceeded with 40% net retention of configuration. Thus competing mechanisms appear to operate, one leading to racemic product, and one to product of retained configuration. The fact that ketone and alcohol cleavages give somewhat different results in dimethyl sulfoxide suggests that the ketone cleavage involves an adduct of some sort between ketone and dimethyl sulfoxide, which on cleavage generates at the front side of an incipient carbanion a proton donor considerably more acidic than dimethyl sulfoxide. A plausible reaction sequence is formulated.

$$C_2H_5\text{—}\overset{*}{C}(CH_3)(C_6H_5)\text{—}C(=O)\text{—}C_6H_5 + \bar{C}H_2SOCH_3 \longrightarrow C_2H_5\text{—}\overset{*}{C}(CH_3)(C_6H_5)\text{—}C(\bar{O})(CH_2SOCH_3)\text{—}C_6H_5 \longrightarrow$$

1,2-Diphenyl-2-methyl-1-butanone

$$C_2H_5\text{—}\overset{*}{C}^{-}(CH_3)(C_6H_5) \quad CH_2COC_6H_5(SOCH_3) \longrightarrow C_2H_5\text{—}\overset{*}{C}(CH_3)(C_6H_5)\text{—}H + CH_3SO\bar{C}HCOC_6H_5$$

2-Phenylbutane

The stereochemical course of the cleavage reactions was found to be somewhat sensitive to the gross structure of the leaving group, but insensitive to the configuration of the leaving group.[10] Table IV records a sample of relevant data. These conclusions apply equally well to the retention and inversion solvents. This general independence of the steric course of the cleavage reaction to the character of the leaving group indicates that intermediates of very similar structures are partitioning between retention and inversion reaction paths. These intermediates differ only slightly when the leaving groups have gross structural differences, and not at all when the leaving groups differ only in configuration. The more bulky leaving groups provide the highest stereospecificities both in retention and inversion solvents.

TABLE IV

Insensitivity of the Stereochemistry of the Cleavage Reactions to the Gross Structure and Configuration of the Leaving Group

$$C_2H_5-\overset{*}{C}(CH_3)(C_6H_5)-\overset{*}{C}(\overset{-}{O}\overset{+}{K})(a)(b) \xrightarrow{ROH} C_2H_5-\overset{*}{C}(CH_3)(C_6H_5)-H$$

Solvent-electrophile	Leaving group substituents a	Leaving group substituents b	Steric result (net)
Dioxane as solvent, *tert*-butyl alcohol as electrophile	CH_3	CH_3	95% Retention
	CH_3	C_2H_5	96% Retention
	CH_3	C_6H_5	96% Retention
	H	C_6H_5	82% Retention
	C_6H_5	H	81% Retention
	C_6H_5	*tert*-BuO	74% Retention
Ethylene or diethylene glycol as solvent and electrophile	CH_3	CH_3	52% Inversion
	CH_3	C_2H_5	48% Inversion
	CH_3	C_6H_5	55% Inversion
	H	C_6H_5	42% Inversion
	C_6H_5	H	41% Inversion
	C_6H_5	$O(CH_2CH_2O)_2H$	38% Inversion

Table V records the results obtained when the structure of one substituent at the seat of substitution is varied from $(CH_3)_2N$[5] to C_2H_5[11] to CH_3O[3,11] to H.[4] In the retention solvent, *tert*-butyl alcohol, stereospecificity decreased somewhat with decreasing bulk of the substituent, but not dramatically. With all substituents, dimethyl sulfoxide as solvent produced racemic product. In ethylene or diethylene glycol as solvent, all substituents gave inversion. Small differences, but no trends are visible in the data.

Cleavage reactions carried out in deuterated and non-deuterated *tert*-butyl alcohol and ethylene glycol gave the same stereochemical results (Table VI).[4] No isotope effects on the partitioning of starting material between the three possible stereochemical paths are evident. Had any one of the three stereochemical processes involved a mechanism in which the carbon–carbon bond was broken and a carbon–hydrogen bond was made in the same transition state, and a competing process

TABLE V

Insensitivity of the Stereochemistry of the Cleavage Reactions to the Substituents at the Seat of Substitution

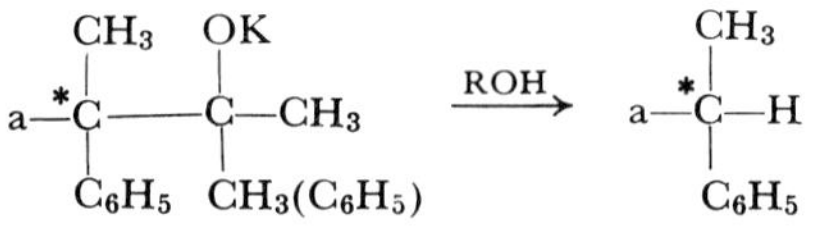

$$\text{a}-\overset{*}{\underset{C_6H_5}{\overset{CH_3}{C}}}-\underset{CH_3(C_6H_5)}{\overset{OK}{C}}-CH_3 \xrightarrow{\text{ROH}} \text{a}-\overset{*}{\underset{C_6H_5}{\overset{CH_3}{C}}}-H$$

Solvent	a	Steric result (net)
tert-BuOH	$(CH_3)_2N$	98% Retention
	C_2H_5	90% Retention
	CH_3O	84% Retention
	H	~71% Retention
$(CH_3)_2SO$–*tert*-BuOH	$(CH_3)_2N$	99% Racemization
	C_2H_5	100% Racemization
	CH_3O	100% Racemization
$(CH_2OH)_2$	$(CH_3)_2N$	34% Inversion
$(CH_2OH)_2$	C_2H_5	52% Inversion
$(CH_2OH)_2$	CH_3O	41% Inversion
$HO(CH_2CH_2O)_2H$	H	~51% Inversion

TABLE VI

Lack of Isotope Effects on Partitioning of Starting Materials Among the Three Possible Stereochemical Paths

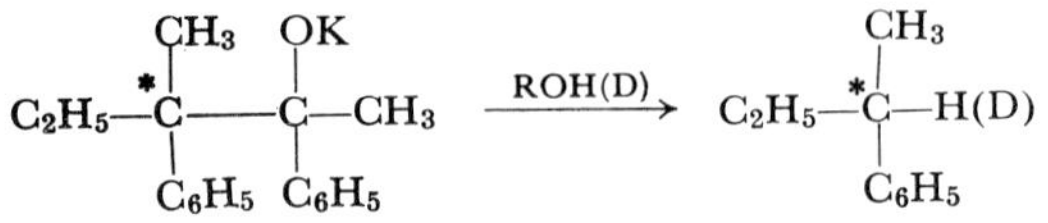

$$C_2H_5-\overset{*}{\underset{C_6H_5}{\overset{CH_3}{C}}}-\underset{C_6H_5}{\overset{OK}{C}}-CH_3 \xrightarrow{\text{ROH(D)}} C_2H_5-\overset{*}{\underset{C_6H_5}{\overset{CH_3}{C}}}-H(D)$$

Solvent	Steric result (net)
tert-BuOH	87% Retention
tert-BuOD	88% Retention
$(CH_2OH)_2$	63% Inversion
$(CH_2OD)_2$	65% Inversion

had involved two different transition states for these acts, this result would be highly improbable.

The question arises whether the leaving group itself can serve as a proton source in those cases in which the carbonyl compound produced has α-hydrogens. To answer this question, 3,4-dimethyl-4-phenyl-3-hexanol was titrated with phenylpotassium in benzene so that all hydroxyl groups were converted to potassium alkoxide groups. The cleavage reaction occurred with 73% net retention, which indicates that the 2-butanone (leaving group) must have served as the proton source.[11]

$$C_2H_5\overset{*}{C}(CH_3)(C_6H_5)-C(\overset{-}{O}\overset{+}{K})(CH_3)-C_2H_5 \xrightarrow{\text{No ROH}} C_2H_5\overset{*}{C}(CH_3)(C_6H_5)-H + CH_2{=}C(OK)-C_2H_5$$

73% Retention

$$C_2H_5\overset{*}{C}(CH_3)(C_6H_5)-C(\overset{-}{O}\overset{+}{K})(C_6H_5)-C_6H_5 \xrightarrow[\text{Diethylene-glycol } (0.1\ M)]{\text{Dioxane}} C_2H_5\overset{*}{C}(CH_3)(C_6H_5)-H + (C_6H_5)_2CO$$

99% Retention

This reaction was less stereospecific than a second one in which the benzene solution was 0.8 *M* in *tert*-butyl alcohol (93% net retention).[12] In a separate experiment, it was demonstrated that 2-phenyl-2-butylpotassium gave 2-phenylbutane when treated with 2-butanone.[2] That the retention mechanism is not dependent on an internal proton source is shown by the fact that 99% net retention was observed when diphenylketone (no protons available) served as leaving group.[16]

The source of the electrophile was clearly demonstrated in the following experiments.[4] The system which gave 2-phenylbutane and acetophenone as product was cleaved with deuterated *tert*-butyl alcohol and ethylene glycol as solvents. Acetophenone in the leaving process might act in competition with the solvent as proton donor. Examination of the 2-phenylbutane produced for deuterium indicated that essentially all of the electrophile had come from the alcoholic medium and none from the acetophenone. This result was obtained both in the retention and inversion solvents, and unequivocally identifies the proton source.

$$C_2H_5\overset{*}{C}(CH_3)(C_6H_5)-C(\overset{-}{O}\overset{+}{K})(C_6H_5)-CH_3 \xrightarrow[(CH_2OD)_2]{tert\text{-BuOD or}} C_2H_5\overset{*}{C}(CH_3)(C_6H_5)-D + CH_3COC_6H_5$$

88% Retention in *tert*-BuOD
65% Inversion in $(CH_2OD)_2$

[16] D. J. Cram, A. Langemann, W. Lwowski, and K. R. Kopecky, *J. Am. Chem. Soc.*, **81**, 5760 (1959).

Other experiments were conducted with systems in which a hydroxyl or amino group attached to the seat of substitution were potential proton donors.[5] The results of Table VII indicate that in dimethyl sulfoxide as

TABLE VII

Effect of Internal Proton Sources on Stereochemical Course of Cleavage Reactions

$$CH_3-\underset{C_6H_5}{\overset{a}{\overset{|}{\underset{|}{C^*}}}}-\underset{CH_3}{\overset{\overset{-\ +}{OK}}{\overset{|}{\underset{|}{C}}}}-CH_3 \xrightarrow{ROH} CH_3-\underset{C_6H_5}{\overset{a}{\overset{|}{\underset{|}{C^*}}}}-H$$

Solvent	Temp. (°C)	Substituent *a*	Steric course (net)
tert-BuOH	141	OH	50% Retention
$(CH_3)_2SO$, *tert*-BuOH	100	OH	100% Racemization
$(CH_2OH)_2$	210	OH	41% Retention
tert-BuOH	142	NH_2	15% Inversion
tert-BuOH	193	NH_2	16% Retention
$(CH_3)_2SO$, *tert*-BuOH	25	NH_2	100% Racemization

solvent, the normal result of 100% racemization was observed. However in *tert*-butyl alcohol or ethylene glycol the results were unusual. With a hydroxyl group at the seat of substitution, about the same amount of retention (40–50%) was observed in both *tert*-butyl alcohol and ethylene glycol. With an amino group at the seat of substitution, in *tert*-butyl alcohol the steric course could be varied from 15% inversion to 16% retention by simply changing the temperature. In other systems, temperature played a minor role. Side reactions prevented the amine system from being examined in ethylene glycol.

Clearly the hydroxyl and amino groups play a role, probably one of providing an internal source of protons. If the hydroxyl group serves as a ligand of a potassium ion, or is hydrogen bonded to the alkoxide anion,

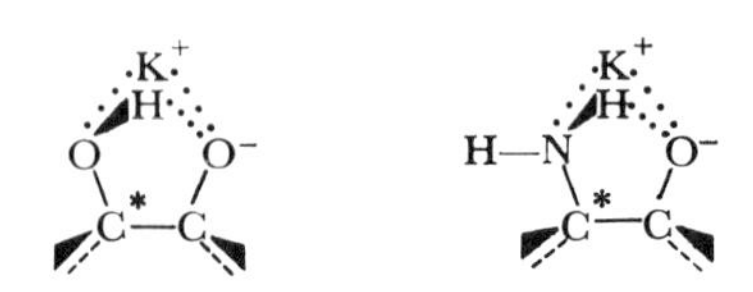

an internal source of protons is available at the front face of the seat of substitution. Thus net retention might be expected. The amino group has two hydrogens, one of which may be oriented at the front and the other at the back of the seat of substitution. As a consequence, a delicate balance between retention and inversion mechanisms is expected.

Carbanions as Intermediates in the Cleavage Reactions

The large body of experimental data of the previous section is *uniquely explained* by mechanisms that involve carbanions, or ion pairs that contain carbanions, as discrete intermediates. The salient facts associated with each of the three types of stereochemical results that must be mechanistically accommodated are as follows:

Retention Mechanism: Solvents of low dielectric constant (low ion pair dissociation power) give high retention of configuration mixed with small amounts of racemization. The stereochemical result is insensitive to the concentration, isotopic type, or acidity of the electrophiles over a large range of pK_a values. The stereochemical result is completely insensitive to the *configuration* of the leaving group, and varies only in a secondary way with the gross structure of the leaving group. The amount of retention changes in a minor way when the cation of the base is varied between the alkali metals, but is reduced to complete racemization when a quaternary ammonium cation is employed. The amount of retention changes only in a minor way when substituents are varied that do not stabilize carbanions and that are attached to the seat of substitution. Retention is observed in the complete absence of proton donors in the medium if the leaving group contains a proton donor, but if proton donors are present in the solvent, these serve as the electrophiles. Attachment of proton-donating substituents to the seat of substitution affects the stereochemical outcome of reaction depending on the number of protons available and their orientation.

Racemization Mechanism: Dimethyl sulfoxide, a non-proton-donating solvent ($pK_a \sim 31$)[14] of high dielectric constant ($\epsilon = 49^{2°e}$) gives complete racemization with all substrates except ketones, which cleave by a special mechanism. Complete racemization is also produced with quaternary ammonium bases in solvents of low dielectric constant.

Inversion Mechanism: Solvents of high dielectric constant which are good proton donors give moderately high inversion mixed with racemization. Appropriate mixtures of proton-donating retention solvents and dimethyl sulfoxide give low amounts of inversion. The stereochemical result is completely insensitive to the *configuration* of the leaving group, and varies only in a secondary way with the gross structure of the leaving group. The stereochemical result varies in only a minor way when the cation of the base is varied between the various alkali metals and quaternary ammonium groups. The amount of inversion changes only in a minor way when substituents are varied which are attached to the seat of substitution and which do not stabilize carbanions (e.g., methoxyl group). Attachment of a hydroxyl to the seat of substitution produces a retention result even in what is normally an inversion solvent.

The mechanisms which satisfactorily explain the facts are summarized in Figure 1.[11] In this scheme, the alkoxide exists as a solvated intimate

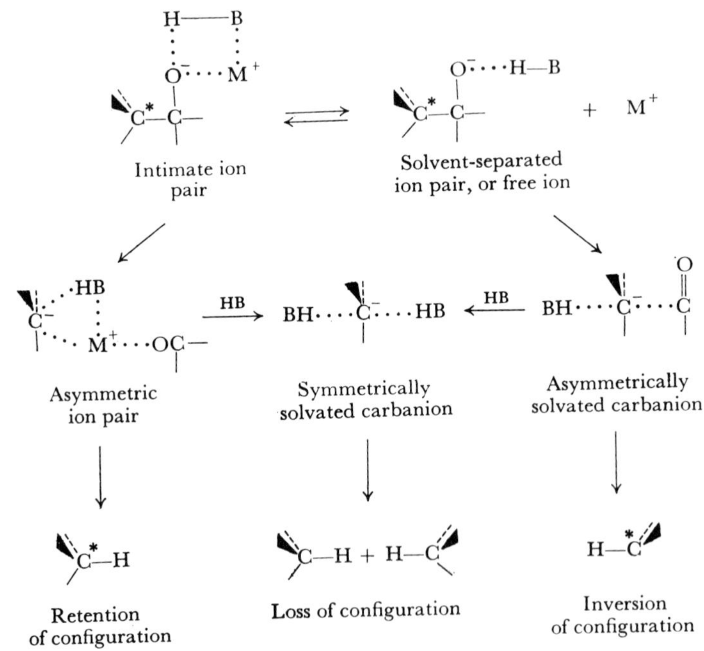

FIG. 1. Mechanisms for electrophilic substitution at saturated carbon with carbon as leaving group.

ion pair in equilibrium with solvent-separated, or free, ions. When M^+ is a metal ion, bound to that ion as ligands are proton donors (HB) from the medium. In nondissociating solvents, the intimate ion pair cleaves to form a metal carbanide ion pair solvated at the front face by proton donors and the leaving group. This species partitions between direct collapse to give product of retained configuration, and dissociation to give symmetrically solvated free carbanion, which collapses to give racemic product. Even if the metal carbanide ion pair is solvated by proton donors at its back face, proton capture from this side is inhibited since a *product-separated ion pair* would be produced, and separation of charge in a low dielectric constant medium is a high-energy process. The retention mechanism is dependent on the ability of the metal cation to gather proton donors as ligands, and on the rotation of the metal cation with its ligands occurring faster than the ion pair dissociates. Substitution of a quaternary ammonium ion for the metal cation results in an ion pair which lasts long enough to dissociate and give racemic product.

In solvents of high dielectric constant, either solvent-separated or free alkoxide ions cleave to give carbanions solvated at the front by the leaving group, and at the back by solvent. If the medium is proton donating, proton capture occurs from the back face of the anion while the front is shielded by the leaving group, and inverted product is produced. This process competes with that in which the leaving group is replaced at the front face of the anion by a solvent molecule, which gives a symmetrically solvated free anion, which in turn goes to racemic product.

With dimethyl sulfoxide as solvent, the solvent-separated ion pair or free anion cleaves to give carbanion solvated at the back face by dimethyl sulfoxide, and at the front by the leaving group. This species lasts long enough to pass into a symmetric environment, and racemic material is produced.

The rates of proton capture by the carbanions must be extremely fast. The difference in pK_a between the products (carbon acids) and the proton donors is probably between 10 and 20 units.

The similarities between these mechanisms and those of the base-catalyzed hydrogen–deuterium exchange reactions discussed in Chapter III are striking. The retention mechanisms for the exchange reaction were associated with media of low dielectric constants, and one of these depended on rotation of a metal cation with its ligands within an ion pair occurring faster than ion pair dissociation. Substitution of a quaternary ammonium ion for the metal cation resulted in complete racemization. Likewise, hydrogen–deuterium exchange occurred with inversion in proton-donating solvents of high dielectric constant. The inversion mechanism depended on asymmetric solvation of the free carbanion. Racemization was associated with dimethyl sulfoxide as solvent in the hydrogen–deuterium exchange experiments. The carbanion was solvated at the back by dimethyl sulfoxide molecules, and lasted long enough to pass into a symmetrical environment and give racemic product. The striking parallel in the spectra of stereochemical results and of reaction conditions for carbon and hydrogen as leaving groups lends considerable support to the mechanistic interpretations.

Competition between Anionic and Radical Cleavage Reactions

When diphenylketone was the leaving group, competition between anionic and radical cleavage of alkoxides was encountered.[16] When optically pure 2-methyl-1,1,2-triphenyl-1-butanol (XIII) was titrated with phenyllithium in dry benzene and heated at 60°, a 35% yield of a mixture of (±)- and *meso*-3,4-dimethyl-3,4-diphenylhexane (XIV) was isolated, along with a 24% yield of *racemic* starting material. In a second experiment, alcohol XIII was heated with dioxane-*tert*-butyl alcohol–lithium *tert*-butoxide. In the early part of the reaction, the medium

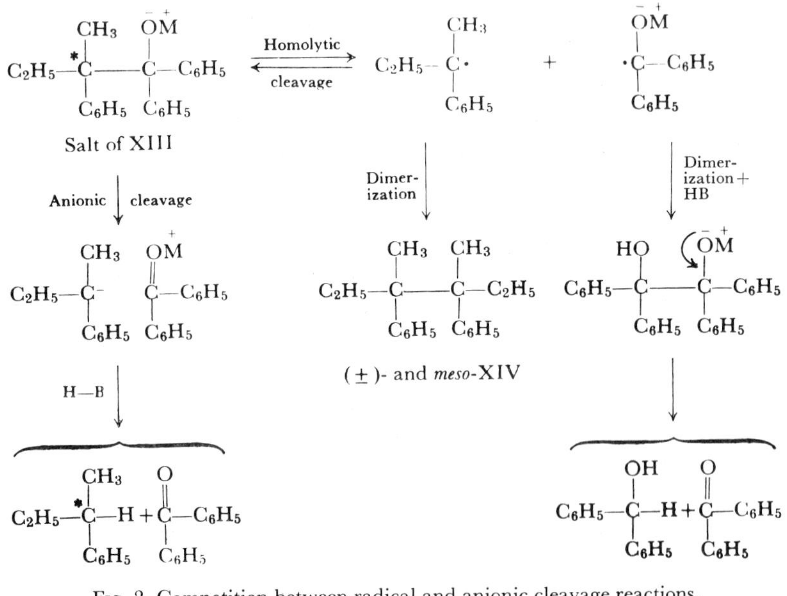

FIG. 2. Competition between radical and anionic cleavage reactions.

turned green, then yellow, then colorless. The products were 25% 2-phenylbutane, 61% hydrocarbon XIV, 66% benzophenone, and 27% benzhydrol. In a third experiment, it was demonstrated that tetraphenylethylene glycol cleaves under the same conditions to give benzhydrol and benzophenone. Figure 2 provides an explanation for these results.

In this scheme, the metal alkoxide of alcohol XIII undergoes competitive homolytic and anionic cleavage, the former being reversible. Homolytic cleavage gives the 2-phenyl-2-butyl radical and the ketyl of benzophenone, which were responsible for the observed color. The 2-phenyl-2-butyl radical loses its configuration, and its recombination with the ketyl gives racemic starting material. Dimerization of the 2-phenyl-2-butyl radicals gives a mixture of racemic and *meso* hydrocarbon XIV, whereas dimerization of the ketyl gives tetraphenylethylene glycol salt. The latter undergoes anionic cleavage to benzhydrol and benzophenone. The original homolytic cleavage competes with the anionic cleavage, which provides 2-phenylbutane and benzophenone.

As might be expected, the 2-phenylbutane obtained by anionic cleavage was of lower optical activity than would have been obtained had the homolytic cleavage been irreversible. An examination was made of the optical purity of the 2-phenylbutane produced as a function of time when alcohol XIII was heated (91°) in dioxane, 0.1 *M*, in diethylene glycol, 0.05 *M*, in potassium glycoxide. Early in the reaction, 2-phenylbutane was produced which was 94% optically pure (retention) whereas at the end of the reaction, the 2-phenylbutane was 81% optically pure. A plot of optical purity vs. time gave a curve which when extrapolated to zero time gave 2-phenylbutane of essentially complete optical purity.[16] These results reflect the racemization of the starting material, which competed with the anionic cleavage reaction. It was demonstrated by others[17] that the 2-phenyl-2-butyl radical does not abstract hydrogen from 2-phenylbutane at a rate which can compete with that of dimerization of the radical. Determination of the yields of the various products led to estimates of the relative rates of the competing processes. The rate of combination of two 2-phenyl-2-butyl radicals must have been about the same as the rate of recombination of 2-phenyl-2-butyl and benzophenone ketyl radicals. The ratio of rates of the heterolytic to homolytic cleavage was estimated to be 1.8 in the above medium.

Salt	$k_{\text{heterolytic}}/k_{\text{homolytic}}$
tert-BuOK	2.2
tert-BuONa	1.6
tert-BuOLi	0.05

In another series of experiments carried out with XIII in dioxane, the ratio of rate constants, $k_{\text{heterolytic}}/k_{\text{homolytic}}$, was found to vary by a factor of about 40 when the base was changed from potassium to sodium to

[17] E. L. Eliel, P. H. Wilken, F. T. Lang, and S. H. Wilen, *J. Am. Chem. Soc.*, **80**, 3303 (1958).

lithium *tert*-butoxide.[16] This result is explained in terms of the amount of covalent character in the O—M bond of the starting alkoxide. The degree of covalent character should vary with the metal in the order, Li > Na > K. In the heterolytic cleavage both the C—C and O—M bonds are broken, whereas in the homolytic cleavage only the C—C bond is broken. Thus the more covalent the O—M bond, the slower the heterolytic cleavage is expected to be relative to that of the homolytic variety.

Cleavage of optically pure alcohol XIII in diethylene glycol–potassium glycoxide for a period almost long enough to complete the reaction led to a 6% recovery of optically pure starting material, and to 2-phenylbutane of 36% optical purity (inversion). In this medium, starting material does not seem to be regenerated from its derived radicals. Possibly the 2-phenyl-2-butyl radical is trapped by the potassium glycoxide molecules supplying hydrogen atoms to the radical to give racemic 2-phenylbutane.[16] The optical purity of the 2-phenylbutane obtained (36%) would have been considerably higher had this alternate radical route for its production been absent.

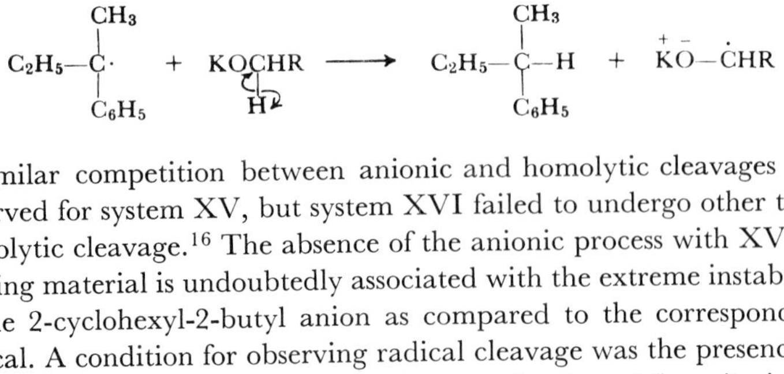

$$C_2H_5-\overset{\displaystyle CH_3}{\underset{\displaystyle C_6H_5}{\overset{|}{\underset{|}{C}}}}\cdot \;+\; KOCHR(H) \longrightarrow C_2H_5-\overset{\displaystyle CH_3}{\underset{\displaystyle C_6H_5}{\overset{|}{\underset{|}{C}}}}-H \;+\; \overset{+}{K}\overset{-}{O}-\dot{C}HR$$

Similar competition between anionic and homolytic cleavages was observed for system XV, but system XVI failed to undergo other than homolytic cleavage.[16] The absence of the anionic process with XVI as starting material is undoubtedly associated with the extreme instability of the 2-cyclohexyl-2-butyl anion as compared to the corresponding radical. A condition for observing radical cleavage was the presence of two phenyls in the leaving group, since application of the criteria for

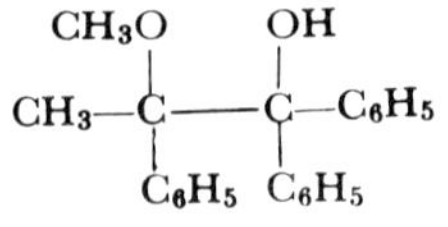

$$CH_3-\overset{\displaystyle CH_3O}{\underset{\displaystyle C_6H_5}{\overset{|}{\underset{|}{C}}}}-\overset{\displaystyle OH}{\underset{\displaystyle C_6H_5}{\overset{|}{\underset{|}{C}}}}-C_6H_5 \qquad\qquad C_2H_5-\overset{\displaystyle CH_3}{\underset{\displaystyle c\text{-}C_6H_{11}}{\overset{|}{\underset{|}{C}}}}-\overset{\displaystyle OH}{\underset{\displaystyle C_6H_5}{\overset{|}{\underset{|}{C}}}}-C_6H_5$$

XV 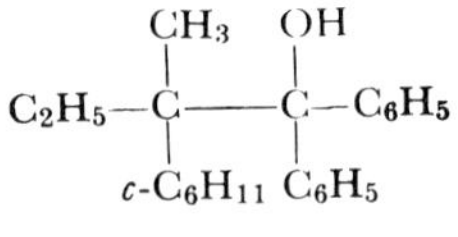 XVI

radical formation (color, racemized recovered starting material, radical dimer formation) to other systems discussed in this chapter gave negative results. Apparently, the homolytic cleavage is promoted by the formation of the relatively stable benzophenone ketyl.

OTHER LEAVING GROUPS IN ELECTROPHILIC SUBSTITUTION AT SATURATED CARBON

The previous section has been concerned with carbanions generated with carbon as the leaving group. In this section, nitrogen and oxygen as the leaving group in the SE reaction are discussed, and then a comparison of the four kinds of leaving groups is made.

Nitrogen as Leaving Group

Study of nitrogen as a leaving group seemed feasible since the latter stages of the Wolff-Kishner reaction probably involve loss of nitrogen from RN_2^- to give a carbanion, which subsequently captures a proton from solvent.[18] The possibility of breaking into the late stages of a Wolff-Kishner reduction was suggested by two analogies. In the McFadyen-Stevens reaction,[19] an arenesulfonhydrazide is treated with base to give ultimately an aldehyde and nitrogen. Treatment of sulfonamides of arylhydrazine with base has in some cases provided the aromatic hydrocarbon.[20] Another approach was suggested by the fact that arylhydrazines can be readily oxidized to give aromatic hydrocarbons.[21]

Wolff-Kishner Reduction:

$$R_2C{=}N{-}NH{-}H + \bar{B} \longrightarrow [R_2C{\cdots}N{\cdots}N{-}H]^- + HB \longrightarrow$$

$$R_2CH{-}N{=}N{-}H + \bar{B} \longrightarrow R_2CH{-}N{=}\bar{N} + HB \longrightarrow$$

$$R_2\bar{C}H + N_2 + HB \longrightarrow R_2CH_2 + N_2 + \bar{B}$$

McFadyen-Stevens Reaction:

$$Ar{-}C({=}O){-}N(H){-}NH{-}SO_2Ar + B \longrightarrow Ar{-}C({=}O){-}N{=}N{-}H + HB + HO_2SAr$$

$$Ar{-}C({=}O){-}N{=}N{-}H + \bar{B} \longrightarrow Ar{-}C({=}O){-}N{=}\bar{N} + HB \longrightarrow Ar{-}C({=}O){-}H + N_2 + \bar{B}$$

[18] (a) W. Seibert, *Ber.*, **80**, 494 (1947); (b) *ibid.*, **81**, 266 (1948); (c) H. H. Symant, H. F. Harnsberger, T. J. Butler, and W. P. Barie, *J. Am. Chem. Soc.*, **74**, 2724 (1952).

[19] J. S. McFadyen and T. S. Stevens, *J. Chem. Soc.*, p. 584 (1936).

[20] R. Escales, *Ber.*, **18**, 893 (1885).

[21] L. Kalb and O. Gross, *Ber.*, **59**, 727 (1926).

Arylhydrazine Oxidation:

$$\text{Ar—NH—NH}_2+[\text{O}] \longrightarrow \text{Ar—N=N—H}+\text{H}_2\text{O}$$

$$\text{Ar—N=N—H}+\bar{\text{B}} \longrightarrow \text{Ar—N=}\bar{\text{N}}+\text{HB} \longrightarrow \text{Ar—H}+\text{N}_2+\bar{\text{B}}$$

Accordingly, optically active 2-phenyl-2-butylhydrazine (XVII), its sulfonamide (XVIII), and the sulfonamide of 2-phenyl-2-butylamine (XIX)[22] were prepared.[23] The maximum rotations were determined, as well as their configurations relative to that of 2-phenylbutane. Treatment of hydrazine XVII with either basic potassium periodate or basic bromine gave 2-phenylbutane, as did treatment of sulfonamide XVIII with base, or XIX with base and hydroxylamine-*O*-sulfonic acid.[22]

$$\overset{*}{\text{R}} = \text{C}_2\text{H}_5\text{—}\overset{*}{\text{C}}(\text{CH}_3)(\text{C}_6\text{H}_5) \qquad \text{Ts} = \text{SO}_2\text{C}_6\text{H}_4\text{CH}_3\text{-}p$$

$$\text{R} = \text{C}_2\text{H}_5\text{—}\overset{*}{\text{C}}(\text{CH}_3)(\text{C}_6\text{H}_5) \qquad \text{Ts} = \text{SO}_2\text{C}_6\text{H}_4\text{CH}_3\text{-}p$$

$\overset{*}{\text{R}}\text{—NH—NH}_2 + [\text{O}]$ (XVII)

$\overset{*}{\text{R}}\text{—NH—NHTs} + \bar{\text{B}}$ (XVIII)

$\overset{*}{\text{R}}\text{—N(Ts)—NH}_2 + \bar{\text{B}}$ from $(\overset{*}{\text{R}}\text{—N(Ts)Na} + \text{NH}_2\text{OS}\bar{\text{O}}_3)$ (Na salt of XIX)

$$\longrightarrow \overset{*}{\text{R}}\text{—N=N—H} \xrightarrow{\text{B}^-} \overset{*}{\text{R}}\text{—N=}\bar{\text{N}} \longrightarrow \overset{*}{\text{R}}^- \xrightarrow{\text{HB}} \overset{*}{\text{R}}\text{—H}$$

An alkyl diimide

When the three reactions were carried out under conditions as close to one another as possible (optically active starting materials in 10% ethanol–90% water, 1.7 *M*, in potassium hydroxide), 2-phenylbutane of the same optical purity (within experimental error) was produced. The reactions were found to go with $32 \pm 6\%$ retention (net) of configuration. In many other media, XVII and XVIII were converted to 2-phenylbutane, in each case the two reactions occurring with the same

[22] A. Nickon and A. Sinz, *J. Am. Chem. Soc.*, **82**, 753 (1960).
[23] D. J. Cram and J. S. Bradshaw, *J. Am. Chem. Soc.*, **85**, 1108 (1963).

stereospecificity and in the same steric direction. The results ranged from $79 \pm 2\%$ retention to 100% racemization, and covered the solvents *tert*-butyl alcohol, ethanol, methanol, water, and dimethyl sulfoxide. These results provide strong evidence that at least one intermediate is common to all three reactions, and this intermediate partitions between racemic and active product. The alkyldiimide is such an intermediate.[23] The formation of alkyl diimide resembles the formation of diimide itself as a reaction intermediate by oxidizing hydrazine, or through elimination reactions.[24]

The presence of a non-base-catalyzed reaction which led to completely racemic 2-phenylbutane was demonstrated for both direct cleavage of sulfonamide XVIII and the oxidative cleavage of hydrazine XVII. This reaction was interpreted as involving homolytic cleavage of the alkyl diimide by either of the two paths formulated. This reaction

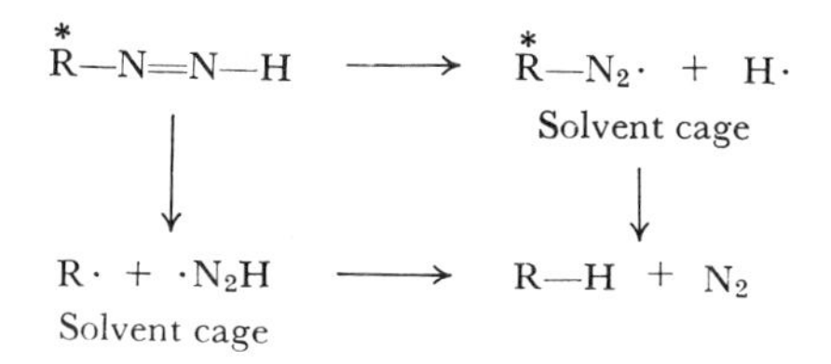

could be suppressed in favor of the stereospecific base-catalyzed anionic cleavage reactions by addition of enough base, at least in all solvents except water.[23] Unless specified otherwise, the reactions discussed below were carried out in the presence of enough base to eliminate the homolytic cleavage reaction.

As in the case of carbon as leaving group, a rough correlation exists between the dielectric constant and the stereochemical course of the substitution reaction with nitrogen as leaving group (see Table VIII). The results range from 78% net retention (*tert*-butyl alcohol) to 32% net inversion (water). However, the scale is shifted toward the retention side with nitrogen as leaving group. Both dimethyl sulfoxide and methanol are retention solvents. With carbon as leaving group, dimethyl sulfoxide was a racemizing and methanol an inversion solvent. The propensity of the alkyl diimide for a retention mechanism is probably associated with the presence of an acidic proton in the leaving group itself that shifts the whole scale toward retention (see below).

[24] (a) S. Hunig, H. R. Muller, and W. Thier, *Tetrahedron Letters*, p. 353 (1961); (b) E. J. Corey, W. L. Mock, and D. J. Pasto, *ibid.*, p. 347 (1961); (c) F. Aylward and M. Savistova, *Chem. & Ind. (London)*, pp. 404, 433 (1961); (d) E. E. van Tamelen, R. S. Dewey, and R. J. Timmons, *J. Am. Chem. Soc.*, **83**, 3725 (1961).

TABLE VIII

Correlation Between Steric Course of Base-Catalyzed Cleavage Reaction and Dielectric Constant of Solvent

$$\overset{*}{R} = C_2H_5-\overset{*}{C}(CH_3)(C_6H_5)$$

Run no.	Starting material	Solvent	$\epsilon^{°C}$	Steric course (net)
1	$\overset{*}{R}NHNHTs + tert\text{-}BuOK$	*tert*-BuOH	11^{19}	78% Retention
2	$\overset{*}{R}NHNH_2 + KIO_3 + tert\text{-}BuOK$	*tert*-BuOH	11^{19}	80% Retention
3	$\overset{*}{R}NHNH_2 + I_2 + tert\text{-}BuOK$	*tert*-BuOH	11^{19}	78% Retention
4	$\overset{*}{R}NHNHTs + n\text{-}BuOK$	*n*-BuOH	18^{25}	68% Retention
5	$\overset{*}{R}NHNHTs + C_2H_5OK$	C_2H_5OH	27^{20}	60% Retention
6	$\overset{*}{R}NHNH_2 + KIO_3 + C_2H_5OK$	C_2H_5OH	27^{20}	55% Retention
7	$\overset{*}{R}NHNHTs + CH_3OK$	CH_3OH	34^{18}	44% Retention
8	$\overset{*}{R}NHNH_2 + KIO_3 + CH_3OK$	CH_3OH	34^{18}	42% Retention
9	$\overset{*}{R}NHNHTs + (CH_3)_3COK$	$(CH_3)_2SO$	49^{20}	44% Retention
10	$\overset{*}{R}NHNH_2 + KIO_3 + (CH_3)_3COK$	$(CH_3)_2SO$	49^{20}	39% Retention
11	$\overset{*}{R}NHNHTs + KOH$	H_2O	80^{20}	5% Inversion
12	$\overset{*}{R}NHNH_2 + KIO_3 + KOH$	H_2O	80^{20}	32% Inversion
13	$\overset{*}{R}NHNH_2 + Br_2 + KOH$	H_2O	80^{20}	33% Inversion

The mechanisms envisioned for explaining the stereochemistry of the cleavage reaction of the alkyl diimide are detailed in Figure 3. In solvents of low dielectric constant, an ion pair is the active form of the base, and at least 2 moles of proton donor are oriented at the front face of the carbanion generated by loss of nitrogen. In solvents of higher dielectric constant, such as methanol or dimethyl sulfoxide, dissociated anion is the active form of the base. Abstraction of a proton from nitrogen gives a mole of alcohol oriented at the front face of the carbanion generated by loss of nitrogen. The activation energy for the formation of the anion is very low, and little if any solvation at the back face of the carbanion is needed for breaking the carbon–nitrogen bond. Hence frontside capture of the carbanion is preferred because of favorable orientation of proton donors at the front. Net inversion is observed in water because random orientation of hydroxyl groups at the back side of the carbanion provides a better statistical chance of backside capture than at the front, which momentarily is shielded by the mole of nitrogen generated.

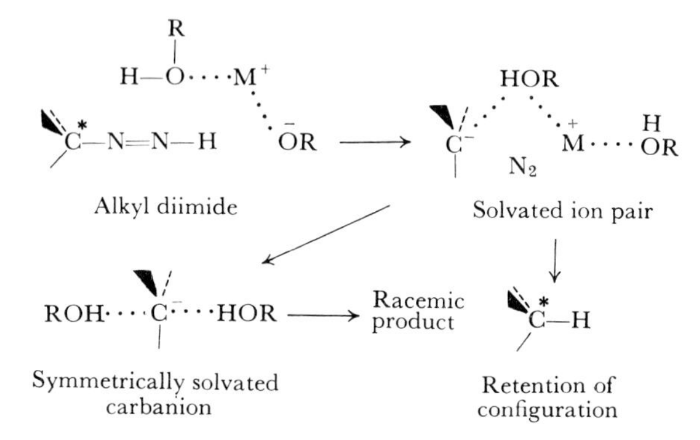

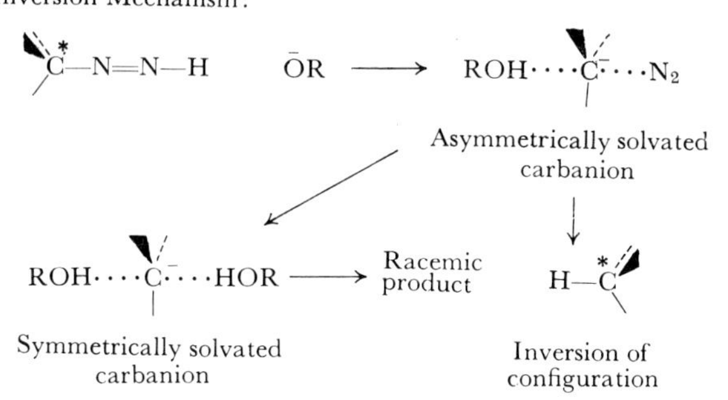

FIG. 3. Mechanisms for electrophilic substitution at saturated carbon with nitrogen as leaving group.

Comparison of runs 11–13 of Table VIII indicates that in water as solvent different amounts of inversion are obtained depending on whether the hydrazine (XVII) is oxidized, or the tosylhydrazide (XVIII) cleaved. Even more anomalous results are obtained when the concentration of potassium hydroxide is increased from the 0.3 *M* concentration used in runs 11–13 to 9.7 *M*. In this base-enriched medium, oxidation of hydrazine XVII with potassium iodate gave

12% net inversion, whereas oxidation with bromine gave 77% retention, and cleavage of tosylhydrazide XVIII gave 70% retention. Here is evidence of two stereospecific mechanisms, one leading to inversion and the other to retention in the same medium.

A feasible explanation for these interesting results is as follows. In water, the tosylhydrazide (XVIII) can undergo either a 1,2- or a 1,1-elimination reaction, the former being favored at low, and the latter at high base concentrations. In the 1,2-elimination, an alkyl diimide is formed in which the proton is several bond distances from the seat of substitution, and in water this species gives low inversion. The 1,1-elimination gives an azamine, whose proton is considerably closer to the seat of substitution, and which gives high retention even in water. The fact that potassium iodate gave net inversion suggests that alkyl diimide was the intermediate, but with bromine, the azamine was formed, probably by 1,1-elimination of HBr from $\overset{*}{R}NHNHBr$.

$$\overset{*}{R}-N(H)-N(H)-Ts + KOH \rightleftharpoons \overset{*}{R}-N(H)-\underline{N}^{-}-Ts + K^{+} + H_2O$$

XVIII → Conjugate base

XVIII $\xrightarrow{-TsH}$ $\overset{*}{R}-N{=}N-H$ (Alkyl diimide) $\xrightarrow{H_2O \mid KOH}$ $H-\overset{*}{R}$ — 32% Inversion (net)

Conjugate base $\xrightarrow{-Ts^{-}}$ $\overset{*}{R}-\overset{+}{N}(H){=}\overset{-}{N}$ (Alkylazamine) $\xrightarrow{H_2O \mid KOH}$ $\overset{*}{R}-H$ — 70% Retention (net)

Some analogy for the formulation of alkylazamine as an intermediate is found in conversions of compounds XX and XXI to XXII,[25c] and other similar reactions in which alkylazamines have been postulated.[25]

$$\left.\begin{array}{l}(C_6H_5CH_2)_2NNH_2 \xrightarrow{[O]} \\ \quad XX \\ (C_6H_5CH_2)_2NNHSO_2Ar \xrightarrow{Base} \\ \quad XXI\end{array}\right\} (C_6H_5CH_2)_2\overset{+}{N}{=}\overset{-}{N} \xrightarrow{-N_2} C_6H_5CH_2CH_2C_6H_5$$

An azamine → XXII

[25] (a) J. Kenner and E. C. Knight, *Ber.*, **69**, 341 (1936); (b) L. A. Carpino, *J. Am. Chem. Soc.*, **79**, 4427 (1957); (c) P. Carter and T. S. Stevens, *J. Chem. Soc.*, p. 1743 (1961).

Oxygen as Leaving Group

Few reactions are known in which oxygen serves as leaving group in electrophilic substitution at saturated carbon. Two kinds of reactions have been used to study the stereochemistry, each of which involves production of 2-phenylbutane (II) as product.[26] In the first, 2-benzyl-oxy-2-phenylbutane (XXIII) when subjected to the action of very strong bases undergoes cleavage. In the second, 2-phenyl-2-butanol (XXIV) or 2-methoxy-2-phenylbutane (XXV) is treated with alkali metals in the presence of proton donors, and again cleavage occurs. As in the previous investigations, the relative configurations of the three starting materials and products were determined,[3,8] as well as their maximum rotations.[27,28]

$$C_2H_5-\overset{*}{C}(CH_3)(C_6H_5)-O-CH(H)C_6H_5 + \bar{B} \longrightarrow C_2H_5-\overset{*}{C}(CH_3)(C_6H_5)-H + HB + C_6H_5CHO$$

XXIII → II

$$C_2H_5-\overset{*}{C}(CH_3)(C_6H_5)-OH + 2M\cdot + HB \longrightarrow C_2H_5-\overset{*}{C}(CH_3)(C_6H_5)-H + MOH + MB$$

XXIV → II

$$C_2H_5-\overset{*}{C}(CH_3)(C_6H_5)-OCH_3 + 2M\cdot + BH \longrightarrow C_2H_5-\overset{*}{C}(CH_3)(C_6H_5)-H + MOCH_3 + MB$$

XXV → II

Unfortunately, benzyl ether XXIII underwent cleavage only under conditions that racemized the 2-phenylbutane after it was formed (*N*-methylaniline, potassium *N*-methylanilide at 180°). The reaction was shown to go with a minimum of 29% retention (net). The hydrocarbon was swept out of the reaction mixture as formed with a stream of nitrogen, but some racemization undoubtedly occurred during this separation. This result is interpreted in Figure 4. The relation of this reaction to the base-catalyzed rearrangement of ethers (Wittig rearrangement) is discussed in Chapter VI.

[26] D. J. Cram, C. A. Kingsbury, and A. Langemann, *J. Am. Chem. Soc.*, **81,** 5785 (1959).
[27] D. J. Cram, *J. Am. Chem. Soc.*, **74,** 2150 (1952).
[28] (a) H. H. Zeiss, *J. Am. Chem. Soc.*, **73,** 2391, 3154 (1953); (b) A. Davies, J. Kenyon, and L. Salome, *J. Chem. Soc.*, p. 3148 (1957).

In the mechanism outlined, the base as a solvated ion pair abstracts a proton from the α-position of the benzyl ether at the same time the C—O bond breaks. A solvated asymmetric ion pair is formed, which collapses by proton capture from the front face of the carbanion. Proton capture from the back face would produce a product-separated ion pair. The low dielectric constant of *N*-methylaniline makes the latter a slower process than the former, which does not require charge separation. To the extent that ion pairs dissociate, symmetrically solvated carbanions are formed, which give racemic material.

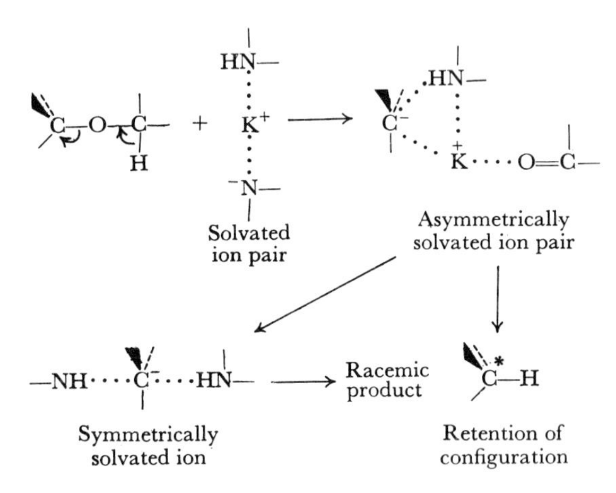

FIG. 4. Mechanism for electrophilic substitution at saturated carbon with oxygen as leaving group.

Table IX records the results of the reductive cleavages of the benzyl alcohol XXIV and the ether XXV with potassium and lithium metals in various alcohols as solvents. Although these reactions had the appearance of being heterogeneous, there is a distinct possibility that the metal dissolved before reaction occurred. The results indicate that 2-phenyl-2-butanol with potassium metal gave about the same amount of inversion (18–22%) in the four alcohol solvents, and that with lithium gave the same amount of retention (9%) in two alcohol solvents. With 2-methoxy-2-phenylbutane as substrate, potassium gave 31% and 14% inversion in *tert*- and *n*-butyl alcohol respectively. Lithium failed to reduce the ether.

Too little is understood about these reactions to provide much mechanistic guidance.[29] However, it is likely that one electron enters

[29] A. J. Birch, *Quart. Rev.* (*London*), **4**, 69 (1950).

TABLE IX

Stereochemical Course of the Reductive Cleavage of 2-Phenyl-2-butanol and 2-Methoxy-2-phenylbutane

Substrate	Solvent	Metal	Steric course (net)
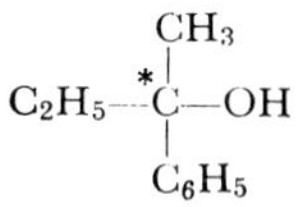 $C_2H_5-\overset{*}{C}(CH_3)(C_6H_5)-OH$	*tert*-BuOH	K	21% Inversion
	$(CH_3)_2CHOH$	K	22% Inversion
	n-BuOH	K	19% Inversion
	C_2H_5OH	K	18% Inversion
	tert-BuOH	Li	9% Retention
	n-BuOH	Li	9% Retention
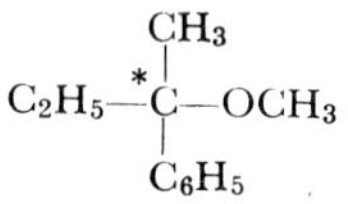 $C_2H_5-\overset{*}{C}(CH_3)(C_6H_5)-OCH_3$	*tert*-BuOH	K	31% Inversion
	n-BuOH	K	14% Inversion

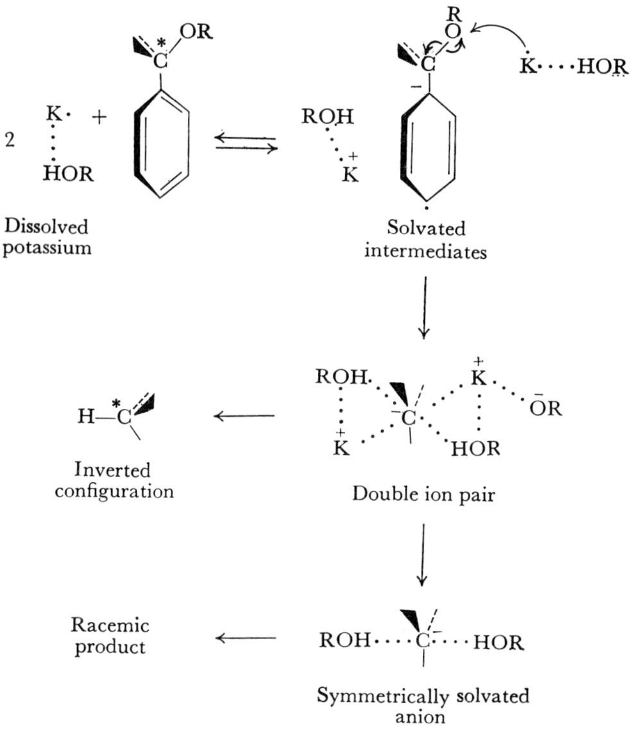

Dissolved potassium

Solvated intermediates

Inverted configuration

Double ion pair

Racemic product

Symmetrically solvated anion

FIG. 5. Mechanism for reductive cleavage of 2-phenyl-2-butanol and 2-methoxy-2-phenylbutane to 2-phenylbutane.

the benzene ring to give a radical anion, and that this species in turn reacts with a second electron at the benzyl position. In this second reaction the carbanion is generated, as well as a metal hydroxide or methoxide. Thus two negative charges are generated, and negative charge is more concentrated at the front than at the back face of the carbanion. The two metal ions are probably on opposite faces of the carbanion in order to minimize charge repulsion. With potassium, net inversion is observed. The potassium ion with its solvation shell orients proton donors at the back face of the carbanion, and the potassium methoxide or hydroxide formed at the front face orients proton donors there. If proton capture occurs at the back face, two potassium alkoxide or hydroxide ion pairs are formed. But if proton capture at the front face should occur, one ion pair and one product-separated ion pair are formed, and this process is unfavorable in solvents of low dielectric constant. As a result, net inversion is the result in solvents of low dielectric constant. This scheme is outlined in Figure 5.

The fact that lithium gives low net retention with conditions under which potassium gives inversion (2-phenyl-2-butanol as substrate) may indicate that the hydroxyl leaving group is acting as the proton source. Lithium hydroxide should be a better proton donor than potassium hydroxide.

Comparison of Leaving Groups

An overall comparison of the stereochemical course of the SE_1 reaction as the leaving group is changed is now possible. Deuterium, carbon, nitrogen, and oxygen have all served as leaving groups in producing the 2-phenyl-2-butyl anion, and proton donors have served as electrophiles. In *tert*-butyl alcohol, deuterium as leaving group gave 87% retention,[30] carbon gave about 95% retention, nitrogen about 80% retention. With *N*-methylaniline as solvent, oxygen gave a minimum of 29% retention. These results have been interpreted in terms of the metal ion of the base rotating with its ligands at the front face of the carbanion, and donating a proton from the same side. In dimethyl sulfoxide, both deuterium[30] and carbon as leaving groups provided complete racemization for the steric course of the reaction. This result pointed to a free anion as the basic catalyst. The carbanion generated was solvated front and back by the non-acidic dimethyl sulfoxide molecules, and this symmetrically solvated carbanion gave racemic product. In dimethyl sulfoxide, with nitrogen as leaving group, the proton associated with the leaving group served as proton donor at the front, and net retention was

[30] D. J. Cram, C. A. Kingsbury, and B. Rickborn, *J. Am. Chem. Soc.*, **83**, 3688 (1961).

observed. In glycol, methanol, or water as solvent, deuterium,[30] carbon, or nitrogen (only in water) gave from 33 to 65% inversion. In these ionizing media, free anions were the basic catalyst, the front face of the

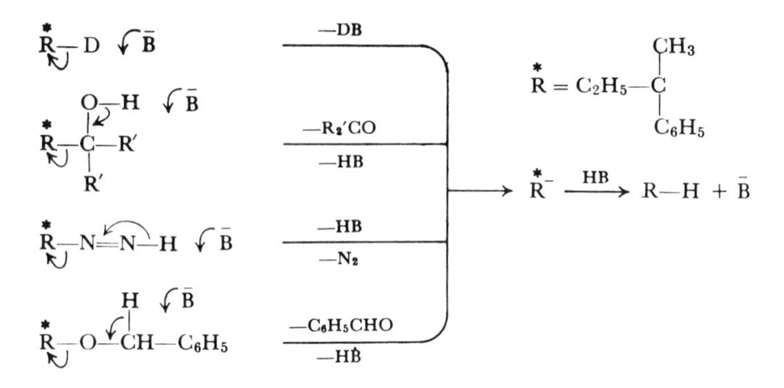

carbanion was shielded by the leaving group, and the back face was hydrogen bonded to solvent. Proton capture at the back face of the carbanion gave net inversion as the steric result. The stereospecific mechanisms competed with racemizing mechanisms in all cases. In the latter, the carbanion lasted long enough to pass into a symmetrical environment.

A second interesting comparison involves the production of 2-phenylbutyronitrile with carbon[6] and deuterium[31] as leaving groups. In nondissociating solvents such as *tert*-butyl alcohol, the nitrile is produced by decarboxylation of the ammonium salt of 2-cyano-2-phenylbutanoic acid with 10% retention.[6] Ammonia-catalyzed hydrogen–deuterium exchange also occurs with net retention.[31] Both mechanisms involve formation of ammonium carbanide ion pairs, which collapse to product faster than they dissociate. The same reactions in dimethyl sulfoxide provide complete racemization as the steric result. In this medium, ion pair dissociation is faster than proton capture, the carbanion passes into a symmetrical environment, and racemization results. In glycol as solvent with any base, decarboxylation occurs with 12% net inversion, and hydrogen–deuterium exchange gives a comparable steric result. In this solvent, the carboxylate anion is the species which cleaves; the carbanion produced is hydrogen bonded at the back face by solvent and shielded at the front by the leaving group. The result is net inversion. In the hydrogen–deuterium exchange free glycoxide anion is the base, a carbanion is produced hydrogen bonded at the front by the leaving

[31] D. J. Cram and L. Gosser, *J. Am. Chem. Soc.*, **86**, 5457 (1964).

group and at the back by solvent. Front side deuterium capture gives back starting material, whereas back side proton capture gives inverted material.

$$C_2H_5-\overset{*}{C}(C{\equiv}N)(C_6H_5)-CO_2^-M^+ + HB \longrightarrow C_2H_5-\overset{*}{C}(C{\equiv}N)(C_6H_5)-H + MB + CO_2$$

$$C_2H_5-\overset{*}{C}(C{\equiv}N)(C_6H_5)-D + :B + HB \longrightarrow C_2H_5-\overset{*}{C}(C{\equiv}N)(C_6H_5)-H + :B + DB$$

A dramatic parallelism of mechanism is evident in the results taken collectively. A large number of reactions, solvents, and bases have been used in these studies, and the results can be rationalized in terms of a relatively small number of concepts. Such phenomena as asymmetric ion pairs, asymmetric solvation, anion and cation rotation within ion pairs, shielding effects, internal proton sources, ion pair dissociation, and very high rates of proton capture underlie the analogies, and make the results comprehensible.

COMPARISON OF THE STEREOCHEMICAL CAPABILITIES OF CARBANIONS AND CARBONIUM IONS

In 1935, Hughes and Ingold[32] suggested that both the monomolecular nucleophilic and electrophilic substitutions at saturated carbon (SN_1 and SE_1 reactions) might occur with retention of configuration. Experiments conducted since that time have clearly demonstrated that the SN_1 reactions, which involve carbonium ion intermediates, can occur with retention,[33] inversion,[34] or with total racemization.[35] Both cations and anions within ion pairs that contain carbonium ions have been shown capable of undergoing rotation and collapse to the covalent state

[32] E. D. Hughes and C. K. Ingold, *J. Chem. Soc.*, p. 244 (1935).

[33] (a) D. J. Cram, *J. Am. Chem. Soc.*, **75**, 332 (1953) and examples cited; (b) C. E. Boozer and E. S. Lewis, *ibid.*, **75**, 3154 (1953).

[34] (a) J. Kenyon, H. Phillips, and V. P. Pittman, *J. Chem. Soc.*, p. 1072 (1935); (b) H. H. Zeiss, *J. Am. Chem. Soc.*, **75**, 3154 (1953); (c) A. Streitwieser, Jr., "Solvolytic Displacement Reactions," McGraw-Hill, New York, 1962, p. 59.

[35] (a) C. L. Arcus, M. P. Balfe, and J. Kenyon, *J. Chem. Soc.*, p. 485 (1938); (b) M. P. Balfe, A. A. Evans, J. Kenyon, and K. N. Nandi, *ibid.*, p. 803 (1946).

Retention Mechanisms

S_N1: Occurs when a complex leaving group carries its own nucleophile, as in the reactions of chlorosulfites in nondissociating solvents.

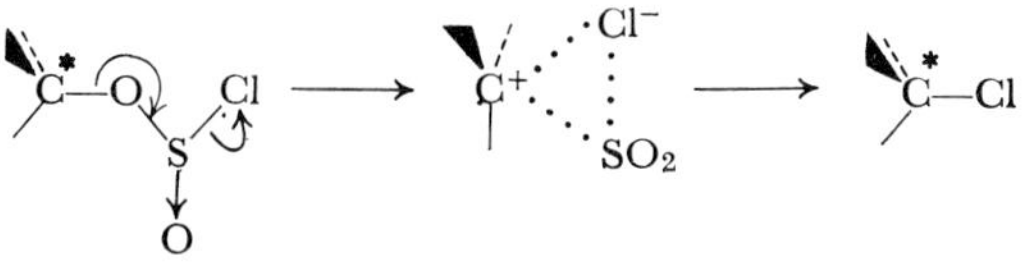

Alkyl chlorosulfite — Asymmetric ion pair — Retained configuration

S_E1: Occurs when complex leaving group carries its own electrophile, as in reactions of alkoxides in nondissociating solvents.

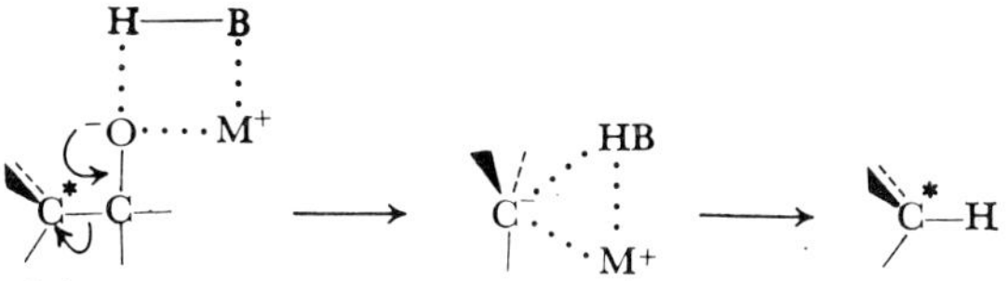

Solvated metal alkoxide — Asymmetric ion pair — Retained configuration

Racemization Mechanisms

S_N1: Occurs in relatively nonnucleophilic but strongly dissociating solvents in which relatively long-lived carbonium ions react with external nucleophiles.

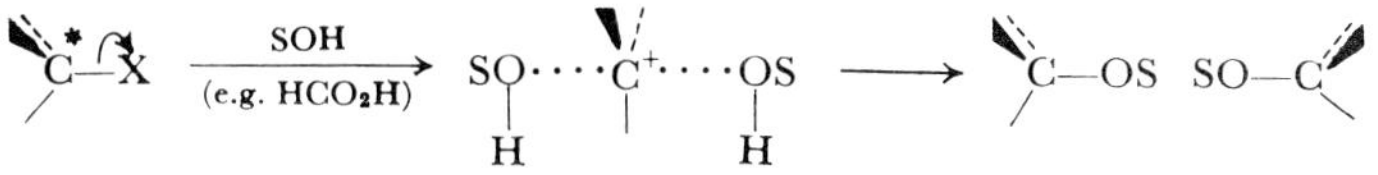

SOH (e.g. HCO_2H)

Alkyl halides, tosylates, etc. — Symmetrically solvated carbonium ion — Racemic product

S_E1: Occurs in non-proton-donating but strongly dissociating solvents in which relatively long-lived carbanions react with external electrophiles.

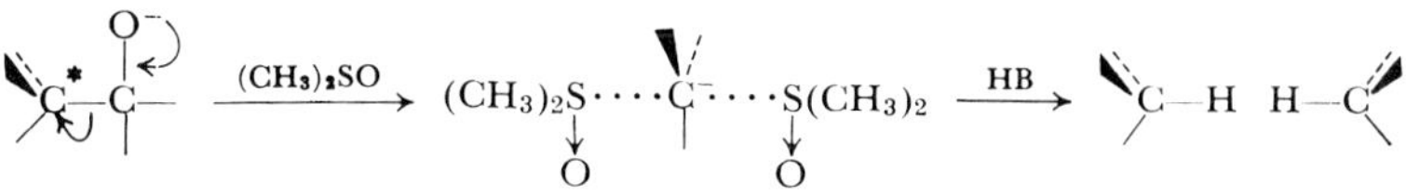

$(CH_3)_2SO$ — $(CH_3)_2S \cdots C^- \cdots S(CH_3)_2$ — HB

Alkoxide anion — Symmetrically solvated carbanion — Racemic product

FIG. 6. *(Continued next page.)*

Inversion Mechanisms

S_N1: Occurs in strongly nucleophilic and ionizing solvents in which short-lived carbonium ions react with external nucleophiles.

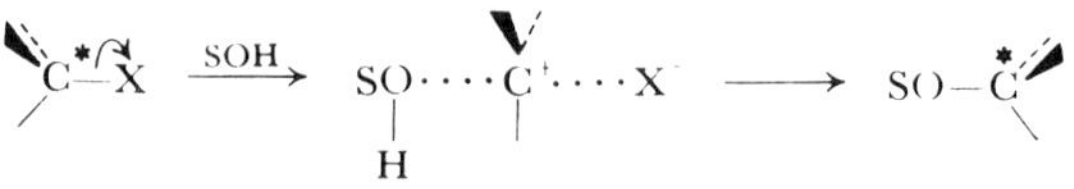

Alkyl halides, tosylates, etc.	Asymmetric ion pair	Inverted configuration

S_E1: Occurs in strongly electrophilic and dissociating solvents in which short-lived carbanions react with external electrophiles.

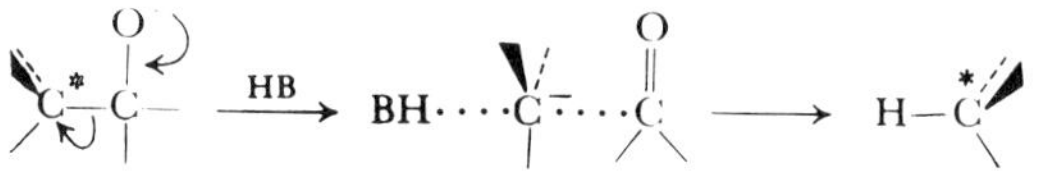

Alkoxide anion	Asymmetrically solvated carbanion	Inverted configuration

Rotations within Ion Pairs

Those that involve carbonium ions:

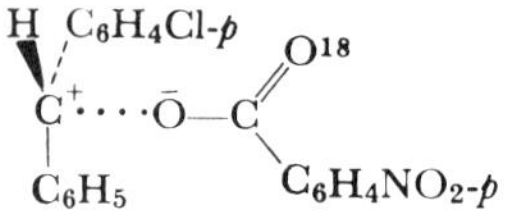

Carboxylate ion rotation within the ion pair is faster than ion pair solvolysis in 90% acetone–10% water.[36c]

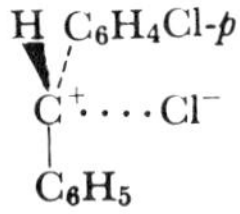

Carbonium ion rotation within the ion pair is faster than exchange reaction with radioactive chloride or acetolysis.[36a]

Those that involve carbanions:

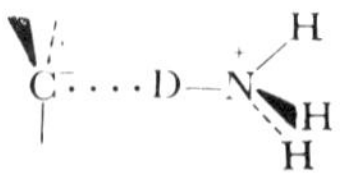

Ammonium ion rotation within the ion pair is faster than ion pair dissociation with 2-(*N*,*N*-dimethylacetamido)-9-methylfluorene system[37] in *tert*-butyl alcohol (see Chapter III).

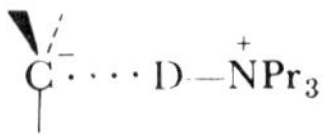

Carbanion rotation within the ion pair is faster than exchange reaction with 2-phenylbutyronitrile system[31] in *tert*-butyl alcohol (see Chapter III).

FIG. 6. Comparison of stereochemical capabilities of carbonium ions and carbanions.

at rates faster than those of ion pair dissociation.[36] Similarly, both cations and anions within ion pairs that contain carbanions have been shown capable of undergoing rotation and collapse to the covalent state at rates faster than those of ion pair dissociation.[31, 37]

The same spectrum of results has been observed to apply to carbanions and carbonium ions. The patterns of stereochemical behavior of the two electronic varieties of carbon are remarkably similar, as is demonstrated in the comparisons of Figure 6.

[36] (a) S. Winstein, J. S. Gall, M. Hojo, and S. Smith, *J. Am. Chem. Soc.*, **82**, 1010 (1960); (b) A. Illiceto, A. Fava, U. Muzzucato, and U. Rossetto, *ibid.*, **83**, 2729 (1961); (c) H. L. Goering, R. G. Brody, and J. F. Levy, *ibid.*, **85**, 3059 (1963).

[37] D. J. Cram and L. Gosser, *J. Am. Chem. Soc.*, **86**, 2950 (1964).

CHAPTER V

Isomerization by Proton Transfer in Unsaturated Systems

Many similarities are found among the unsaturated rearrangements of carbonium ions, carbon radicals, and carbanions. This relationship reflects the ability of such unsaturated substituents as the vinyl, ethynyl, and aryl groups to distribute positive charge, negative charge, and unpaired electrons when attached to carbon in the appropriate valence state. Unlike most of the radical and cationic rearrangements in unsaturated systems, those of the anionic variety for the most part are isomerizations. which involve simply a reshuffling of hydrogen atoms and of double bonds. These rearrangements, which are base catalyzed and involve carbanionic intermediates, are the subject of this chapter. Other anionic rearrangements are discussed in Chapter VI.

The three kinds of rearrangement that are treated here are put in general form in Equations (1), (2), and (3). All are base catalyzed, and all involve ambident or polydent anions as intermediates. A number of interesting and fundamental questions arise concerning these reactions. (1) To what extent and under what circumstances are these rearrangements intramolecular? (2) Do allylic anions exhibit *cis-trans*-isomerism? (3) How do alkyl substituents affect the rates of allylic rearrangement of the simple alkenes? (4) How do substituents affect the positions of equilibrium between isomeric olefins? (5) What structural and environmental features control the collapse ratios of ambident or polydent

$$(1)\quad -\overset{|}{C}{=}\overset{|}{C}-\underset{H}{\overset{|}{C}}- \; + \; :B \rightleftarrows \underbrace{-\overset{|}{C}\cdots\overset{|}{C}\cdots\overset{|}{C}-}_{-} \; + \; HB \rightleftarrows -\underset{H}{\overset{|}{C}}-\overset{|}{C}{=}\overset{|}{C}- \; + \; :B$$

$$(2)\quad -C{\equiv}C-\underset{H}{\overset{|}{C}}- \; + \; :B \rightleftarrows \underbrace{-C\overset{\cdots}{=}C\cdots\overset{|}{C}-}_{-} \; + \; HB \rightleftarrows \underset{H}{>}C{=}C{=}C< \; + \; :B$$

anions? Although of intrinsic interest, the answers to these questions have implications for the behavior of ambident or polydent anions toward electrophiles other than proton donors.

(3) [cyclohexadiene with exocyclic C= and H] + :B ⟶ [delocalized anion]$^-$ + HB ⟶ [benzene with —C—H] + :B

INTRAMOLECULARITY

From observations of low or negative isotope effects in base-catalyzed hydrogen isotope exchanges of carbon acids,[1a] rate-equilibrium correlations for dissociation of carbon acids[1b] (see Chapter I), and stereochemical studies,[2] it was concluded that $k_{-1} \gg k_2$ in Equation (4) under certain circumstances. This conclusion suggested that if the carbanion

$$(4)\quad -\overset{|}{\underset{|}{C}}-H + :B \underset{k_{-1}}{\overset{k_1}{\rightleftharpoons}} -\overset{|}{\underset{|}{C}}^{-}\cdots HB \xrightarrow[DB]{k_2} -\overset{|}{\underset{|}{C}}^{-}\cdots DB \longrightarrow -\overset{|}{\underset{|}{C}}-D + :B$$

generated were allylic, proton capture at the site distant from the original covalent site might occur faster than isotopic exchange. Accordingly, compound I was found to isomerize in *tert*-butyl alcohol-*O*-*d*–potassium *tert*-butoxide to II with 54% intramolecularity.[3] Subsequently, other examples of base-catalyzed intramolecular proton transfers in other allylic systems were reported.[4]

System I possesses good qualifications for study of the base-catalyzed allylic proton transfer reaction. The reaction occurs under mild enough conditions to allow a variety of solvents and bases to be employed. The equilibrium constants between conjugated (II and III) and uncon-

[1] (a) D. J. Cram, D. A. Scott, and W. D. Nielsen, *J. Am. Chem. Soc.*, **83**, 3696 (1961); (b) R. Stewart, J. P. O'Donnell, D. J. Cram, and B. Rickborn, *Tetrahedron*, **18**, 917 (1962).

[2] D. J. Cram and L. Gosser, *J. Am. Chem. Soc.*, **85**, 3890 (1963).

[3] D. J. Cram and R. T. Uyeda, *J. Am. Chem. Soc.*, **84**, 4358 (1962).

[4] (a) S. Bank, C. A. Rowe, Jr., and A. Schriesheim, *J. Am. Chem. Soc.*, **85**, 2115 (1963); (b) W. von E. Doering and P. P. Gaspar, *ibid.*, **85**, 3043 (1963); (c) G. Bergson and A. M. Weidler, *Acta Chem. Scand.*, **17**, 862, 1798, 2691, 2724 (1963); (d) R. B. Bates, R. H. Carnighan, and C. E. Staples, *J. Am. Chem. Soc.*, **85**, 3032 (1963); (e) V. A. Mironov, E. V. Sobolev, and A. N. Elizarova, *Tetrahedron*, **19**, 1939 (1963); (f) W. R. Roth, *Tetrahedron Letters*, **17**, 1009 (1964); (g) S. McLean and P. Haynes, *ibid.*, **34**, 2385 (1964).

jugated (I) components of the allylic system very strongly favor the conjugated products. The products once formed do not exchange under

$$\underset{\text{I}}{CH_3-\overset{\overset{C_6H_5}{|}}{\underset{\underset{H}{|}}{C}}-CH{=}CH_2} \xrightarrow{\text{Base}} \underset{\text{II}}{\overset{C_6H_5 \quad\quad H}{\underset{CH_3 \quad\quad CH_3}{C{=}C}}} + \underset{\text{III}}{\overset{C_6H_5 \quad\quad CH_3}{\underset{CH_3 \quad\quad H}{C{=}C}}}$$

conditions for rearrangement. The starting material possesses an asymmetric center, so the stereochemistry of simple exchange can be studied at the same time that the rearrangement is examined. The thermodynamically more stable isomer II dominates in the product by a large factor.[5]

In the rearrangement of I to II, the ratio (% intramol.)/(% intermol.) varied by a maximum factor of about 20 as solvent, base, and position of the isotopic label (substrate vs. solvent) were varied (see Table I).[5] Intramolecularity was highest in nondissociating solvents where metal alkoxides were employed, and was lowered slightly by substitution of a quaternary ammonium for a potassium ion. Not much difference was observed between dimethoxyethane–*tert*-butyl alcohol, and *tert*-butyl alcohol itself as solvent. Ethylene glycol, a strongly dissociating solvent, gave lower intramolecularity, but only by a small amount. Dimethyl sulfoxide–methanol gave the highest intramolecularity observed with deuterated substrate. In the two solvent-base systems in which *trans*-olefin (III) was examined, the degree of intramolecularity approximated that of the *cis*-isomer (II). The remarkable feature of the data is that the variable which produced the greatest change in the ratio of intramolecularity to intermolecularity was the original location of the deuterium. When deuterium was in the medium, higher values of the ratio were observed than when it was in the substrate, the factors ranging from 3 to 10. The results are recorded in Table I.

Two general mechanisms are clearly incompatible with the facts. In the first, each of the two ends of the allylic anion is hydrogen bonded to a hydroxyl group, the benzyl position involving the leaving group, and the terminal position involving a molecule from the medium (see A). Such a mechanism only provides for *intermolecularity*, and not the *intramolecularity*. In the second mechanism, the leaving group hydrogen bonds one face of the allylic carbanion at both ends of the resonating system, and a molecule of solvent of the opposite isotopic type hydrogen-bonds the opposite face in the same way (see B). This mechanism

[5] D. J. Cram and R. T. Uyeda, *J. Am. Chem. Soc.*, **86**, 5466 (1964).

TABLE I

Intramolecularity in Base-Catalyzed Rearrangement of 3-Phenyl-1-butene (I) to cis-2-Phenyl-2-butene (II)

Substrate	Solvent	Base	Atom D in II	(% Intramol.)/(% Intermol.)
I-h	$(CH_3OCH_2)_2$, 5 *M tert*-BuOD	*tert*-BuOK	0.50[a]	1.0
I-d	$(CH_3OCH_2)_2$, 5 *M tert*-BuOH	*tert*-BuOK	0.17	0.2
I-h	*tert*-BuOD	*tert*-BuOK	0.49	1.0
I-d	*tert*-BuOH	*tert*-BuOK	0.23	0.3
I-h	*tert*-BuOD	$(CH_3)_4NOD$	0.55	0.8
I-d	*tert*-BuOH	$(CH_3)_4NOH$	0.06	0.06
I-h	$(CH_2OD)_2$	$DOCH_2CH_2OK$	0.69[b]	0.45
I-d	$(CH_2OH)_2$	$HOCH_2CH_2OK$	0.12	0.14
I-d	91% $(CH_3)_2SO$, 9% CH_3OH	CH_3OK	0.32	0.47

[a] *trans*-2-Phenyl-2-butene contained 0.51 atom of D.
[b] *trans*-2-Phenyl-2-butene contained 0.67 atom of D.

requires the same amount of deuterium in the product irrespective of whether the label was in solvent or substrate.

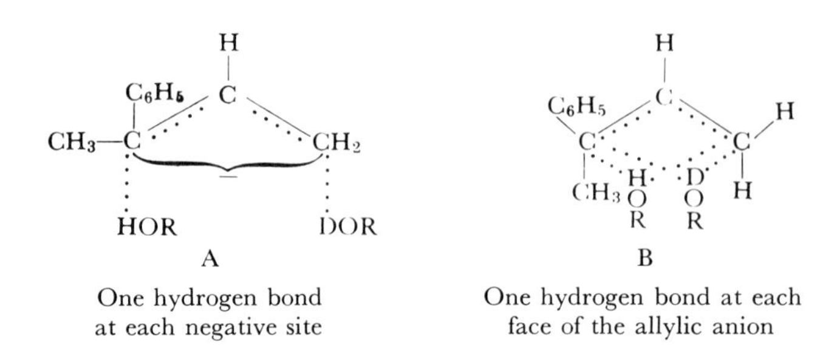

One hydrogen bond at each negative site

One hydrogen bond at each face of the allylic anion

The most likely mechanism involves the leaving group becoming hydrogen bonded at both anionic sites at the same time. To provide the best geometry for this hydrogen bonding, the anionic sites might have a hybridization between sp^2 and sp^3, and the C—C—C bond angle might decrease to somewhat less than 120°. Collapse of this discrete intermediate to rearranged material provides for the intramolecular component of the mechanism. Exchange of the hydrogen-bonded leaving group with molecules of the opposite isotopic type in the medium and collapse of new isotopically labeled, hydrogen-bonded allylic anion to rearranged material provide for the intermolecular mechanism. The presence or absence of a metallic cation with its ligands seems to be relatively unimportant to the operation of the intramolecular mechanism.

An even more delicate probe of mechanism is derived from combining the stereochemical results of hydrogen–deuterium exchange of I with the results of the rearrangement. Although isotopic exchange of I was about 20 times slower in rate than was isomerization, the relative rates of isotopic exchange (rate constant, k_e) and of racemization (rate constant, k_α) were determined in two solvent-base systems. In *tert*-butyl alcohol–potassium *tert*-butoxide, $k_e/k_\alpha > 10$, so exchange occurred with high retention of configuration. In ethylene glycol–potassium ethylene glycoxide, $k_e/k_\alpha = 0.7$, so exchange went with net inversion (see Chapter II). Thus the stereochemistry of the exchange reaction exhibited the same solvent dependence as did those of 2-phenylbutane and 1-phenylmethoxyethane.[6]

The stereochemical and rearrangement data are nicely accommodated by the mechanism of Figure 1. In this scheme, base as an ion pair (in nondissociating solvents) or as a free anion (in dissociating solvents) abstracts a proton from the benzyl carbon atom to form, in one transition state, a carbanion hydrogen bonded to the same hydroxyl group at both the benzyl and methylene carbon. When a metal cation was present as in *tert*-butyl alcohol, exchange occurred at the front face of the carbanion, collapse of which to starting material gave material of retained configuration. Exchange involved rotation of the metal cation and its ligands within the ion pair and collapse to starting material, both processes occurring faster than ion pair dissociation. In the dissociating solvent, ethylene glycol, the asymmetrically solvated anion exchanged with solvent to form an isotopically labeled inverted solvated species, which collapsed to inverted exchanged starting material.

No correlation is visible between the intramolecularity and such factors as solvent acidity, concentration of proton donors (changes were very small), dissociating power of the solvent, or the character of the base. Features such as the existence of the allylic anion as part of an ion pair or in a dissociated state, or coordination of the cation of the ion pair with solvent, seem to play roles subservient even to the isotope effects. These results contrast with those observed for the stereochemistry of the base-catalyzed hydrogen–deuterium exchange reaction at asymmetric carbon. The steric course of the exchange reaction was little affected by the placement of the isotope, and greatly affected

[6] D. J. Cram, C. A. Kingsbury, and B. Rickborn, *J. Am. Chem. Soc.*, **83**, 3688 (1961).

by other factors. Unlike the stereochemical paths, the intramolecular reaction course seems dependent largely on hydrogen bonding.

Schriesheim and co-workers[4a] observed that deuterated 1-pentene isomerized to deuterated 2-pentene at 55° in dimethyl sulfoxide, 0.43 *M* in potassium *tert*-butoxide and 0.43 *M* in *tert*-butyl alcohol. The rate of

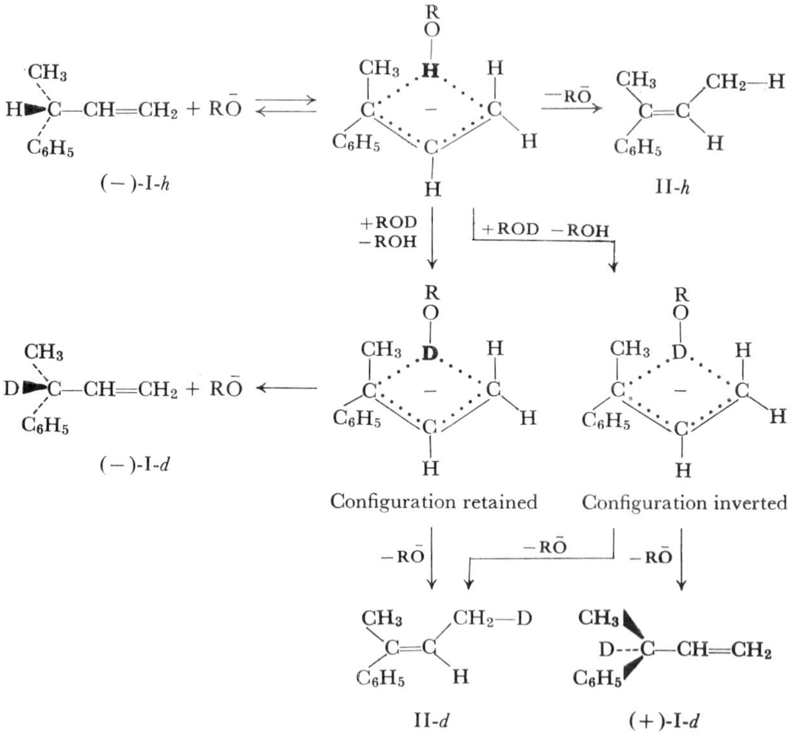

FIG. 1. Mechanism of base-catalyzed exchange and isomerization reactions.

isomerization was about 16 times that of exchange. The higher intramolecularity in this rearrangement as compared to that of the 3-phenyl-1-butene system (I) is probably associated with both the pK_a relationships of the hydrocarbon and the medium and the concentration of the proton pool. On the MSAD scale (see Chapter I) dimethyl sulfoxide has an estimated pK_a of 36, 1-pentene of about 38, and *tert*-butyl alcohol (in dimethyl sulfoxide -0.4 *M* in base and alcohol) of ~ 25. In dimethyl sulfoxide, free alkoxide anion is probably the active basic species. Deuterium abstraction from deuterated 1-pentene produces the carbanion hydrogen bonded to *tert*-butyl alcohol-*O-d*, the latter being much more acidic than dimethyl sulfoxide. The proton capture process involving the carbanion and alcohol is probably extremely rapid (ΔpK_a of 1-pentene and *tert*-butyl alcohol ~ 13), and the concentration of unlabeled *tert*-butyl alcohol in the medium was not enough to produce a large intermolecular component. Reaction between the anion of 1-pentene and dimethyl sulfoxide is probably relatively slow compared with that between anion and alcohol since ΔpK_a of 1-pentene and dimethyl sulfoxide ~ 2.

$$D_2C{=}CD{-}CD_2{-}C_2D_5 + R\bar{O} \longrightarrow D_2C\overset{\overset{D}{|}}{\cdots C \cdots}CDC_2D_5 \cdots D \cdots OR \longrightarrow$$

$$CD_3CD{=}CDC_2D_5$$

The pK_a of 3-phenyl-1-butene is probably about 34, and those of the alcohols of Table I are 16–19. Although the ΔpK_a of the hydrocarbon and proton pool was large enough to give very rapid proton capture, the concentration of proton pool was very high. As a result the intra- and intermolecular processes were more competitive. Reduction of the concentration of methanol in dimethyl sulfoxide in the last run of Table I probably would have increased intramolecularity.

In systems where many hydrogen bonding sites are available, intramolecularity appears to be dependent on ion pairing. Doering and Gaspar[4b] observed that 1,1-dideuterocycloheptatriene in dimethyl sulfoxide–triethylcarbinol–potassium triethylcarboxide underwent isotope exchange-isomerization with little intramolecularity. However, in triethylcarbinol–potassium triethylcarboxide, intramolecular isomerization exceeded intermolecular isomerization by a factor of 12.[4b] Unlike the simple allylic systems, the cycloheptatrienide anion has seven hydrogen bonding sites, and one alcohol molecule could at most hydrogen-bond to only two or three of them at one time. In the nondissociating solvent, triethylcarbinol, the potassium ion could be close to two adjacent charge-distributing sites at the most. In such an intermediate, proton capture should occur only at these sites, since charge separation in a solvent of low dielectric constant would result should proton capture occur at a distant site. Thus intramolecularity is observed. In dimethyl sulfoxide (high dielectric constant), a dissociated carbanion is probably the intermediate, the seven hydrogen-bonding sites become equivalent, and intermolecularity predominates. The random character of the deuterium distribution observed in the product obtained in triethylcarbinol indicates that the potassium ion and its deuterated ligand moved around the ring faster than deuterium was captured by the carbanion.

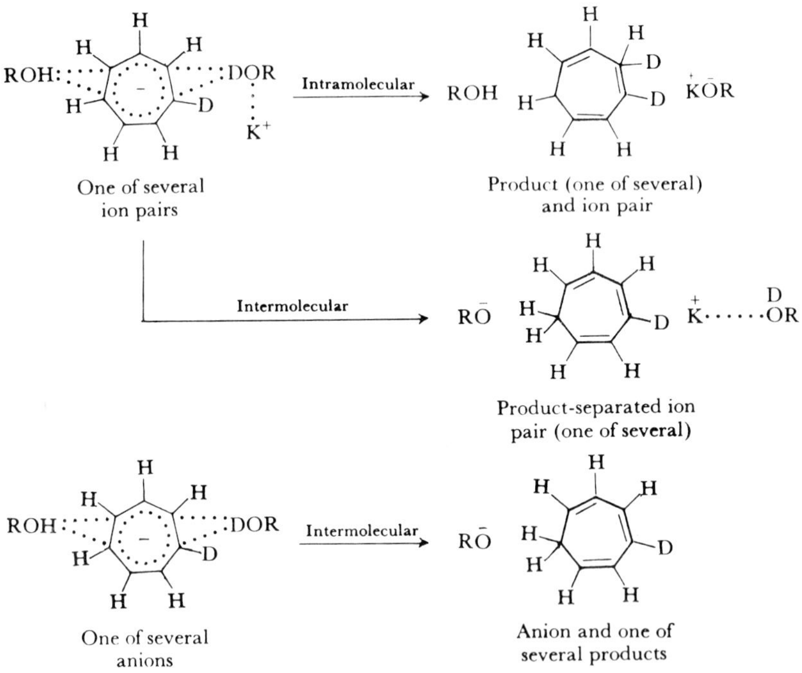

ROH: H H H H H D :DOR K$^+$ —Intramolecular→ ROH H H H D D H H H $\overset{+}{K}\bar{O}R$

One of several ion pairs

Product (one of several) and ion pair

—Intermolecular→ $R\bar{O}$ H H H H H D H H $\overset{+}{K}$······OR D

Product-separated ion pair (one of several)

ROH: H H H H H D :DOR —Intermolecular→ $R\bar{O}$ H H H H H D H H

One of several anions

Anion and one of several products

Mironov *et al.*[4e] observed that 5-deuterocyclopentadiene when heated to 60° gives a mixture of starting material, 1-deuterocyclopentadiene, and 2-deuterocyclopentadiene in the proportion of about 1 : 1 : 1 without formation of noticeable amounts of non- or di-deuterated cyclopentadienes. These results suggest that an intramolecular equilibration occurred in which the proton or its isotope wandered randomly around the π-system of the cyclopentadienide system without appreciable transfer between molecules.

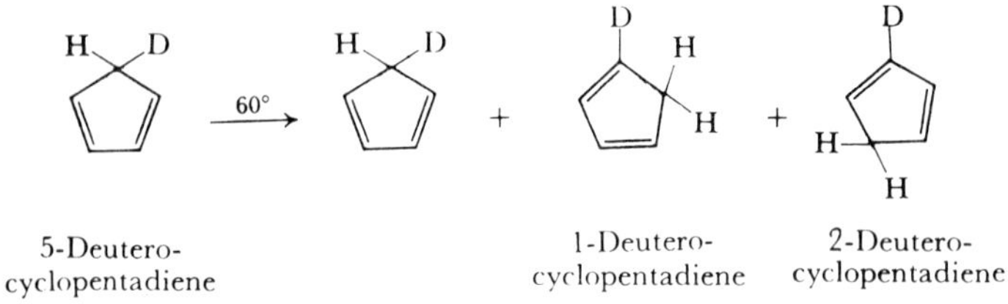

H D —60°→ H D + D H H + D H H

5-Deutero-cyclopentadiene

1-Deutero-cyclopentadiene

2-Deutero-cyclopentadiene

Roth[4f] made similar observations both in the liquid and gas phase with deuterated cyclopentadiene and indene, and concluded that the isomerizations were intramolecular. McLean and Haynes[4g] found that 5-methylcyclopentadiene rearranged at room temperature in carbon tetrachloride or in homogeneous form. When carried out in dimethoxyethane containing deuterium oxide, no deuterium incorporation was observed. The isomerization rate was dramatically accelerated by addition of bases.

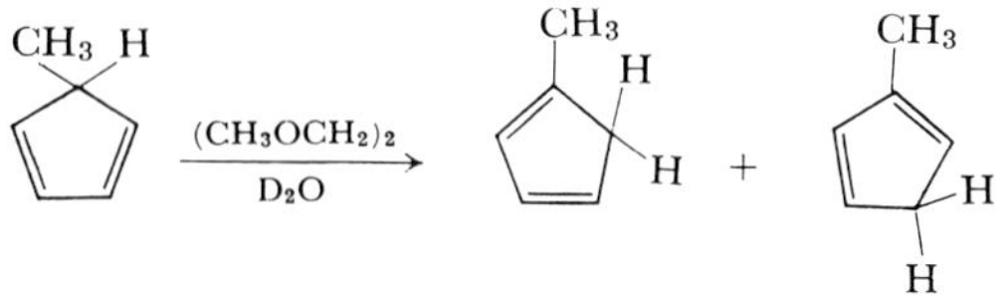

$\xrightarrow[D_2O]{(CH_3OCH_2)_2}$

5-Methylcyclopentadiene → 1-Methylcyclopentadiene + 2-Methylcyclopentadiene

Bergson and Weidler[4c] found that optically active 1-methylindene underwent triethylamine-catalyzed rearrangement to 3-methylindene in pyridine. The rates of racemization and of isomerization were equal. When the reaction was conducted in the presence of 5 moles of deuterium oxide (per mole of starting material) the product contained a negligible amount of deuterium, a fact which indicates that under the conditions used, the reaction was intramolecular.

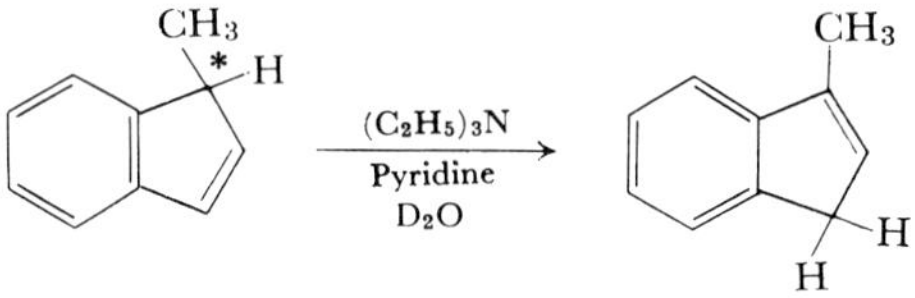

$\xrightarrow[\text{Pyridine},\ D_2O]{(C_2H_5)_3N}$

3-Methylindene → 1-Methylindene

These authors also observed that with triethylenediamine as base and pyridine as solvent, optically active 3-methyl-1-*tert*-butylindene gave 1-methyl-3-*tert*-butylindene with very high stereospecificity. Similar results were observed when an isopropyl group was substituted for the *tert*-butyl group. These results constitute an elegant demonstration of 1,3-asymmetric induction, and demonstrate that only one face of the π-system of electrons was involved in the proton transfer, which undoubtedly involved *tert*-ammonium carbanide ion pairs as intermediates.

In dimethyl sulfoxide as solvent and *tert*-amine as base, 1-isopropyl-3-methylindene (optically active) racemized and isomerized at the same rate. This result indicates that the ion pairs formed dissociated faster than proton transfer occurred. In *tert*-butyl alcohol–potassium *tert*-butoxide, the racemization rate exceeded the isomerization rate by

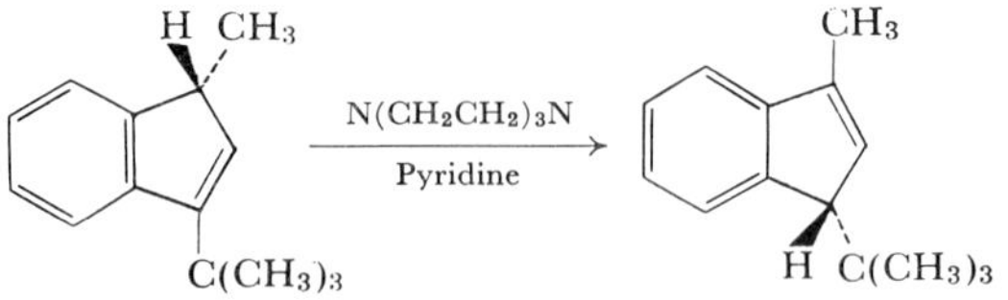

H CH_3 $\xrightarrow[\text{Pyridine}]{N(CH_2CH_2)_3N}$ CH_3

$C(CH_3)_3$ H $C(CH_3)_3$

3-Methyl-1-*tert*-butylindene 1-Methyl-3-*tert*-butylindene

about a factor of 10. The pK_a of the indene system is probably around 20, and that of *tert*-butyl alcohol is 19. As a result, proton capture by carbanion is probably slower than dissociation. Proton capture by the allylic anion appears to occur at the least hindered site, and hence racemization is faster than isomerization.

These examples of intramolecular allylic proton transfers catalyzed by base have their counterparts in similar reactions catalyzed by enzyme systems. Enough examples have been encountered to make the phenomena appear general.[7]

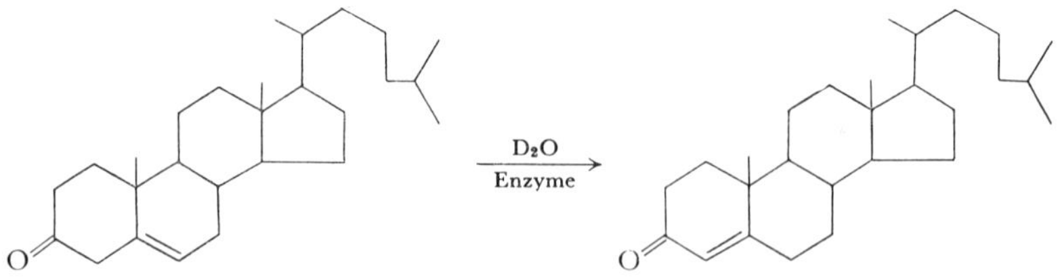

O $\xrightarrow[\text{Enzyme}]{D_2O}$ O

No deuterium

$CH_3—\underset{CH_3}{\underset{|}{C}}=CH—CH_2—O—PO_3PO_3H \xrightarrow[\text{Enzyme}]{T_2O} CH_2=\underset{CH_3}{\underset{|}{C}}—CH_2—CH_2—O—PO_3PO_3H$

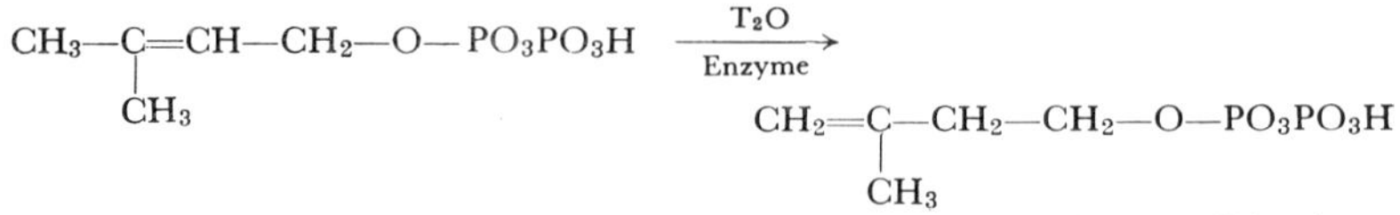

2% of T expected from equilibration of exchangeable positions

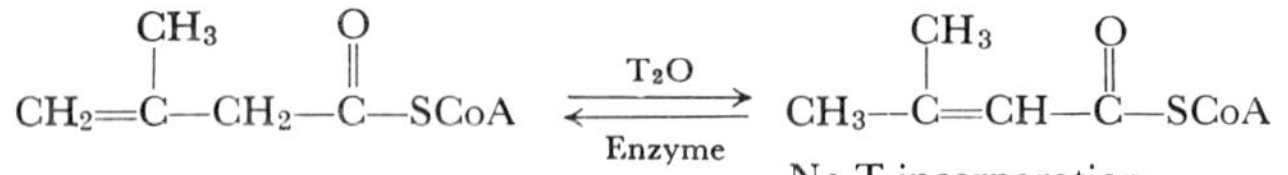

$CH_2=\overset{CH_3}{\overset{|}{C}}—CH_2—\overset{O}{\overset{\|}{C}}—SCoA \underset{\text{Enzyme}}{\overset{T_2O}{\rightleftarrows}} CH_3—\overset{CH_3}{\overset{|}{C}}=CH—\overset{O}{\overset{\|}{C}}—SCoA$

No T incorporation

The observed competition between intramolecular and intermolecular allylic 1,3-proton transfers of carbanion chemistry recalls similar phenomena found in allylic 1,3-chloride ion transfers of carbonium ion

[7] (a) F. S. Kawahara and P. Talalay, *J. Biol. Chem.*, **235**, PC 1 (1960); (b) B. W. Agranoff, H. Eggerer, U. Henning, and F. Lynen, *ibid.*, **235**, 326 (1960); (c) H. C. Rilling and M. J. Coon, *ibid.*, **235**, 3087 (1960).

chemistry.[8] During the acetolysis of α,α-dimethylallyl chloride, intramolecular isomerization to γ,γ-dimethylallyl chloride occurred at rates of the same order of magnitude as the solvolysis rates. A bridged allylic chloride was formulated (C) whose geometry resembles that of the protium allylic anion discussed above (D).

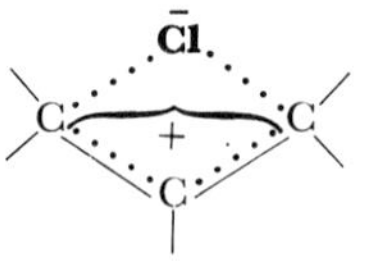

C: bridged allyl cation with $\bar{Cl}$ bridging the terminal carbons (+ charge)

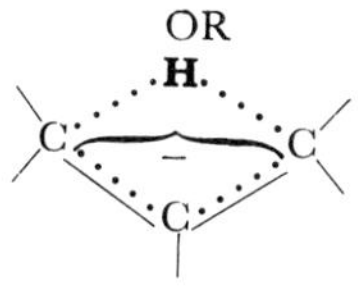

D: bridged allyl anion with H bridging the terminal carbons (− charge), OR above H, D on central carbon

The classical evidence for the existence of a bimolecular mechanism of prototropy was based on a comparison of rate processes intrinsic to the ethoxide anion–catalyzed methyleneazomethine rearrangement of IV to V in ethanol-*O-d*, or dioxane–ethanol-*O-d*.[9] With k_i as the initial rate constant for isomerization, k_e the initial rate constant for introduction of deuterium into the system as a whole, and k_α the rate constant for loss of optical activity, an equality of values of these rate constants was taken as evidence for the absence of a carbanionic intermediate in this rearrangement, and the operation of a "bimolecular mechanism" for the proton transfer.[9] In the isomerization of IVa, b, and c, $k_i = k_\alpha$,[10] and for the isomerization of IVc, $k_i = k_\alpha = k_e$.[10,11]

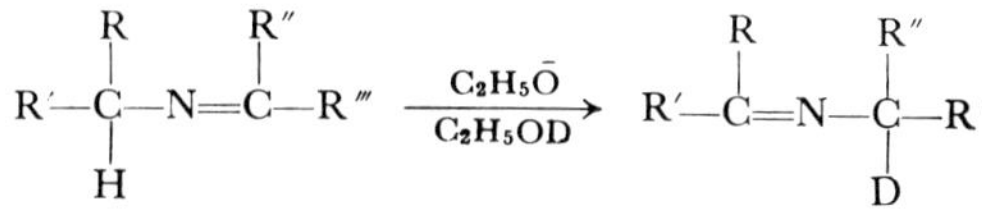

$$\underset{\text{IV}}{R'\text{—}C(R)(H)\text{—}N{=}C(R'')\text{—}R'''} \xrightarrow[C_2H_5OD]{C_2H_5\bar{O}} \underset{\text{V}}{R'\text{—}C(R){=}N\text{—}C(R'')(D)\text{—}R'''}$$

System	R	R′	R″	R‴
a	C_6H_5	CH_3	C_6H_5	*p*-ClC_6H_4
b	C_6H_5	CH_3	C_6H_5	C_6H_5
c	*p*-$C_6H_5C_6H_4$	C_6H_5	C_6H_5	H

[8] W. G. Young, S. Winstein, and H. L. Goering, *J. Am. Chem. Soc.*, **73**, 1958 (1951).

[9] C. K. Ingold, "Structure and Mechanism in Organic Chemistry," Cornell Univ. Press, Ithaca, New York, 1953, pp. 572–574.

[10] (a) C. K. Ingold and C. L. Wilson, *J. Chem. Soc.*, p. 1493 (1933); (b) C. K. Ingold and C. L. Wilson, *ibid.*, p. 93 (1934); (c) S. K. Hsu, C. K. Ingold, and C. L. Wilson, *ibid.*, p. 1774 (1935).

[11] R. P. Ossorio and E. D. Hughes, *J. Chem. Soc.*, p. 426 (1952).

Bimolecular (one-stage) mechanism of prototropy:

$$\text{B:}\ \ {-}\underset{\text{H}}{\overset{|}{\text{C}}}{-}\text{N}{=}\overset{|}{\text{C}}{-}\ \ \text{D—B} \longrightarrow {-}\underset{\text{B}\cdots\text{H}}{\overset{|}{\text{C}}}\cdots\text{N}\cdots\underset{\text{D}\cdots\text{B}}{\overset{|}{\text{C}}}{-} \longrightarrow \text{B—H}\ \ {-}\overset{|}{\text{C}}{=}\text{N}{-}\underset{\text{D}}{\overset{|}{\text{C}}}{-}\ \ \text{:B}$$

Transition state.

A reexamination of the methyleneazomethine rearrangement demonstrated that carbanions do indeed intervene as intermediates, and that the reaction resembles that of its carbon analog.[12a] The value of $K_{eq.}$ (VI/V) was found to be 1.2. After 8% isomerization of optically active V to VI in one-to-one dioxane–ethylene glycol-*O-d* (potassium ethylene glycoxide) at 100°, recovered starting material exhibited little or no racemization or isotopic exchange. These observations are compatible with $k_i = k_\alpha = k_e$ for V→VI, and are in harmony with those made by Ingold and co-workers[10] with regard to the behavior of IVa, b, and c.

$$\underset{\text{V}}{CH_3{-}\overset{*}{\underset{H}{\overset{C_6H_5}{C}}}{-}N{=}\overset{C_6H_4Cl\text{-}p}{C}{-}C_6H_4Cl\text{-}p} \longrightarrow \underset{\text{VI}}{CH_3{-}\overset{C_6H_5}{C}{=}N{-}\underset{H}{\overset{C_6H_4Cl\text{-}p}{C}}{-}C_6H_4Cl\text{-}p}$$

When VI was allowed to isomerize 10% to V under the same conditions, recovered starting material (VI) was found to have undergone > 95% isotopic exchange in the benzhydryl position. When *tert*-butyl alcohol-*O-d*, potassium *tert*-butoxide at 75° was substituted for ethylene glycol, and isomerization was 2.5%, the recovered starting material (VI) had undergone 38% isotopic exchange at the benzhydryl position. A value of $k_e'/k_i = 19$ was calculated from the data for *tert*-butyl alcohol, where k_e' is the rate constant for introduction of deuterium *into starting material only*. The results of the run in dioxane–ethylene glycol indicate that $k_e'/k_i > 10$ in this solvent as well. If the isomerizations were entirely intermolecular, these values would equal the ratios of rate constants for collapse of a carbanion intermediate to give VI and V, respectively (k_a/k_b). Presence of an intramolecular component in the isomerization would increase the value of this "collapse ratio."

$$\text{VI-}h \xrightarrow{\text{B:}} \underbrace{CH_3{-}\overset{C_6H_5}{C}\cdots N\cdots\overset{C_6H_4Cl\text{-}p}{C}}_{-}{-}C_6H_4Cl\text{-}p \begin{array}{l} \xrightarrow{k_a} \text{VI-}d \\ \xrightarrow{k_b} \text{V-}d \end{array}$$

[12] (a) D. J. Cram and R. D. Guthrie, *J. Am. Chem. Soc.*, **87**, 397 (1965); (b) D. J. Cram and R. D. Guthrie, unpublished results.

The high value of this collapse ratio indicates why $k_i = k_\alpha = k_e$ for V → VI in spite of the intervention of a carbanion as an intermediate. The very close structural similarity of this system to those of Ingold and co-workers[10] leaves little doubt that carbanions intervened in their systems. The fact that these investigators observed $k_i = k_\alpha$ and $k_i = k_\alpha = k_e$ reflects a collapse ratio that favored product, as was observed for V → VI.

A system was designed (VII) which favored carbanion collapse to the asymmetric center of the α-phenylethylamino component of the azomethylene system.[12a] The value of $K_{eq.}$ (VIII/VII) was found to be 14.9 at 100°. When optically active VII was allowed to undergo 17% isomerization in *tert*-butyl alcohol-*O-d* (potassium *tert*-butoxide as base) at 75°, recovered starting material had undergone 57% isotopic exchange but only 3% racemization. The rearranged product (VIII) had undergone 62% exchange of one atom of hydrogen for deuterium in the methylene position. If k_α' is defined as the rate constant for racemization of the starting material (k_α was the rate constant for loss of optical activity due to both racemization and isomerization), and k_e' is the rate constant for introduction of deuterium into the starting material only, then $k_e'/k_\alpha' = 28$, and $k_e'/k_i \sim 4.7$. The amount of starting material regenerated from carbanion at 17% conversion is not large enough to affect this value substantially.

$$CH_3-\overset{*}{C}(C_6H_5)(H)-N{=}CH-C(CH_3)_3 \xrightarrow{tert\text{-BuOK}} CH_3-C(C_6H_5)\cdots N\cdots CH-C(CH_3)_3^{\,-}$$

VII-*h* — Aza-allylic anion

$$\xrightarrow{tert\text{-BuOD}} CH_3-\overset{*}{C}(C_6H_5)(D)-N{=}CH-C(CH_3)_3 \quad \text{(VII-}d\text{)}$$

$$\xrightarrow{tert\text{-BuOD}} CH_3-C(C_6H_5){=}N-CH(H\ (D))-C(CH_3)_3 \quad \text{(VIII-}h\text{ and VIII-}d\text{)}$$

VII-*d* — VIII-*h* and VIII-*d*

The value of k_e'/k_i approximates the ratio of rate constants for collapse of a carbanion intermediate to give VII and VIII (k_a'/k_b') if all carbanion collapse to VII involved isotopic exchange. The intramolecularity for conversion of VII to VIII is calculated from the exchange observed in VIII to be a minimum of 38%. If the observed exchange of VII before isomerization is taken into account, the intramolecularity becomes roughly 50%. Correction of the collapse ratio for this effect gives $(k_a' + k_c')/k_b' \sim 7$ (see formulation for definitions of rate constants).

The course of the conversion of optically active VII-*h* to VIII was also examined at 25° in dimethyl sulfoxide-d_6, 3.1 M in methanol-O-d and 0.30 M in potassium methoxide. In two experiments, isomerization

$$\text{VII-}h + \bar{\text{B}} \underset{k_c'}{\rightleftarrows} \underbrace{\text{C}_6\text{H}_5\text{—}\overset{\text{CH}_3}{\overset{|}{\text{C}}}\text{....N....CH}}_{-}\text{—C(CH}_3)_3 + \text{HB} \xrightarrow[\text{BD}]{k_a'} \text{VII-}d$$

$$\xrightarrow[\text{BD}]{k_b'} \text{VIII-}d$$

was allowed to go 14 and 23%. Recovered VII showed 17 and 29% isotopic exchange and 18 and 33% racemization, respectively. Values of $k_e'/k_i = 1.24$ and 1.3 and $k_e'/k_\alpha' = 0.91$ and 0.84 were calculated, respectively. Product VIII recovered from the run carried to 23% isomerization contained 80% of one atom of deuterium in its methylene group. If exchange of starting material before isomerization is considered, the intramolecular component in isomerization is calculated as about 25%, and $k_e'/k_i' \sim 1.6$. The same isomerization in the same medium carried 25% of the way at 75° gave $k_e'/k_i \sim 1.4$ and $k_e'/k_\alpha' \sim 0.90$. At 14% isomerization at 75°, the product (VIII) contained 84% of one atom of deuterium in its methylene group. At the higher temperature, reaction occurred with significantly lower intramolecularity. Under the conditions of the last two runs, VIII showed no significant isotopic exchange in its methylene position in control experiments. Thus the stereochemical course of exchange of VII changed from high retention to racemization when dimethyl sulfoxide–methanol was substituted for *tert*-butyl alcohol, and the collapse ratio of the carbanion changed by a factor of about 4 in the direction of favoring VIII.

In all solvents, the collapse ratio for VII → VIII favors proton capture at the benzyl carbon of the intermediate carbanion. As a consequence, the rate constants for loss of optical activity (k_α), for isomerization (k_i), and for isotopic exchange of the system as a whole (k_e) are far different from one another. The high value of k_e'/k_α' (28) in *tert*-butyl alcohol (nondissociating solvent) indicates that exchange of VII occurs with high retention of configuration. For 3-phenyl-1-butene (I) [5] in the same solvent, $k_e'/k_\alpha' > 10$. The intramolecularity for the isomerization of VII to VIII ($\sim 50\%$) approximates that observed for the isomerization of 3-phenyl-1-butene (I).

Clearly carbanions intervene as intermediates in these isomerization, racemization, and exchange reactions, and the claim for a bimolecular mechanism of prototropy [10] is without foundation.

Jacobs and Dankner [13] observed that certain aryl acetylenes were

[13] T. L. Jacobs and D. Dankner, *J. Org. Chem.*, **22**, 1424 (1957).

isomerized to allenes when treated with basic alumina. This rearrangement resembles that of the allylic rearrangement except that the three carbon atoms involved in the proton transfers are linear, and farther apart than in the allylic rearrangement. In spite of this difference, isomerization of IX to X with potassium alkoxide bases in deuterated alcohols proceeded intramolecularly to an extent comparable to that observed in the 3-phenyl-1-butene system (see Table II).[14] That this

TABLE II

Intramolecularity in the Base-Catalyzed Rearrangement of 1,3,3-Triphenylpropyne (IX) *to Triphenylallene* (X)

Substrate	Solvent	Base	T (°C)	% Intra-molecularity
IX-h	*tert*-BuOD	*tert*-BuOK	30	22
IX-d	CH_3OH	CH_3OK	30	19
IX-d	$(CH_3)_2SO$–1.6 M *tert*-BuOH	$N(CH_2CH_2)_3N$	30	88
IX-d	$(CH_3)_2SO$–3.9 M CH_3OH	$N(CH_2CH_2)_3N$	30	88
IX-d	$(CH_3)_2SO$–3.9 M CH_3OH[a]	$N(CH_2CH_2)_3N$	30	85
IX-d	$(CH_3)_2SO$–3.9 M CH_3OH[a]	$(CH_2)_5NH$	30	58

[a] Solution was 0.14 M in $N(CH_2CH_2)_3NHI$.

isomerization should have an intramolecular component was suggested by the isoracemization of nitrile XI[15] (see Chapter II). This intramolecular racemization in the presence of external proton donor was

$$\text{Ar—CH}_2\text{—C}\equiv\text{C—C}_6\text{H}_5 \xrightarrow[\text{alumina}]{\text{Basic}} \text{Ar—CH=C=CH—C}_6\text{H}_5$$

$$\underset{\text{XI (Optically active)}}{\text{C}_2\text{H}_5\text{—}\overset{\text{H}}{\overset{|}{\underset{\text{C}_6\text{H}_5}{\underset{|}{\text{C}^*}}}}\text{—C}\equiv\text{N}} \xrightarrow{\text{N(C}_3\text{H}_7)_3} \text{C}_2\text{H}_5\text{—}\underset{\text{C}_6\text{H}_5}{\underset{|}{\text{C}}}\text{=C=N}\diagup^{\text{H}}$$

$$\downarrow (\text{C}_2\text{H}_7)_3\text{N}$$

$$\underset{\text{XI (Racemic)}}{\text{C}_6\text{H}_5\text{—}\overset{\text{C}_2\text{H}_5}{\overset{|}{\underset{\text{H}}{\underset{|}{\text{C}}}}}\text{—C}\equiv\text{N}} \xleftarrow{(\text{C}_3\text{H}_7)_3\text{N}} \text{C}_6\text{H}_5\text{—}\overset{\text{C}_2\text{H}_5}{\overset{|}{\text{C}}}\text{=C=N}\diagdown_{\text{H}}$$

[14] D. J. Cram, F. Willey, H. P. Fischer, and D. A. Scott, *J. Am. Chem. Soc.*, **86**, 5370 (1964).

[15] D. J. Cram and L. Gosser, *J. Am. Chem. Soc.*, **86**, 2950 (1964).

interpreted as involving a "conducted tour mechanism," parts of which involved the intramolecular proton transfer from carbon to nitrogen and back to carbon. The fact that IX isomerizes to X with high intramolecularity under conditions of the isoracemization (tripropylamine-*tert*-butyl alcohol) lends strong support to the conducted tour mechanism. The mechanism proposed for the isomerization of IX to X is found in Figure 2.

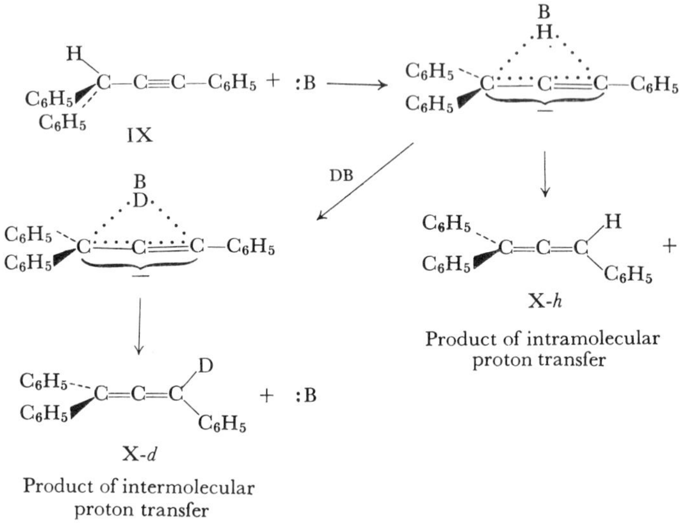

IX

DB

X-*h*

Product of intramolecular proton transfer

X-*d*

Product of intermolecular proton transfer

FIG. 2. Intramolecular proton transfer in the acetylene-to-allene rearrangement.

The intramolecular base-catalyzed isomerizations considered thus far have involved 1,3-proton migrations. An example of a 1,5-proton migration is found in the base-catalyzed rearrangement of XII to give XIII.[14] The fact that a benzene ring is generated from a triene makes this rearrangement particularly facile. It was conveniently conducted at 75° in triethylcarbinol with tripropylamine as catalyst, which makes it comparable in kinetic acidity to nitrofluorene system XIV, with an estimated pK_a of 18.[15]

Table III records the results of rearrangement of XII in a variety of deuterated solvents and with a variety of bases.[14] The pattern of results resembles those observed with the 1,3-proton rearrangement observed in the 3-phenyl-1-butene rearrangement (see Table I). With a variety of solvents ranging from ethylene glycol to tetrahydrofuran–10% water to dimethyl sulfoxide–10% methanol to methanol to *tert*-butyl alcohol and with metal alkoxides as bases, the percent intramolecularity varied only

TABLE III

Intramolecularity in a Base-Catalyzed 1,5-Proton Rearrangement of Triene XII *to Triarylmethane* XIII

Run no.	Solvent	Base	T (°C)	% Intra-molecular	$k_{intramol.}/k_{intermol.}$
1	$DOCH_2CH_2OD$	$DOCH_2CH_2OK$	55	17	0.20
2	$(CH_2)_4O$–10% D_2O	DONa	25	34	0.51
3	$(CD_3)_2SO$–10% CH_3OD	CH_3OK	25	40	0.67
4	CH_3OD	CH_3ONa	25	47	0.88
5	*tert*-BuOD	*tert*-BuOK	25	50	1.0
6	$(C_2H_5)_3COD$	$(C_3H_7)_3N$	75	98	49
7	$(C_2H_5)_3COD$ [a]	$(C_3H_7)_3N$	75	97	32
8	$(C_2H_5)_3COD$ [b]	$(C_3H_7)_3N$	75	98	49

[a] Solution was 0.1 M in $(C_4H_9)_4NI$.
[b] Solution was 0.1 M in $(C_3H_7)_3NDI$.

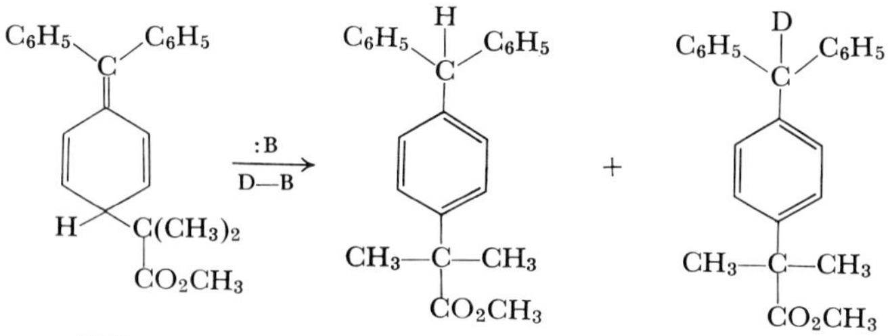

XII $\xrightarrow[D—B]{:B}$ XIII-*h* + XIII-*d*

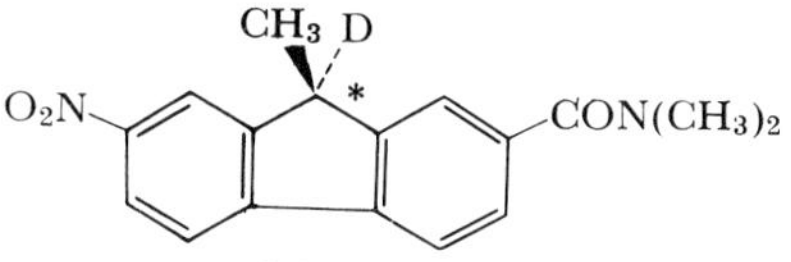

XIV

from a low of 17% to a high of 50% (runs 1–5). The ratio of rate constants, $k_{intramol.}/k_{intermol.}$ increased by a factor of 5 in passing from ethylene glycol (run 1) to *tert*-butyl alcohol (run 5). A dramatic increase in intramolecularity was observed (98%) when tripropylamine was employed as base and triethylcarbinol as solvent (run 6). When reactions were run in the presence of tetrabutylammonium iodide (0.1 M) or tripropylammonium iodide (0.1 M), results were the same within experimental error ($k_{intramol.}/k_{intermol.} \cong 50$). Thus in passing from *tert*-butyl alcohol–potassium *tert*-butoxide to triethylcarbinol–tripropylamine, the ratio of rate constants ($k_{intramol.}/k_{intermol.}$) increased by a factor of about 50. This increase in intramolecularity is undoubtedly associated with the change in charge type of the base rather than the minor change in solvent.

The intramolecular components in these 1,5-rearrangements can be visualized as occurring either by two consecutive 1,3-rearrangements, or by a less specific 1,5-movement of the proton by the molecule of base. The former could consist of two 1,3-hydrogen bonding stages, or could have a covalent intermediate, in which two hydrogens are attached to the *o*-position of the incipient benzene ring. The latter possibility might be detected with appropriate labeling techniques.

The almost exclusive intramolecularity of the rearrangement with tripropylamine in triethylcarbinol-*O-d* ($k_{intramol.}/k_{intermol.} \sim 50$) is explained in Figure 3. The same principles are involved should a direct

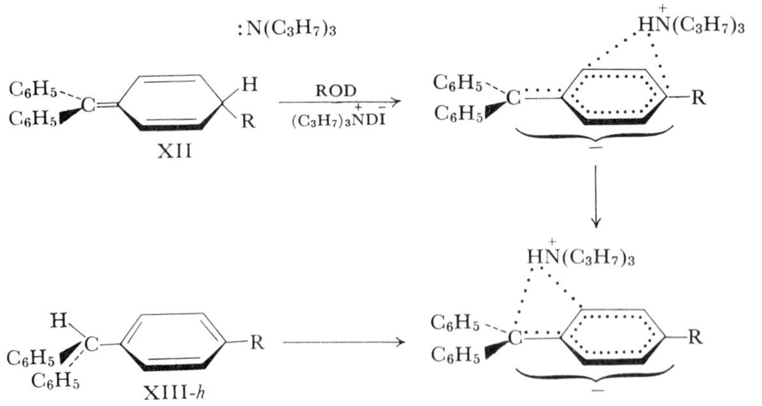

FIG. 3. Mechanism of 1,5-intramolecular proton migration with tripropylamine as catalyst ($R = C(CH_3)_2CO_2CH_3$).

1,5-shift occur or a covalent intermediate intervene. In the mechanism of Figure 3, two hydrogen-bonded ion pairs are formulated as intermediates. Although the carbanion is undoubtedly hydrogen bonded at many other sites, deuteron capture at these sites does not occur since such a process would lead to a *product-separated ion pair in a solvent of low dielectric constant*. Proton capture from the tripropylammonium group gives two neutral molecules. The difference in pK_a between XII and the tripropylammonium ion is probably about 8 pK_a units, and sufficient to make the proton capture process faster than ion pair dissociation.

Although *tert*-butyl alcohol–potassium *tert*-butoxide also possesses a low dielectric constant, the potassium ion has both the leaving group and a deuterated molecule from solvent as ligands, and simple rotation of the potassium ion with its ligands places the deuterated hydroxyl group in a position to react with the carbanion. This process lowers the intramolecularity. The even lower intramolecularity in methanol and ethylene glycol is probably associated with free anions acting as the basic catalyst, solvent molecules becoming hydrogen bonded at distant sites, and deuterium capture competing with proton capture. Another factor that tends to lower the intramolecularity when alkoxide ions or ion pairs serve as bases is the fact that the pK_a's of the proton donors and that of the starting material are comparable. As a consequence the carbanion probably lives longer, and allows more time for solvent to take the place of the leaving group.

The high intramolecularity exhibited by the rearrangement of XII with tripropylamine as base lends substantial support to the "conducted tour mechanism" for isoracemization of nitrofluorene system XIV.[15] The rearrangement of XII as visualized in Figure 3 occurs across the face of a benzyl anion, and this conversion is half of the isoracemization process for system XIV (see Chapter III). If indeed the rearrangement involves only one face of the aromatic nucleus, the tripropylamine-catalyzed reaction should be very highly stereospecific, and would be an example of 1,5-asymmetric induction of high specificity.

GEOMETRIC STABILITY OF ALLYLIC ANIONS

Both allylic cations[16] and radicals[17] have been demonstrated to exhibit geometric stability. Although the configurational stability of allylic organometallic compounds has received some attention,[18] only one study has been made of allylic anions generated by proton abstraction.[19] The potassium *tert*-butoxide–catalyzed isomerization and hydrogen isotope exchange reactions of *cis*- and *trans*-α-methylstilbene (*cis*-XV and *trans*-XV) and α-benzylstyrene (XVI) were studied. The rates at which each olefin underwent isotopic exchange and the rates at which two of the olefins went to the equilibrium mixture were determined, as well as the amounts of all three olefins in the equilibrium

[16] (a) W. G. Young, S. H. Sharman, and S. Winstein, *J. Am. Chem. Soc.*, **82**, 1376 (1960); (b) J. H. Brewster and H. O. Bayer, *J. Org. Chem.*, **29**, 105 (1964).

[17] C. Walling and W. Thaler, *J. Am. Chem. Soc.*, **83**, 3877 (1961).

[18] (a) R. H. DeWolfe and W. G. Young, *Chem. Rev.*, **56**, 753 (1956); (b) P. D. Sleezer, S. Winstein, and W. G. Young, *J. Am. Chem. Soc.*, **85**, 1890 (1963); (c) E. J. Lampher, *ibid.*, **79**, 5578 (1957); (d) W. O. Haag and H. Pines, *ibid.*, **82**, 387 (1960).

[19] D. H. Hunter and D. J. Cram, *J. Am. Chem. Soc.*, **86**, 5478 (1964).

mixture. From these data, the following facts emerged. (1) The *cis*- and *trans*-olefins isomerize into one another through α-benzylstyrene (XVI) as an intermediate, but do not isomerize detectably into one another directly. (2) The *cis*- and *trans*-allylic anions serve as intermediates in the isomerization and exchange reactions, and maintain their geometric integrity in the process. (3) The collapse of the *cis*-allylic anion to *cis*-XV and XVI favored *cis*-XV by a factor of 6.5, whereas the collapse of the *trans*-allylic anion to *trans*-XV and XVI favored *trans*-XV by a factor of 8.8. (4) The equilibrium mixture of the three olefins at 75° in *tert*-butyl alcohol contained 19.5% *cis*-XV, 78.5 *trans*-XV, and 2% XVI. (5) In the kinetically controlled isomerization of XVI to *cis*- and *trans*-XV in *tert*-butyl alcohol, the latter predominated over the former by a factor of 11. (6) The rearrangement of α-benzylstyrene (XVI) to *cis*-XV proceeded with 55% intramolecularity and to *trans*-XV with 36% intramolecularity.

$$(CH_3)(C_6H_5)C{=}C(H)(C_6H_5) \underset{}{\overset{:B}{\rightleftharpoons}} [CH_2\cdots C(C_6H_5)\cdots C(H)(C_6H_5)]^- + HB \overset{:B}{\rightleftharpoons} CH_2{=}C(C_6H_5){-}CH_2{-}C_6H_5$$

cis-XV   *cis*-Carbanion   XVI

$$(CH_3)(C_6H_5)C{=}C(C_6H_5)(H) \overset{:B}{\rightleftharpoons} [CH_2\cdots C(C_6H_5)\cdots C(C_6H_5)(H)]^- + HB \overset{:B}{\rightleftharpoons} CH_2{=}C(C_6H_5){-}CH_2{-}C_6H_5$$

trans-XV   *trans*-Carbanion

An interesting question allied to that of the geometric stability of the allyl anion is concerned with the relation between the kinetically and thermodynamically controlled ratio of *cis-trans*-isomers produced in a carbanionic allylic rearrangement. Evidence has accumulated that isomers tend to predominate in kinetically controlled processes in which two hydrogens are *cis* to one another.[20] Examples are listed.

A detailed kinetic analysis of this effect was made by Schriesheim and Rowe[20f] as applied to the isomerization of 1-alkenes to 2-alkenes. At 55°

[20] (a) W. O. Haag and H. Pines, *J. Am. Chem. Soc.*, **83**, 1701 (1961); (b) T. J. Prosser, *ibid.*, **83**, 1701 (1961); (c) C. C. Price and W. H. Snyder, *ibid.*, **83**, 1773 (1961); (d) A. Schriesheim, J. E. Hofmann, and C. A. Rowe, Jr., *ibid.*, **83**, 3731 (1961); (e) C. C. Price and W. H. Snyder, *Tetrahedron Letters*, **2**, 69 (1962) [an exception is noted in this article]; (f) A. Schriesheim and C. A. Rowe, Jr., *Tetrahedron Letters*, **10**, 405 (1962); (g) M. D. Carr, J. R. P. Clark, and M. C. Whiting, *Proc. Chem. Soc.*, p. 333 (1963); (h) P. L. Nichols, S. F. Herb, and R. W. Riemenschneider, *J. Am. Chem. Soc.*, **73**, 247 (1951).

in dimethyl sulfoxide–potassium *tert*-butoxide, ratios of *cis*- to *trans*-2-alkenes produced by 1-alkenes were extrapolated to zero time. These

$$CH_3CH_2CH{=}CH_2 \xrightarrow[Al_2O_3]{\text{Na on}} \underset{CH_3\quad\quad CH_3}{\overset{H\quad\quad H}{C{=}C}}$$ Predominant isomer[20a]

$$ROCH_2CH{=}CH_2 \xrightarrow{tert\text{-BuOK}} \underset{RO\quad\quad CH_3}{\overset{H\quad\quad H}{C{=}C}}$$ Predominant isomer[20b, c]

$$RCH_2CH{=}CH_2 \xrightarrow[(CH_3)_2SO]{tert\text{-BuOK}} \underset{R\quad\quad CH_3}{\overset{H\quad\quad H}{C{=}C}}$$ Predominant isomer [20d, f, g]

ratios reflect kinetic control of products, and were compared with the ratios at $t = \infty$, which reflect thermodynamic control of products. Table IV contains the results.

TABLE IV

Comparison of cis- and trans-2-Alkenes Produced by Base-Catalyzed Isomerization of 1-Alkenes in Kinetic and Thermodynamically Controlled Processes[20f]

Start. material	$(cis/trans)\ t \to 0$	$(cis/trans)\ t \to \infty$	$\frac{(cis/trans)\ t \to 0}{(cis/trans)\ t \to \infty}$
$CH_2{=}CH{-}CH_2{-}CH_3$	47.4	0.25	190
$CH_2{=}CH{-}CH_2{-}CH_2{-}CH_3$	10.8	0.23	47
$CH_2{=}CH{-}CH_2{-}CH(CH_3)_2$	3.2	0.23	14
$CH_2{=}CH{-}CH_2{-}C(CH_3)_3$	0.25	<0.001	>250

Clearly the kinetically controlled processes produced a ratio of olefins favorable to the thermodynamically unstable isomer. This generalization does not apply to isomerization of α-benzylstyrene to *cis*- and *trans*-α-methylstilbene[19] or to isomerization of *N*,*N*-dimethylallylamine to *N*,*N*-dimethylpropenylamine.[20e]

Whiting and co-workers[20g] observed that *trans*-3-octene in ethylenediamine–lithium 2-aminoethylamide at 25° isomerized initially to give essentially only *cis*-2-octene. Under the same conditions, *cis*-3-octene initially gave *trans*-4-octene as the essentially exclusive product.

In those cases where kinetic control leads to the less stable isomer, Price and Snyder[20e] pointed to a possible explanation which assumes a ground state conformational control of products. It is assumed that that conformation is favored in the starting material which places β-hydrogens *cis* to the double bond, and thus allows the π-electrons of the double

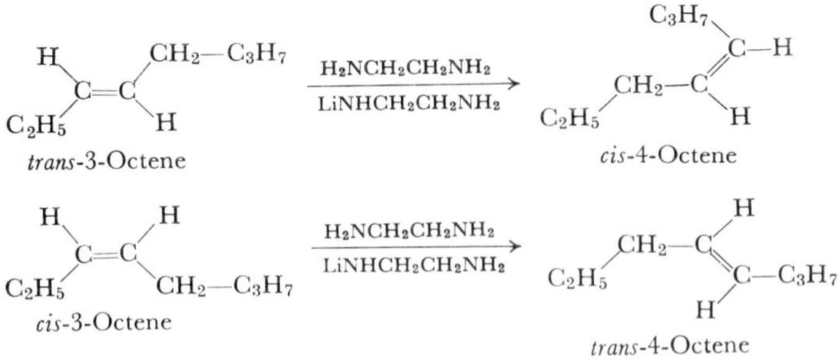

$$\text{H}(\text{C}_2\text{H}_5)\text{C}{=}\text{C}(\text{H})\text{CH}_2{-}\text{C}_3\text{H}_7 \xrightarrow[\text{LiNHCH}_2\text{CH}_2\text{NH}_2]{\text{H}_2\text{NCH}_2\text{CH}_2\text{NH}_2} \text{C}_2\text{H}_5{-}\text{CH}_2{-}\text{C(H)}{=}\text{C(H)}{-}\text{C}_3\text{H}_7$$

trans-3-Octene → *cis*-4-Octene

$$\text{H}(\text{C}_2\text{H}_5)\text{C}{=}\text{C}(\text{H})\text{CH}_2{-}\text{C}_3\text{H}_7 \xrightarrow[\text{LiNHCH}_2\text{CH}_2\text{NH}_2]{\text{H}_2\text{NCH}_2\text{CH}_2\text{NH}_2} \text{C}_2\text{H}_5{-}\text{CH}_2{-}\text{C(H)}{=}\text{C(H)}{-}\text{C}_3\text{H}_7$$

cis-3-Octene → *trans*-4-Octene

bond to interact with two of the hydrogens of the β-carbon. The conformation presumed favored by 2-pentene is formulated.

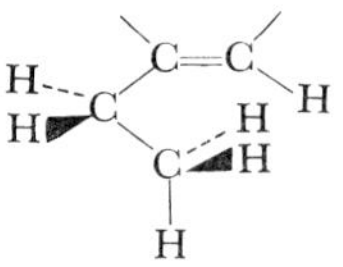

Ground state conformation presumed favored for 1-butene

Another possible explanation focuses attention on a balance between eclipsing effects and steric inhibition of solvation in the transition state for proton abstraction. These two steric effects operate in the opposite direction. In general, the more internally compressed a molecule by internal eclipsing of the nonreactive parts of the molecule, the more exposed is the reactive site to attacking reagents and solvating solvent molecules. The less internally compressed and the more dispersed the nonreactive parts of a molecule, the more hindered is the reactive site to attacking reagents and solvating solvent molecules. In the transition states of those rearrangements leading to the *cis*-products, carbanion and metal ion solvation may be more important sterically than the nonbonded interactions within the hydrocarbon portions of the molecules.

HYDROCARBON SUBSTITUENT EFFECTS ON THE RATES OF ALLYLIC REARRANGEMENTS OF ALKENES

The rates of isomerization of a number of 1-alkenes were determined by Schriesheim and co-workers.[21] The reactions were conducted in

[21] (a) A. Schriesheim and C. A. Rowe, Jr., *J. Am. Chem. Soc.*, **84**, 3161 (1962); (b) A. Schriesheim, C. A. Rowe, Jr., and L. Naslund, *ibid.*, **85**, 2111 (1963).

dimethyl sulfoxide–potassium *tert*-butoxide at 55°, and were followed to 30–50% conversion. Although the products once formed frequently underwent further rearrangement, under the conditions used these processes were slow enough not to interfere with the initial allylic rearrangement. The nonterminal, more highly substituted olefins except when highly hindered were more stable than the starting materials. The relative rates are listed in Table V, and are corrected for the number of hydrogens on the α-carbon available for abstraction.

TABLE V

Hydrocarbon Substituent Effects on the Relative Rates of Isomerization of Terminal Alkenes at 55° in Dimethyl Sulfoxide–Potassium tert-Butoxide[21b]

R Group	Group I $CH_2{=}CH{-}CH_2{-}R$	Group II $CH_2{=}C(CH_3){-}CH_2{-}R$	Group III $CH_2{=}CH{-}CH(CH_3){-}R$	Group IV $CH_2{=}C(R){-}CH_2{-}CH_3$
CH_3	1	1	1	1
H	—	—	4.13	6.5
C_2H_5	0.57	0.43	0.54	0.58
C_3H_7	0.55	0.50	—	—
i-C_3H_7	0.17	0.12	0.15	—
tert-C_4H_9	0.0074	0.0085	—	—
$CH_2{=}CH$	20×10^5	—	—	—
C_6H_5	9.4×10^5	—	—	—

Within experimental error, Equation (1) applies to the kinetic data of the four groups of alkenes. The ranking of the alkyl group substituents is independent of the position of the alkyl substituent on the allyl system.

$$\left(\log \frac{k_{\mathrm{I}}^{\mathrm{R}}}{k_{\mathrm{I}}^{\mathrm{CH_3}}}\right) = \left(\log \frac{k_{\mathrm{II}}^{\mathrm{R}}}{k_{\mathrm{II}}^{\mathrm{CH_3}}}\right) = \left(\log \frac{k_{\mathrm{III}}^{\mathrm{R}}}{k_{\mathrm{III}}^{\mathrm{CH_3}}}\right) = \left(\log \frac{k_{\mathrm{IV}}^{\mathrm{R}}}{k_{\mathrm{IV}}^{\mathrm{CH_3}}}\right) = \left(\log \frac{k^{\mathrm{R}}}{k^{\mathrm{CH_3}}}\right)_{\mathrm{av.}} \quad (1)$$

Divinyl methane and allylbenzene were both about 6 powers of 10 faster than 1-butene, and methyl is about 2 powers of 10 more acidifying than *tert*-butyl. The groups fall in the following order with respect to their ability to enhance the rate of isomerization, the range covering over 7 powers of 10:

$$CH_2{=}CH \geqslant C_6H_5 > H > CH_3 > C_2H_5 > C_3H_7 > i\text{-}C_3H_7 > tert\text{-}C_4H_9$$

In Figure 4, the values of $(\log k^{R}/k^{CH_3})_{av.}$ are plotted against the values of Taft's σ^* constants, which measure the polar effect of groups.[22] Although hydrogen, methyl, ethyl, and propyl describe a good straight line, isopropyl and *tert*-butyl fall far below the line, the magnitude of the deviation increasing with the bulk of the substituent. Here steric effects are clearly visible: steric inhibition of coplanarity of the transition state for anion formation, and steric inhibition of solvation of the anion.

In a different study,[23] the rates of isomerization of the methylenecycloalkanes were determined in dimethyl sulfoxide–potassium *tert*-butoxide.[23] The rates relative to that of methylenecyclohexane at 55°

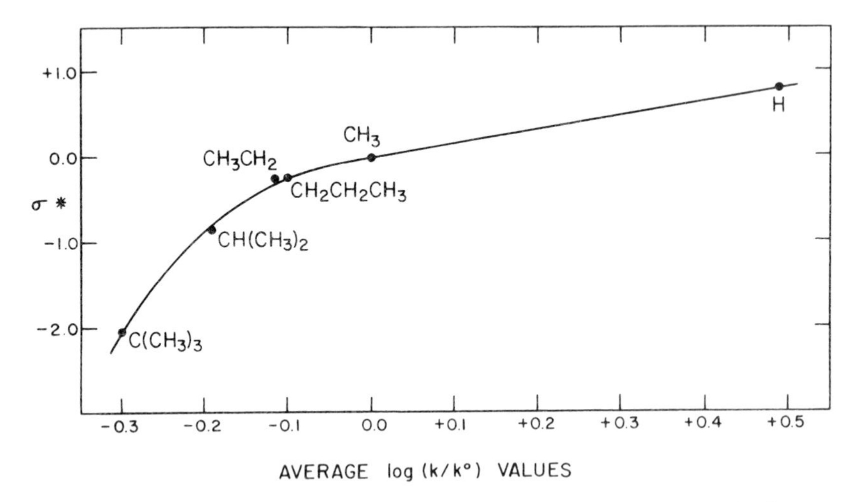

FIG. 4. Plot of σ^* against average log (k/k°) values for rates of alkene isomerizations.

are recorded in Table VI, along with the activation parameters. Two open-chain terminal alkenes are included for comparison.[21] The rates of base-catalyzed bromination in water at 0° of the corresponding cycloalkanones were also measured. The fact that a plot of $\log k_{\text{isomerization}}$ vs. $\log k_{\text{bromination}}$ for the various sized ring compounds proved to be linear was interpreted as demonstrating that proton abstraction by base was rate determining for each reaction, and that in each olefin-ketone pair, the same stereoelectronic effects governed proton abstraction from both alkene and ketone.

The rates of isomerization cover a range of 3 powers of 10, with

[22] R. W. Taft, Jr., *in* "Steric Effects in Organic Chemistry" (M.-S. Newman, ed.), Wiley, New York, 1956, Chapter 13.

[23] A. Schriesheim, R. J. Muller, and C. A. Rowe, Jr., *J. Am. Chem. Soc.*, **85**, 3164 (1962).

TABLE VI

Relative Rates (per Reactive Hydrogen) of Base-Catalyzed Isomerization of Methylenecycloalkanes to Methylcycloalkenes in Dimethyl Sulfoxide[23]

Compound	Relative rate	$\Delta H^‡$ (Kcal./mole)	$\Delta S^‡$ (e.u.)
Methylenecyclobutane ($=CH_2$)	1070	13.3	−27.3
Methylenecyclopentane ($=CH_2$)	454	13.3	−17.0
Methylenecyclohexane ($=CH_2$)	1	27.1	0.7
Methylenecycloheptane ($=CH_2$)	5.8	19.7	−18.0
Methylenecyclooctane ($=CH_2$)	17	16.8	−25.4
$(CH_3CH_2)_2C{=}CH_2$	35	—	—
$CH_3CH_2CH{=}CH_2$	385	16.6	−22.3

methylenecyclobutane exhibiting the fastest and methylenecyclohexane the slowest rate. This difference was interpreted as reflecting a transition state for proton abstraction which was largely carbanionic in character, and in which the three carbons of the allylic system and their attached substituents approached a planar configuration. Such a configuration maximizes electron delocalization, and stabilizes the transition state.[24] The cyclobutyl and cyclopentyl systems in their ground states are not far from the geometries of their transition states, except that in each, two pairs of hydrogens become uneclipsed in passing to the transition state. The latter effect is probably more important in the reaction of methylene-cyclobutane, and is the major factor in the rate acceleration. Angle strain effects predict that the cyclobutyl should be slower than the cyclopentyl

[24] E. J. Corey, *J. Am. Chem. Soc.*, **76**, 175 (1954).

system, since more angle strain is generated by confinement of an sp^2 carbon in a four-membered than in a five-membered ring.[25] Although angle strain is certainly playing a role, that role is not dominant. The interplay of eclipsing and angle strain effects underlies the other rate factors in Table VI.

The heats of activation vary by over 10 Kcal./mole, and the entropy of activation by 28 e.u. as the size of the ring systems is changed. For methylenecyclobutane isomerization, the very low heat of activation is somewhat compensated by the high negative entropy of activation. The latter indicates a rigid transition state with a high order of solvent and base organization. All of the other systems are much closer in $\Delta S^{\ddagger}$ values to methylenecyclobutane than to methylenecyclohexane, for which $\Delta S^{\ddagger} \sim 0$. In the transition state for this olefin, the allylic anion appears to be less well developed, charge less delocalized, and solvent less rigidly bound.

OLEFIN EQUILIBRIA

Equilibration of isomeric olefins has a longer history than most topics of this book, a history that is mingled with the evolution of the modern theory of tautomerism.[26]

Equilibration of the five 2-methylpentenes in dimethyl sulfoxide–potassium *tert*-butoxide at 55° provided a mixture whose composition is close to that predicted from thermodynamic data[21a] (Table VII). These data are representative of a large body of results that indicate that successive substitution of alkyl groups (not highly branched) for hydrogens of ethylene tends to give more stable structures than their less substituted isomers. This generalization grows out of the merging of the electron-releasing inductive effect of the alkyl groups, and the electron-withdrawing effect of the vinyl groups. With *tert*-butyl and like groups, steric effects are superimposed, and the generalization can break down, particularly with *cis*-isomers.

The data of Table VII are instructive (the percentage calculated is probably more reliable for the alkenes present in small amounts). In the equilibrium mixture, 2-methyl-2-pentene > 2-methyl-1-pentene > 4-methyl-1-pentene by factors of 80 to 10 to 1, respectively. Each additional alkyl group substituted for hydrogen on ethylene increases the stability relative to its isomers by a factor of 5–10.

[25] H. C. Brown, J. H. Brewster, and H. Shechter, *J. Am. Chem. Soc.*, **76**, 467 (1954).
[26] C. K. Ingold, "Structure and Mechanism in Organic Chemistry," Cornell Univ. Press, Ithaca, New York, 1953, Chapter X.

TABLE VII

Composition of Equilibrium Mixture of 2-Methylpentenes in Dimethyl Sulfoxide at 55° [21a]

Compound	% Observed	% Calculated
$CH_2{=}C(CH_3){-}CH_2{-}CH_2{-}CH_3$	11.3	11.0
$CH_3{-}C(CH_3){=}CH{-}CH_2{-}CH_3$	80.0	80.7
$CH_3{-}CH(CH_3){-}C(H){=}C(H){-}CH_3$ (H above C, H below C)	7.2	5.5
$CH_3{-}CH(CH_3){-}C(H){=}C(H){-}CH_3$ (both H below)	1.2	1.8
$CH_3{-}CH(CH_3){-}CH_2{-}CH{=}CH_2$	0.3	1.0

Conjugative effects on equilibria of vinyl and aryl groups with the carbon–carbon double bond are even more important. For example, at equilibrium propenylbenzene dominates over allylbenzene by an estimated factor of 3 to 4 powers of 10,[27a,b] and 1,3-pentadiene over 1,4-pentadiene by a factor of greater than 600 at 25°.[27c,d] However, when incorporated in ring systems, the conjugative effect is combined with enforced conformational effects, and the importance of the conjugative effect becomes less predictable.[27e]

Conjugative effects on equilibria of carbonyl and cyano groups with carbon–carbon double bonds have been studied extensively by Kon, Linstead, and co-workers.[28] Systems such as those listed have been examined, and the following generalizations emerged. (1) If the γ-positions of unsaturated acids, esters, ketones, or nitriles are unsubstituted, equilibria between the α,β- and β,γ-compounds very strongly

[27] (a) A. G. Catchpole, E. D. Hughes, and C. K. Ingold, *J. Chem. Soc.*, p. 8 (1948); (b) L. Pauling, "Nature of the Chemical Bond," Cornell Univ. Press, Ithaca, New York, 1960, p. 196; (c) J. E. Kilpatrick, C. W. Beckett, E. J. Prosen, K. S. Pitzer, and F. D. Rossini, *J. Res. Natl. Bur. Std.*, **42**, 225 (1949); (d) J. B. Conant and G. B. Kistiakowsky, *Chem. Rev.*, **20**, 181 (1937); (e) R. B. Bates, R. H. Carnighan, and C. E. Staples, *J. Am. Chem. Soc.*, **85**, 3030 (1963).

[28] Reviewed by (a) J. W. Baker, "Tautomerism," Routledge, London, 1934, Chapter 9; (b) C. K. Ingold, "Structure and Mechanism in Organic Chemistry," Cornell Univ. Press, Ithaca, New York, 1953, p. 562.

favor the conjugated isomer, irrespective of substitution at the β-position. (2) Introduction of alkyl groups (particularly methyl) into the γ-positions shifts the equilibrium toward the unconjugated isomer, which can

$$C_2H_5CH_2CH{=}CHCO_2^- \rightleftarrows C_2H_5CH{=}CHCH_2CO_2^-$$

$$C_2H_5CH_2CH{=}CHCO_2C_2H_5 \rightleftarrows C_2H_5CH{=}CHCH_2CO_2C_2H_5$$

$$CH_3CH_2C(C_2H_5){=}CHCOCH_3 \rightleftarrows CH_3CH{=}C(C_2H_5)CH_2COCH_3$$

$$(CH_3)_2CHCH{=}CHCN \rightleftarrows (CH_3)_2C{=}CHCH_2CN$$

in some cases become the predominant isomer. (3) Substitution of aryl groups for hydrogen in the γ-position makes the β,γ-isomer the more stable by a large factor. (4) Substitution of alkyl groups (particularly methyl) at the α-position favors the α,β-isomer. These generalizations reflect the superposition of the inductive effects of alkyl groups on the conjugative effects of unsaturated groups on isomer stability. A crude ranking of effects in decreasing order of importance gives the following order: aryl or vinyl conjugation > carbonyl or cyano conjugation $\sim$ alkyl group inductive effects. Although fewer data are available, the nitro group would appear to resemble the carbonyl and cyano groups in its effect on equilibria.[29a]

The conjugative effect of the unshared electrons of amines, ethers, and sulfides strongly favors the vinyl over the propenyl forms of these compounds, although in most cases equilibrium constants are not available.[20b, c, e, 29b]

$$CH_2{=}CH{-}CH_2{-}\ddot{A}{-} \rightleftarrows \left\{CH_3{-}CH{=}CH{-}\ddot{A}{-} \longleftrightarrow CH_3{-}\bar{C}H{-}CH{=}\overset{+}{A}{-}\right\}$$

Favored isomer

The effect on olefin equilibria of functional groups centered around second-row elements is particularly interesting. O'Connor and coworkers[30] gathered data for equilibria between allyl and vinyl sulfides, sulfoxides, and sulfones (Table VIII).

[29] (a) Yu. V. Baskov, T. Urbanski, M. Witanowski, and L. Stefaniak, *Tetrahedron*, **20**, 1519 (1964); (b) D. S. Tarbell and W. E. Lovett, *J. Am. Chem. Soc.*, **78**, 2259 (1956).

[30] (a) D. E. O'Connor and W. I. Lyness, *J. Am. Chem. Soc.*, **85**, 3044 (1963); (b) D. E. O'Connor and C. D. Broaddus, *ibid.*, **86**, 2267 (1964); (c) D. E. O'Connor and W. I. Lyness, *ibid.*, **86**, 3840 (1964).

TABLE VIII

Composition of Equilibrium Mixtures for Unsaturated Sulfides, Sulfoxides, and Sulfones

$$\underset{3}{RCH_2}—\underset{2}{CH}=\underset{1}{CH}—X \underset{\text{Base}}{\overset{K}{\rightleftharpoons}} \underset{3}{RCH}=\underset{2}{CH}—\underset{1}{CH_2}—X$$

Vinyl form Allyl form

R	X	K
H	SCH_3	<0.01
H	$SOCH_3$	0.25
H	SO_2CH_3	0.80
C_3H_7	SCH_3	0.5
C_3H_7	$SOCH_3$	24
CH_3	$SOCH_3$	32
C_3H_7	SO_2CH_3	>99

The data of Table VIII indicate that substitution of an alkyl group for a hydrogen in the 3-position of the unsaturated systems increases the value of the equilibrium constant by factors that range from 40 to 100. This same kind of effect has been observed in the olefin equilibria discussed previously. The data also indicate that $CH_3S > CH_3SO > CH_3SO_2$ in ability to stabilize the vinyl form of the olefin compared to the allyl form. The equilibrium constant changes by a maximum factor of about 200 as these groups are changed.

Variation of R from methyl to propyl has little effect on K for the sulfoxide system, and presumably for the other systems as well. For those systems in which R is propyl, the values of the equilibrium constants in effect measure the relative abilities of X—CH_2 and X to interact favorably with the carbon–carbon double bond. Stabilization effects due to sulfur-electron-pair delocalization into the double bond should decrease in the order $CH_3S > CH_3SO > CH_3SO_2$. This is the same order observed for the ability of X to stabilize the double bond. This effect, although undoubtedly present, fails to explain the fact that $CH_3SOCH_2 \gg CH_3SO$ and $CH_3SO_2CH_2 \gg CH_3SO_2$ in ability to stabilize the respective olefins. Overlap between the *p*-orbitals of the double bond and the *d*-orbitals of sulfur fails to explain the data, since such overlap should stabilize the vinyl relative to the allyl form in each pair, and the opposite is observed.

The major factor responsible for these orders is probably the inductive effect. The order of electron-withdrawing tendency is $CH_3SO_2 >$

$CH_3SO > CH_3S$, and for each pair, $X > XCH_2$. Just as the carbon–carbon double bond is stabilized by the electron-releasing inductive effect of alkyl groups, it should be destabilized by the electron-withdrawing effect of electronegative groups.[30c] Groups that possess strong electron-withdrawing properties tend to form bonds to carbon which are richest in *p*-character, since *p*-orbitals are more extended than *s*-orbitals. The vinyl group forms bonds with sp^2 and the allyl with sp^3 orbitals, the latter being richer in *p*-character. Thus the inductive effect seems mainly responsible for the positions of equilibria in these olefins, with electron pair–π interactions superimposed.

COLLAPSE RATIOS FOR ALLYLIC ANIONS

Uncertainties in detailed interpretations of kinetic results arise from the presence of "back reactions" in most proton abstraction processes (see Chapter III). Observed rate constants for proton abstraction are frequently composite, since the proton recapture process by the carbanion is very fast, and competes with other reactions of the carbanion. In terms of the allylic rearrangement of Equation (2), k_{-1} might be of comparable or even greater value than k_{-2}, in which case a knowledge of

(2) $-\underset{\text{H}}{\underset{|}{\overset{|}{C}}}=\overset{|}{C}-\overset{|}{C}-$ + :B $\underset{k_{-1}}{\overset{k_1}{\rightleftarrows}}$ [C⋯C⋯C allylic anion ⋯ HB] $\underset{k_{-2}}{\overset{k_2}{\rightleftarrows}}$ $-\underset{\text{H}}{\underset{|}{\overset{|}{C}}}-\overset{|}{C}=\overset{|}{C}-$ + :B

$$k_{\text{obs.}} = \frac{k_1 \times k_{-2}}{k_{-1} + k_{-2}}$$

k_{-2}/k_{-1} (the collapse ratio) is indispensable to an understanding of mechanism. When knowledge of the relative values of k_{-1} and k_{-2} is absent, the hope is that $k_{-2} \gg k_{-1}$, or that $k_{-2}/(k_{-1}+k_{-2})$ remains constant when the rates of rearrangement of several systems are compared. Thus the questions of what factors control collapse ratios and how these might be determined are of prime importance.

For many years, the Hughes-Ingold rule seemed to satisfactorily summarize the facts known about collapse ratios.[31] This rule states,[32]

[31] A. G. Catchpole, E. D. Hughes, and C. K. Ingold, *J. Chem. Soc.*, p. 11 (1948).

[32] C. K. Ingold, "Structure and Mechanism in Organic Chemistry," Cornell Univ. Press, Ithaca, New York, 1953, p. 565.

"when a proton is supplied by acids to the mesomeric anion of weakly ionizing tautomers of markedly unequal stability, then the tautomer which is most quickly formed is the thermodynamically least stable: It is also the tautomer from which the proton is lost most quickly to bases." This generalization is particularly successful when unsaturated carbanions are stabilized by charge distribution onto electronegative elements. For example,[32] when a metal salt of phenylnitromethane is acidified, the unstable aci-nitro tautomer is first formed, which then isomerizes to the more stable nitro form. The aci-form (oxygen acid) is more acidic than the nitro form (carbon acid), and loses its proton faster.

In another example,[33] the sodium salt of ethyl cyclopentenylmalonate

$$\left[C_6H_5CH \cdots \overset{+}{N} \begin{smallmatrix} \cdot O \\ \cdot O \end{smallmatrix}\right]^- \overset{+}{M} \xrightarrow{\text{Fast}} C_6H_5CH{=}N(\rightarrow O)OH \quad \text{Aci-form}$$

$$\xrightarrow{\text{Slow}} C_6H_5CH_2{-}N(\rightarrow O){=}O \quad \text{Nitro form}$$

Salt of phenyl-nitromethane

was acidified with benzoic acid in ether–petroleum ether. The thermodynamically unstable unconjugated isomer was initially produced as the predominant component.

$$\text{(cyclopentenyl anion)}{\cdots}\bar{C}(CO_2C_2H_5)_2\ \overset{+}{Na} \xrightarrow{C_6H_5CO_2H} \text{(cyclopent-1-enyl)}{-}CH(CO_2C_2H_5)_2 + \text{(cyclopentylidene)}{=}C(CO_2C_2H_5)_2$$

Isomer initially produced

More stable isomer

The rule appears to apply to the cyclohexadiene system.[27e] At 95° in *tert*-amyl alcohol–potassium *tert*-amyloxide, an equilibrium mixture of cyclohexadienes favors the conjugated 1,3-isomer by a factor of 2.2 over the unconjugated 1,4-isomer. In the same medium deuterated, low conversion of the 1,4- into the 1,3-isomer, or of the 1,3- into the 1,4-isomer, gave yields of the mono-deuterated 1,4- and 1,3-isomers in the ratio of 8 to 1. The fact that this ratio was the same with each starting

[33] W. E. Hugh and C. A. R. Kon, *J. Chem. Soc.*, p. 775 (1930).

material indicated a common allylic anion as intermediate. The less stable isomer was produced 8 times faster than the stable isomer by protonation of this carbanion. Thus $k_{-2}/k_{-1} = 8$, and $k_1/k_2 \simeq 0.05$.

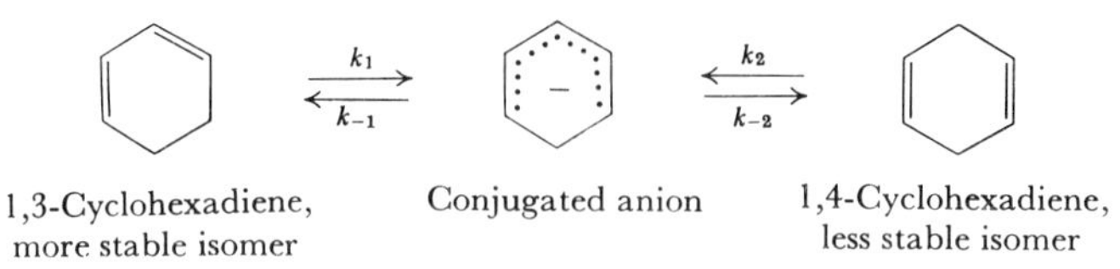

Other examples of application of this rule are found in the steroid field.[34]

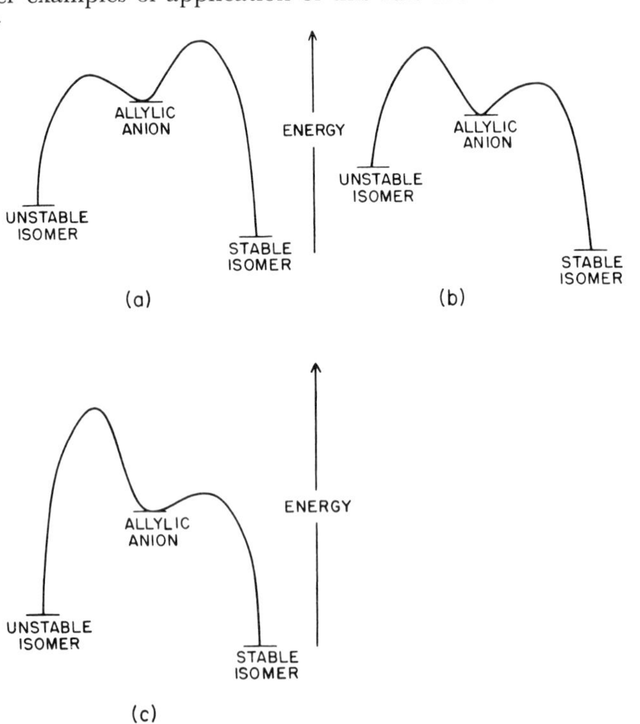

FIG. 5. Activation energy–reaction coordinate profiles for formation and collapse of allylic and related anions.

Three basic activation energy–reaction coordinate profiles for formation and collapse of allylic and related anions are shown in Figure 5. The first of these (a) embodies the Hughes-Ingold rule, of which abundant examples are available. In the second (b), the allylic anion collapses

[34] S. K. Malhotra and H. J. Ringold, *J. Am. Chem. Soc.*, **85**, 1538 (1963).

preferentially to give the thermodynamically more stable isomer, but the less stable isomer is still the more kinetically acidic. Thus (b) reflects exceptions to the first part of the rule, but not the second part. In the third profile (c), the allylic anion collapses preferentially to the thermodynamically more stable isomer, which is also the most kinetically acidic. This third diagram violates both parts of the Hughes-Ingold rule.

Several examples of energy profile (b) are available. One involves the base-catalyzed conversion of 3-phenyl-1-butene (I) to 2-phenyl-2-butene (II). At equilibrium in *tert*-butyl alcohol–potassium *tert*-butoxide

$$\underset{\text{I}}{CH_3-\overset{\overset{C_6H_5}{|}}{\underset{\underset{H}{|}}{C}}-CH{=}CH_2} \underset{k_{-1}}{\overset{k_1}{\rightleftharpoons}} \left[\overset{C_6H_5}{\underset{CH_3}{C}} \cdots \overset{H}{C} \cdots \overset{H}{\underset{H}{C}}\right]^{-} \underset{k_{-2}}{\overset{k_2}{\rightleftharpoons}} \underset{\text{II}}{\overset{C_6H_5}{\underset{CH_3}{C}}{=}\overset{H}{\underset{CH_3}{C}}}$$

at 75°, isomer II dominates over I by a factor too large to be easily measured.[35] In many deuterated solvents and bases, the rate of introduction of deuterium into the terminal methyl group of II was about 20 times faster than into the benzyl position of I during the isomerization of I to II.[5] Clearly, the allylic anion produces the most thermodynamically stable isomer, with $k_{-2}/k_{-1} \sim 20$. The results are in direct contradiction to the first part of the Hughes-Ingold rule. Comparison of the rate constants of isomerization of I with those for isotopic exchange of II indicated that k_1/k_2 equaled 1 to 2 powers of 10. Thus the least stable isomer was the faster to lose a proton, and the second part of the rule applies. Since $K = k_1 k_{-2}/k_2 k_{-1}$, K must be very high valued ($\sim 10^3$ to 10^4).

In contrast to the phenylbutene system (I and II) is the behavior exhibited by VII and VIII in *tert*-butyl alcohol–potassium *tert*-butoxide.[12] The value of the equilibrium constant for VII$\rightleftharpoons$VIII is about 15. The value of $k_{-1}/k_{-2} \sim 7$; thus the collapse ratio of the aza-allylic anion favors the thermodynamically unstable isomer. The rate of exchange of VII at the benzyl position proved to be faster than the rate of exchange of VIII, and $k_1/k_2 > 1$. Thus this aza-allylic system clearly obeys the Hughes-Ingold rule.

A second system that conforms to the energy profile (b) of Figure 5 involves isomerizations of XVI to *cis*- and *trans*-XV and the reverse

[35] D. J. Cram and R. T. Uyeda, unpublished results.

reaction in potassium *tert*-butoxide–*tert*-butyl alcohol.[19] The composition of the equilibrium mixture at 75° is indicated by the percentages

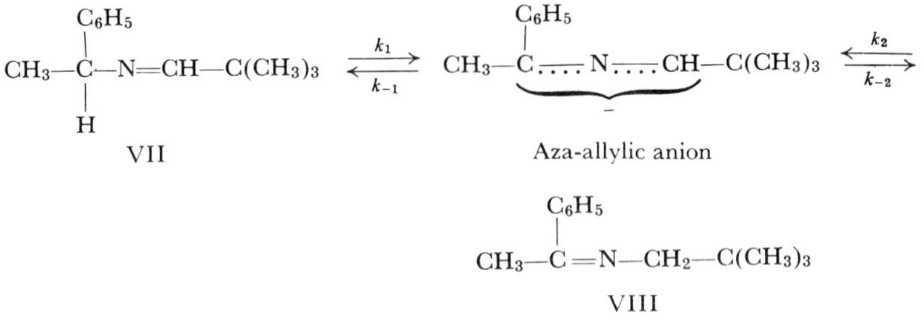

$$\underset{\text{VII}}{CH_3\text{—}\overset{\overset{C_6H_5}{|}}{\underset{\underset{H}{|}}{C}}\text{—}N\text{=}CH\text{—}C(CH_3)_3} \underset{k_{-1}}{\overset{k_1}{\rightleftarrows}} \underset{\text{Aza-allylic anion}}{CH_3\text{—}\overset{\overset{C_6H_5}{|}}{C}\cdots N\cdots CH\text{—}C(CH_3)_3} \underset{k_{-2}}{\overset{k_2}{\leftrightarrows}} \underset{\text{VIII}}{CH_3\text{—}\overset{\overset{C_6H_5}{|}}{C}\text{=}N\text{—}CH_2\text{—}C(CH_3)_3}$$

underneath the formulas. Through combinations of kinetics of isomerization, the thermodynamic data, and kinetics of deuterium incorporation, the following ratios of rate constants were calculated (rate constant are defined in the formulation):

$$\frac{k_1}{k_2} = 4.5 \qquad \frac{k_{-2}}{k_{-1}} = 8.8 \qquad \frac{k_3}{k_4} = 1.5 \qquad \frac{k_{-4}}{k_{-3}} = 6.5$$

Both the *cis*- and *trans*-anions give predominantly the more stable isomers upon proton capture, but the least thermodynamically stable

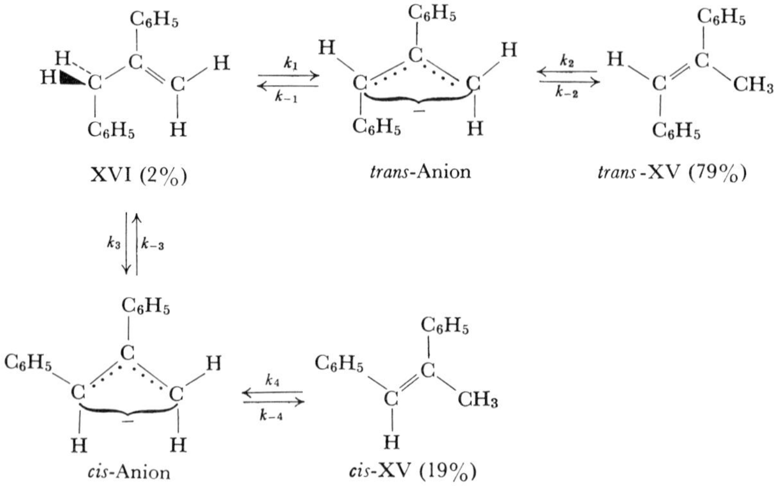

XVI (2%) $\underset{k_{-1}}{\overset{k_1}{\rightleftarrows}}$ *trans*-Anion $\underset{k_{-2}}{\overset{k_2}{\leftrightarrows}}$ *trans*-XV (79%)

XVI $\underset{k_3}{\downarrow}\uparrow_{k_{-3}}$ *cis*-Anion $\underset{k_{-4}}{\overset{k_4}{\leftrightarrows}}$ *cis*-XV (19%)

isomer (XVI) loses its proton to base faster than the more stable isomers (*cis*- and *trans*-XV).

An example of a system that possesses energy profile (c) of Figure 4 involves the four isomeric olefins, XVII–XX.[36] In *tert*-butyl alcohol–potassium *tert*-butoxide at 40°, the four olefins equilibrated to give the

XVII (52%) $\underset{k_{-1}}{\overset{k_1}{\rightleftarrows}}$ E $\underset{k_{-2}}{\overset{k_2}{\leftrightarrows}}$ XVIII (31%)

G F

XX (1%) $\rightleftarrows$ H $\leftrightarrows$ XIX (16%)

composition indicated under the formulas. The initial rate constants for *isomerization* of each olefin decrease in value in the following order, $k_{XVIII} > k_{XVII} > k_{XIX} > k_{XX}$. This reactivity order correlates with the order of decreasing stability of the four allylic anions predicted on the basis of 1,3-steric repulsions within the allylic anion, namely, E > F > G > H. Isomer XX, which is the least thermodynamically stable, is the slowest to undergo isomerization.

The rates at which XVII and XVIII isomerize into one another were much faster than the rates at which these compounds gave XX and XIX.

[36] D. J. Cram and S. W. Ela, unpublished results.

When isomerized in *tert*-butyl alcohol-*O-d*, XVII underwent isotopic exchange at the benzyl position much faster than it isomerized to XVIII. From the data, the collapse ratio (excluding intramolecularity) was calculated to be $k_{-1}/k_{-2} \gtrsim 30$. Thus the allylic anion, E, protonates faster at that position which gives the most thermodynamically stable isomer (XVII), and violates both parts of the Hughes-Ingold rule.

CHAPTER VI

Molecular Rearrangements

Carbanion rearrangements are fewer in number than their carbonium ion counterparts. Positive carbon is unsaturated and exhibits a strong affinity for any groups in its vicinity that contain unshared pairs of electrons. Nitrogen, oxygen, sulfur, and the halogens all have unshared electron pairs, are commonly encountered elements in organic compounds, and frequently enter into carbonium ion rearrangements. The electron-deficient elements, such as boron or aluminum, might be expected to enter into carbanion rearrangements, but studies based on this expectation are only now commencing.[1] Ring-chain anionic, cationic, and radical rearrangements are all known, particularly those that involve three- and four-membered ring compounds. The 1,2-rearrangements of carbonium ions that involve saturated alkyl as migrating group find some analogy in the 1,2-rearrangements of carbanions. Although this parallelism between the ability of the anions and cations of carbon to enter into rearrangements exists, the wealth of examples in carbonium ion chemistry dwarfs those of carbanion chemistry.

This chapter is divided into the following sections: ring-chain anionic rearrangements, 1,2-rearrangements, and rearrangements with 1,3-elimination reaction stages.

RING-CHAIN ANIONIC REARRANGEMENTS

A number of rearrangements that appear to have carbanionic intermediates are known which possess the general characteristics of Equation (1). Examples are known in which $n = 1$ or 2, but not where n is larger

[1] (a) D. S. Matteson and J. O. Waldbillig, *J. Am. Chem. Soc.*, **85**, 1019 (1963); (b) D. J. Pasto, *ibid.*, **86**, 3039 (1964); (c) D. S. Matteson and J. O. Waldbillig, *ibid.*, **86**, 3778 (1964).

valued. The driving force for ring opening probably reflects release of angle strain and, in some systems, the formation of carbanions which are stabilized either by conjugating groups or by substitution of hydrogen for alkyl groups. Ring closure reactions occur only when substituent effects are favorable. Conjugation between a carbanion and a three-membered ring probably provides some stabilization for the ring form of the anion when $n = 1$.

(1)

$(CH_2)_n$ C—C—L (ring) ⟶ [$(CH_2)_n$ C—$\bar{C}$— (ring) ⇅ $(CH_2)_n$ —$\bar{C}$— C═C—] $\xrightarrow{E}$ $(CH_2)_n$ C—C—E (ring)

$(CH_2)_n$ —C(L)— C═C— ⟶ $\xrightarrow{E}$ $(CH_2)_n$ —C(E)— C═C—

In effect, these transformations are electrophilic substitutions, reaction being initiated by carbanion formation, and terminated by capture by an electrophile. Two kinds of leaving groups have been used, nitrogen and metal ions. Proton donors, carbonyl carbon, and halogens have been used as electrophiles. In the following sections, nitrogen and metal ions are discussed as leaving groups.

Nitrogen as Leaving Group

Equations (2) through (6) describe reactions that potentially involve alkyl diimides as intermediates which, in the presence of strong base, appear to generate carbanions. In the absence of base, or in the presence of weak base, alkyl diimides also decompose by a thermal reaction, probably involving a radical cage mechanism (the reactions of Equations (3) and (4) were used to generate the alkyl diimide).[2]

Application of the Wolff-Kishner reaction (Equation (2)) to cyclopropylcarboxaldehyde resulted in methylcyclopropane,[3] and reduction of cyclopropyl methyl ketone gave ethylcyclopropane without any

[2] D. J. Cram and J. S. Bradshaw, *J. Am. Chem. Soc.*, **85**, 1108 (1963).

[3] E. Renk, P. R. Shafer, W. H. Graham, R. H. Mazur, and J. D. Roberts, *J. Am. Chem. Soc.*, **83**, 1987 (1961).

rearrangement.[4] Cyclopropylcarbonyl derivatives containing β-aryl groups undergo some rearrangement during Wolff-Kishner reduction.[5a]

$$(2)\quad >C{=}N{-}NH_2 \xrightarrow{:B} H{-}\overset{|}{C}{-}N{=}N{-}H\ \ddot{B} \xrightarrow{-N_2} H{-}\overset{|}{\underset{|}{C}}^{-} \xrightarrow{HB} H{-}\overset{|}{\underset{|}{C}}{-}H$$

$$(3)\quad -\overset{|}{\underset{|}{C}}{-}NH{-}NH_2+[O] \longrightarrow -\overset{|}{\underset{|}{C}}{-}N{=}N{-}H \xrightarrow{:B} -\overset{|}{\underset{|}{C}}^{-} \xrightarrow{HB} -\overset{|}{\underset{|}{C}}{-}H$$

$$(4)\quad -\overset{|}{\underset{|}{C}}{-}\overset{H}{\overset{|}{N}}{-}NH{-}SO_2Ar \xrightarrow{:B} -C{-}N{=}N{-}H \xrightarrow{:B} -\overset{|}{\underset{|}{C}}^{-} \xrightarrow{HB} -\overset{|}{\underset{|}{C}}{-}H$$

$$(5)\quad -\overset{|}{\underset{|}{C}}{-}\overset{\overset{+}{K}}{\bar{N}}{-}SO_2Ar+H_3\overset{+}{N}{-}O{-}\bar{S}O_3 \longrightarrow -\overset{|}{\underset{|}{C}}{-}\overset{|}{\underset{\underset{NH_2}{|}}{N}}{-}SO_2Ar \xrightarrow{:B}$$

$$-\overset{|}{\underset{|}{C}}{-}N{=}N{-}H \xrightarrow{:B} -\overset{|}{\underset{|}{C}}^{-} \xrightarrow{HB} -\overset{|}{\underset{|}{C}}{-}H$$

$$(6)\quad -\overset{|}{\underset{|}{C}}{-}NH_2+NHF_2 \longrightarrow -\overset{|}{\underset{|}{C}}{-}\overset{H}{\overset{|}{N}}{-}NH{-}F \xrightarrow{:B} -\overset{|}{\underset{|}{C}}{-}N{=}N{-}H \xrightarrow{:B}$$

$$-\overset{|}{\underset{|}{C}}^{-} \xrightarrow{HB} -\overset{|}{\underset{|}{C}}{-}H$$

Use of the Nickon-Sinz reaction[6] (Equation (5)) and the *p*-toluenesulfonamide of cyclopropylcarbinylamine gave 1-butene.[5b] Application of the reaction of Equation (6) to cyclopropylmethylcarbinylamine gave 2-pentene,[5c] and to cyclopropylcarbinylamine, 1-butene.[5b]

Other results have a bearing on the mechanism of alkyl diimide decomposition. Treatment of cinnamylamine with difluoramine gives exclusively the rearranged material, allylbenzene.[5c] In this case, rearrangement occurred in spite of the fact that the thermodynamically unstable isomer was produced. The difluoramine reaction is not carried out in the presence of strong base, and the alkyl diimide decomposition

[4] P. Pomerantz, A. Fookson, T. W. Mears, S. Rothberg, and F. L. Howard, *J. Res. Natl. Bur. Std.*, **52**, 59 (1954).

[5] (a) C. L. Bumgardner and J. P. Freeman, *Tetrahedron Letters*, p. 737 (1964); (b) C. L. Bumgardner, K. J. Martin, and J. P. Freeman, *J. Am. Chem. Soc.*, **85**, 97 (1963); (c) C. L. Bumgardner and J. P. Freeman, *ibid.*, **86**, 2233 (1964).

[6] (a) A. Nickon and A. Sinz, *J. Am. Chem. Soc.*, **82**, 753 (1960); (b) A. Nickon and A. S. Hill, *ibid.*, **86**, 1152 (1964).

might not involve carbanions. Thermal decomposition of the alkyl diimide could give rearranged product by a concerted or by a radical cage reaction. In connection with the last possibility, application of the

$$\text{cyclo-}(CH_2)_2CH{-}C(R){=}O \xrightarrow[NaOH]{NH_2NH_2} \text{cyclo-}(CH_2)_2CH{-}CH(R){-}N{=}N{-}H \longrightarrow \text{cyclo-}(CH_2)_2CH{-}CH_2R$$

R = H or CH_3 R = H or CH_3 R = H or CH_3

$$\text{cyclo-}CH_2CH(C_6H_5)CH{-}CHO \xrightarrow[NaOH]{NH_2NH_2} \text{cyclo-}CH_2CH(C_6H_5)CH{-}CH_2{-}N{=}N{-}H \longrightarrow$$

$$\text{cyclo-}CH_2CH(C_6H_5)CH{-}CH_3 + C_6H_5CH_2CH_2CH{=}CH_2$$

$$\text{cyclo-}(CH_2)_2CH{-}CH_2{-}NHSO_2C_6H_4CH_3\text{-}p + H_2NOSO_3Na \xrightarrow{NaOH}$$

$$\text{cyclo-}(CH_2)_2CH{-}CH_2{-}N{=}N{-}H \longrightarrow CH_3CH_2CH{=}CH_2$$

$$\text{cyclo-}(CH_2)_2CH{-}CH(R){-}NH_2 + NHF_2 \longrightarrow \text{cyclo-}(CH_2)_2CH{-}CH(R){-}N{=}N{-}H \longrightarrow$$

R = H or CH_3 R = H or CH_3

$$CH_3CH_2CH{=}CHR$$

R = H or CH_3

reaction of Equation (4) to the benzenesulfonamide of 2-phenyl-2-butylhydrazine (optically active) in the absence of base gave racemic 2-phenylbutane by what appeared to be a radical cage reaction. In the presence of base, the reaction went partially stereospecifically by an anionic mechanism. The direction and extent of the transformation depended on the exact reaction conditions[2] (see Chapter IV). The cyclopropylcarbinyl radical and related species sometimes undergo rearrangement and sometimes do not, depending on the method of generation and the environment.[7]

$$C_6H_5{-}CH{=}CH{-}CH_2{-}NH_2 + NHF_2 \longrightarrow$$

$$C_6H_5{-}CH{=}CH{-}CH_2{-}N{=}N{-}H \ (\text{cyclic}) \longrightarrow C_6H_5{-}CH_2{-}CH{=}CH_2$$

$$C_2H_5{-}C(CH_3)(C_6H_5){-}NH{-}NH{-}SO_2Ar \longrightarrow \overset{*}{R}{-}N{=}N{-}H \longrightarrow [R\cdot\cdot N_2H] \ (\text{Cage}) \longrightarrow R{-}H \ (\text{Racemic})$$

$$\overset{*}{R}{-}N{=}N{-}H \xrightarrow{B:} R^- \xrightarrow{H-B} \overset{*}{R}{-}H \ (\text{Optically active})$$

It appears that alkyl diimides can decompose by a variety of mechanisms, and when a carbanion does indeed intervene as an intermediate in reactions (2) through (6), its fate depends on its exact environment and lifetime. More results are needed before the mechanistic picture is clear.

Metal Ions as Leaving Groups

Preparation and examination of the properties and reactions of certain organometallic compounds have provided interesting examples of ring-chain rearrangements. Roberts and Mazur[8a] observed that products derived from the Grignard reagent of cyclopropylcarbinyl bromide possessed the allylcarbinyl structure. Furthermore, nuclear magnetic resonance studies of the freshly prepared Grignard reagent demonstrated the structure to be $\gtrsim 99\%$ rearranged.[8b]

[7] (a) R. Breslow, *in* "Molecular Rearrangements" (P. de Mayo, ed.), Wiley (Interscience), New York, 1963, Vol. 1, pp. 291–293; (b) E. S. Huyser and J. D. Taliaferro, *J. Org. Chem.*, **28**, 3442 (1963).

[8] (a) J. D. Roberts and R. H. Mazur, *J. Am. Chem. Soc.*, **73**, 2509 (1951); (b) M. S. Silver, P. R. Shafer, J. E. Nordlander, C. Ruchardt, and J. D. Roberts, *ibid.*, **82**, 2646 (1960); (c) D. J. Patel, C. L. Hamilton, and J. D. Roberts, private communication; (d) K. L. Servis and J. D. Roberts, *J. Am. Chem. Soc.*, **86**, 3773 (1964).

The reversible character of the ring-chain rearrangement was demonstrated by Roberts and co-workers,[8] who started with either C^{14} or deuterium-labeled allylcarbinyl bromide, prepared the Grignard

$$\text{(cyclo-}CH_2CH_2\text{)}CH—CH_2—Br \xrightarrow[\text{Ether}]{\text{Mg}} \text{(cyclo-}CH_2CH_2\text{)}CH—CH_2—MgBr \longrightarrow CH_2(CH=CH_2)—CH_2—MgBr$$

$$\xrightarrow{C_6H_5NCO} CH_2(CH=CH_2)—CH_2—\underset{\underset{O}{\|}}{C}—NH—C_6H_5$$

reagent, and either oxygenated or carbonated the products. Although only open-chain products were obtained, the label was scrambled in such a way as to indicate unequivocally that the Grignard reagent rearranges after its formation and prior to conversion to final products.

$$CH_2=CH—CH_2—CD_2—Br \xrightarrow[\text{Mg}]{\text{Ether}}$$

$$CH_2=CH—CH_2—CD_2—MgBr \xrightarrow{O_2} CH_2=CH—CH_2—CD_2OH$$

$$\updownarrow$$

$$\text{(cyclo-}D_2C—H_2C\text{)}CH—CH_2—MgBr$$

$$\updownarrow$$

$$CH_2=CH—CD_2—CH_2—MgBr \xrightarrow{O_2} CH_2=CH—CD_2—CH_2OH$$

The half-life for equilibration of the isomeric Grignard reagents was 30 hr. at +27° and 40 min. at +55°, corresponding to an activation energy of about 23 Kcal./mole.[8b]

The fact that none of the label was found in the terminal position of the product indicates that a nonclassical bicyclobutanide anion did not intervene as an intermediate in this rearrangement. These results contrast with those of parallel experiments in carbonium ion chemistry,[8d] which point to bicyclobutonium cations as discrete intermediates in solvolyses of compounds such as cyclopropylcarbinyl or allylcarbinyl tosylates. The anionic bridged structure contains four delocalized electrons, whereas the cationic has two. This fact correlates with the $4n+2$ Hückel rule which provides delocalization energies for two electrons ($n = 0$) in an unsaturated cyclic system but none for four.

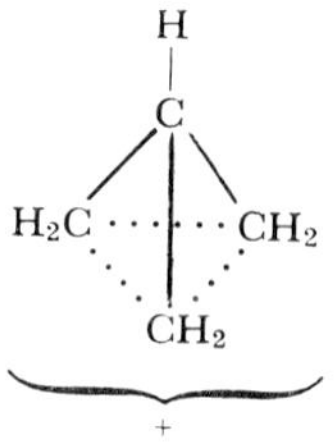

H
C
H_2C CH_2
CH_2
+

Bicyclobutonium ion

H
C
H_2C CH_2
CH_2

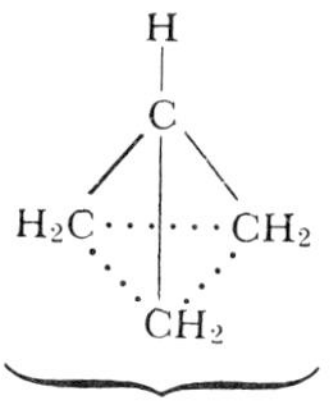

Bicyclobutanide ion

In another study, Roberts and co-workers[8c] prepared the Grignard reagent of cyclopropylcarbinyl bromide in dimethyl ether at $-24°$, and quenched aliquots of the filtered solution with benzoic acid. A mixture of methylcyclopropane and 1-butene was produced, the former decreasing with time from a maximum of 45% to only trace amounts after 48 hr. From the rate of disappearance of the cyclopropylcarbinylmagnesium bromide, the authors calculated a half-life of 126 min. for the reagent, and a free energy of activation of 19 Kcal./mole at $-24°$. The rate-determining step in the isotopic scrambling of allylcarbinylmagnesium bromide was the formation of the cyclopropylcarbinylmagnesium bromide, and at $-24°$ the free energy of activation for this process was 25 Kcal./mole.[8c] From these data, the equilibrium constant

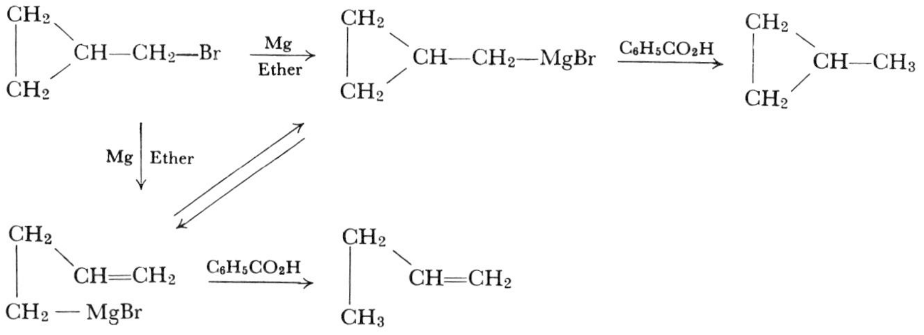

CH_2 / CH_2 > CH—CH_2—Br —(Mg, Ether)→ CH_2 / CH_2 > CH—CH_2—MgBr —($C_6H_5CO_2H$)→ CH_2 / CH_2 > CH—CH_3

Mg | Ether ↓ ⇌

CH_2 > CH=CH_2, CH_2—MgBr —($C_6H_5CO_2H$)→ CH_2 > CH=CH_2, CH_3

between cyclopropylcarbinylmagnesium bromide and allylcarbinylmagnesium bromide was calculated to be 8×10^{-6} at $-24°$, heavily favoring the open-chain form.

When cyclopropylcarbinyl bromide in the presence of traces of trifluoracetic acid or benzoic acid was treated with magnesium in either dimethyl ether at $-24°$ or diethyl ether at $35°$, both methylcyclopropane and 1-butene were produced. In dimethyl ether, about 55% of the hydrocarbon was cyclic and 45% open chain. In diethyl ether at the higher temperature, about 32% of the hydrocarbon was cyclic and 68% open chain. Similar experiments with allylcarbinyl bromide produced 1-butene and only traces of methylcyclopropane. Since the interconversions of the cyclic and open-chain Grignard reagents are much slower processes than that of proton capture, the percentages of the two hydrocarbons produced must represent the percentages of the two Grignard reagents initially generated. Thus in both solvents, the cyclic bromide must produce directly both Grignard reagents. Since the ratio changed with temperature, the two processes must have different activation energies. Substitution of cyclopropylcarbinyl iodide or chloride for the bromide resulted in production of substantially less cyclic hydrocarbon.

$CH_2{=}CH{-}CH{=}CH{-}CH_2{-}CH_2{-}Br$ $\xrightarrow[\text{Ether}]{\text{Mg}}$

$CH_2{=}CH{-}CH{=}CH{-}CH_2{-}CH_2{-}MgBr$

⇅

$CH_2{=}CH{-}CH(MgBr){-}CH_2{-}CH{-}CH_2$ (cyclopropyl ring: CH—CH₂ joined to CH₂)

⇅

$BrMg{-}CH_2{-}CH{=}CH{-}CH_2{-}CH{-}CH_2$ (cyclopropyl ring: CH—CH₂ joined to CH₂)

$\xrightarrow{C_2H_5OH}$

$CH_2{=}CH{-}CH{=}CH{-}CH_2{-}CH_3$

77%

$CH_2{=}CH{-}CH_2{-}CH{-}CH_2{-}CH_2$ (cyclopropyl ring: CH—CH₂ joined to CH₂)

10%

$CH_3{-}CH{=}CH{-}CH{-}CH_2{-}CH_2$ (cyclopropyl ring: CH—CH₂ joined to CH₂)

cis and *trans*

13%

The Grignard reagent prepared from 1-bromo-3,5-hexadiene in diethyl ether gave no cyclic product when treated with carbon dioxide, but did give substantial amounts of cyclic products when treated with oxygen or ethanol.[9] In this more extended conjugated system, apparently the extra vinyl group stabilizes the cyclic forms of the Grignard reagent enough to provide reasonable amounts at equilibrium.

The same worker[9] prepared the Grignard reagent of γ,γ-diphenylallylcarbinyl bromide. With deuterium labeling and quenching experiments similar to those carried out with the allylcarbinyl magnesium bromide,[8b] it was found that equilibration of the α- and β-positions of the Grignard reagent was complete in 5 hr. in ether at 25°. Equilibrated Grignard reagent, when treated with proton donors, gave largely open-chain product. A small amount of cyclic hydrocarbon was shown to be formed at a stage earlier than the point at which proton donor was added. Deuterated proton donor gave undeuterated cyclic product.

$$(C_6H_5)_2C{=}CH{-}CH_2{-}CD_2{-}MgBr \xrightleftharpoons{\text{Ether}} (C_6H_5)_2C{=}CH{-}CD_2{-}CH_2{-}MgBr$$

With this organometallic system, oxygenation of the Grignard reagent gave substantial amounts of cyclic alcohol. Clearly the distribution of alcohols in the product reflects the composition of the equilibrium mixture, because none of the cyclic Grignard reagent could be detected by NMR spectroscopy. Without the two phenyl groups, no cyclic products were obtained.

$$\left[\begin{array}{l}CH_2\\ |\quad \diagdown CH{=}C(C_6H_5)_2\\ CH_2{-}MgBr\end{array} \xrightleftharpoons{\text{Ether}} \begin{array}{l}CH_2\\ |\quad \rangle CH{-}C(C_6H_5)_2\\ CH_2 \qquad\qquad |\\ \qquad\qquad\quad MgBr\end{array}\right] \xrightarrow{HX}$$

$$\underset{80\text{–}86\%}{\begin{array}{l}CH_2\\ |\quad \diagdown CH{=}C(C_6H_5)_2\\ CH_3\end{array}} + \underset{4\text{–}10\%}{\begin{array}{l}CH_2\\ |\quad \rangle CH{-}CH(C_6H_5)_2\\ CH_2\end{array}}$$

In work with organolithium compounds, Lansbury and co-workers[10] demonstrated that cyclopropylcarbinyllithium can be prepared, but that it isomerizes to allylcarbinyllithium at a rate dependent on the amount of ether in the medium. Benzaldehyde, benzophenone, and

[9] M. E. Howden, Ph.D. Thesis, California Institute of Technology, Pasadena, California, 1962.

[10] (a) P. T. Lansbury and V. A. Pattison, *J. Am. Chem. Soc.*, **85**, 1886 (1963); (b) P. T. Lansbury, V. A. Pattison, W. H. Clement, and J. D. Sidler, *ibid.*, **86**, 2247 (1964).

benzoyl chloride as electrophiles gave approximately the same ratios of rearranged and unrearranged products, the amount of rearrangement in 90% hexane–10% ether being only a function of time. Rearranged reagent appeared to be the more stable form in ether, but in tetrahydrofuran, the cyclic form predominated. When prepared from the iodide, cyclopropylcarbinyllithium-α,α-d_2 when treated with benzaldehyde gave cyclopropylcarbinylphenylcarbinol without any deuterium scrambling.

Roberts and Maercker[11] treated cyclopropyldiphenylmethyl methyl ether with sodium–potassium alloy in ether, and the organometallic produced (deep red) was protonated and carbonated. The products were unrearranged. However, when the organometallic was treated with

$$(CH_2)_2CH{-}CH_2{-}I + s\text{-BuLi} \xrightarrow[10\%\ \text{Ether}]{90\%\ \text{Hexane}}$$

$$\left\{ (CH_2)_2CH{-}CH_2{-}Li \rightleftarrows CH_2(Li){-}CH_2{-}CH{=}CH_2 \right\}$$

$$\downarrow C_6H_5CHO \qquad\qquad \downarrow C_6H_5CHO$$

$$(CH_2)_2CH{-}CH_2{-}CH(OH){-}C_6H_5 \qquad\qquad CH_2{=}CH{-}CH_2{-}CH_2{-}CH(OH){-}C_6H_5$$

either lithium bromide or magnesium bromide and then either carbonated or protonated, the products were completely rearranged, open-chain materials. When allowed to stand in deuterated dimethyl sulfoxide–potassium *tert*-butoxide at 20° for 20 hr., benzylcyclopropane became completely deuterated in the benzyl position without undergoing any ring opening. These results are consistent with the expectation that the carbon–potassium bond is ionic, and the conjugated, cyclic form of the ion pair is more stable than the non-conjugated, open form of the ion pair. However, the carbon–magnesium or carbon–lithium bonds possess enough covalent character to make the nonconjugated open form of the organometallic the more stable of the isomers.

[11] J. D. Roberts and A. F. Maercker, private communication.

Equilibria between open-chain and ring organometallic compounds have also been demonstrated for systems that contain four-membered

$$\text{cyclo-}C_3H_5\text{—}C(C_6H_5)_2\text{—}OCH_3 \xrightarrow[\text{Ether}]{\text{Na–K}} \text{cyclo-}C_3H_5\text{—}C(C_6H_5)_2^-K^+ \xrightarrow{CO_2} \text{cyclo-}C_3H_5\text{—}C(C_6H_5)_2\text{—}CO_2H$$

$$\text{cyclo-}C_3H_5\text{—}C(C_6H_5)_2^-K^+ \xrightarrow{\text{LiBr or MgBr}_2} CH_2\text{—Li(MgBr)}\text{—}CH_2\text{—}CH{=}C(C_6H_5)_2$$

$$\text{cyclo-}C_3H_5\text{—}C(C_6H_5)_2^-K^+ \xrightarrow{H_2O} \text{cyclo-}C_3H_5\text{—}CH(C_6H_5)_2$$

$$CH_2\text{—Li(MgBr)}\text{—}CH_2\text{—}CH{=}C(C_6H_5)_2 \xrightarrow[(H_2O)]{CO_2} HO_2C\text{—}CH_2\text{—}CH_2\text{—}CH{=}C(C_6H_5)_2$$

$$\text{cyclo-}C_3H_5\text{—}CH_2\text{—}C_6H_5 + CD_3SOCD_3 \xrightarrow[\text{20°, 20 hr.}]{\textit{tert}\text{-BuOK}} \text{cyclo-}C_3H_5\text{—}CD_2\text{—}C_6H_5$$

rings.[12] The organomercury derivative of 5-chloro-1-hexene was prepared, converted to the lithium derivative, and the organometallic product quenched by treatment with proton donors.

The lithium compound, when refluxed in cyclohexane and quenched, gave > 99% 3-methyl-1-pentene. The magnesium compound prepared directly from the chloride was heated in refluxing tetrahydrofuran, and aliquots were quenched periodically. The proportion of 3-methyl-1-pentene in the mixture of that alkene and 1-hexene increased at a decreasing rate to a maximum of 20%. Apparently with time the organometallic abstracted protons from solvent. Reaction of 5-chloro-1-hexene with sodium in hydrocarbon solvents gave mixtures of the two olefins, with 3-methyl-1-pentene predominating. In this case too, the organometallic apparently abstracted protons from the medium at a rate competitive with rearrangement.

Cyclobutylcarbinyl chloride when treated with sodium in tetradecane gave a 7 to 1 ratio of 1-pentene to methylcyclobutane, the protons arising as a result of reaction of the organometallic with starting chloride (E_2 reaction). Reaction of the same chloride with lithium in benzene gave a ratio of 13 to 1 of 1-pentene to methylcyclobutane, whereas magnesium in tetrahydrofuran at 65° gave greater than 99.8% 1-pentene

[12] (a) E. A. Hill, H. G. Richey, Jr., and T. C. Rees, *J. Org. Chem.*, **28**, 2161 (1963); (b) H. G. Richey, Jr., and E. A. Hill, *J. Am. Chem. Soc.*, **86**, 421 (1964).

when quenched with proton donors after a period of time. However, the amount of rearranged material increased with time, which indicates that the organometallic compound first formed and then rearranged.

$$\begin{array}{l} CH_2{-}CH{=}CH_2 \\ | \\ CH_2{-}CH{-}Cl \\ \quad\;\; | \\ \quad\;\; CH_3 \end{array} \xrightarrow[(2)\ HgCl_2]{(1)\ Mg} \begin{array}{l} CH_2{-}CH{=}CH_2 \\ | \\ CH_2{-}CH{-}HgCl \\ \quad\;\; | \\ \quad\;\; CH_3 \end{array} \xrightarrow{M}$$

5-Chloro-1-hexene

$$\begin{array}{l} CH_2{-}CH{=}CH_2 \\ | \\ CH_2{-}CH{-}M \\ \quad\;\; | \\ \quad\;\; CH_3 \end{array} \rightleftharpoons \begin{array}{l} CH_2{-}CH{-}CH_2{-}M \\ | \qquad\; | \\ CH_2{-}CH{-}CH_3 \end{array} \rightleftharpoons \begin{array}{l} M{-}CH_2 \quad CH{=}CH_2 \\ \quad\;\; | \qquad\;\; | \\ \quad\;\; CH_2{-}CH{-}CH_3 \end{array}$$

$$\downarrow H_2O \qquad\qquad\qquad\qquad \downarrow H_2O$$

$$\begin{array}{l} CH_2{-}CH{=}CH_2 \\ | \\ CH_2{-}CH_2{-}CH_3 \end{array} \qquad\qquad \begin{array}{l} CH_3 \quad CH{=}CH_2 \\ | \qquad\;\; | \\ CH_2{-}CH{-}CH_3 \end{array}$$

1-Hexene 3-Methyl-1-pentene

These results indicate that cyclobutylmethylmetallics form, and rearrange to open-chain isomers in a reaction competitive with protonation from substances in the medium. If no protonation occurs before quenching, then rearrangement is essentially complete.

$$\begin{array}{l} CH_2{-}CH{-}CH_2M \\ | \qquad\; | \\ CH_2{-}CH_2 \end{array} \rightleftharpoons \begin{array}{l} CH_2{-}CH{=}CH_2 \\ | \\ CH_2{-}CH_2M \end{array}$$

$$\downarrow BH \qquad\qquad\qquad \downarrow BH$$

$$\begin{array}{l} CH_2{-}CH{-}CH_3 \\ | \qquad\; | \\ CH_2{-}CH_2 \end{array} \qquad \begin{array}{l} CH_2{-}CH{=}CH_2 \\ | \\ CH_2{-}CH_3 \end{array}$$

Unlike their lower homologs, cyclopentylmethyl and cyclohexylmethyl chlorides when treated with sodium did not produce any open-chain alkenes. Release of strain seems to be a requirement for the rearrangement.[12b]

1,2-REARRANGEMENTS

A number of 1,2-rearrangements are initiated by carbanion formation in much the same way that carbonium ions frequently initiate rearrangement. Both varieties of transformation usually compete with substitution, elimination, and fragmentation reactions. In carbonium ion rearrangements, halogen, nitrogen, oxygen, or carbon usually serve as leaving groups, although rearrangement is sometimes initiated by proton addition to alkenes. In carbanion rearrangements, hydrogen or metals are the usual leaving groups, although some rearrangements are initiated by addition of negative ions to carbonyl groups. In this section, the Stevens, Wittig, aryl, and benzilic acid rearrangements are discussed in turn.

Stevens and Related Rearrangements

The Stevens rearrangement[13] involves migration of an alkyl group from a quaternary ammonium to an adjacent carbanionic center. Reaction is initiated by proton abstraction from a carbon usually activated by both the quaternary ammonium and a carbonyl group. However, phenyls can also serve as activating groups.[14] Similar rearrangements occur with sulfonium salts.[13b]

The usual migrating groups[13] are allyl, 1-phenylethyl, benzhydryl, 9-fluorenyl, 3-phenylpropargyl, and phenacyl, although methyl migrates to a sufficiently basic center.[14] Electron attracting groups substituted in the *m*- or *p*-positions of a migrating benzyl group accelerate the rearrangement[13b, 15] in the following order: $NO_2 > I, Br, Cl > CH_3, H > OCH_3$. Substituents in the *p*-position of the phenacyl group, which activate the proton-leaving group, have a rate effect of the opposite type, electron-providing groups being the most activating: $CH_3O > CH_3 > H, Cl, Br, I > NO_2$.[13c] Thus rearrangement is facilitated by withdrawing electrons from the migration origin, and by localizing electrons at the migration terminus.

Joint rearrangement of two compounds with similar isomerization rates did not result in "cross-bred products,"[15] even when detection of such a product was facilitated by radioactive carbon labeling experiments.[16] When carried out with an asymmetric α-phenylethyl as

[13] (a) T. S. Stevens, E. M. Creighton, A. B. Gordon, and M. MacNicol, *J. Chem. Soc.*, p. 3193 (1928); (b) T. Thomson and T. S. Stevens, *ibid.*, p. 55 (1932); (c) J. L. Dunn and T. S. Stevens, *ibid.*, p. 1926 (1934) and intervening papers.

[14] (a) G. Wittig, R. Mangold, and G. Felletschin, *Ann.*, **580**, 116 (1948); (b) G. Wittig, *Angew. Chem.*, **63**, 15 (1951); *ibid.*, **66**, 10 (1954).

[15] T. S. Stevens, *J. Chem. Soc.*, p. 2107 (1930).

[16] R. A. W. Johnstone and T. S. Stevens, *J. Chem. Soc.*, p. 4487 (1955).

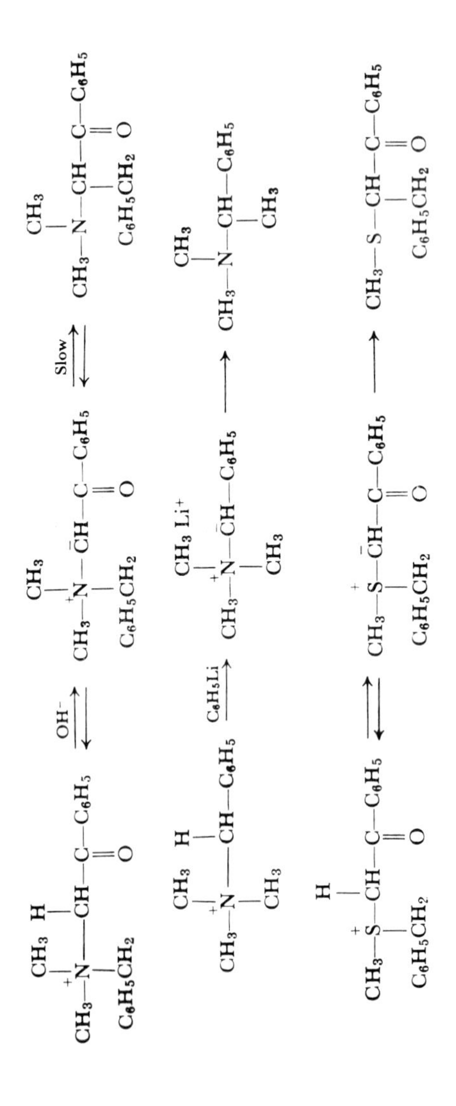

migrating group, rearrangement occurred with 97% preservation of optical activity,[17] and with retention of configuration.[18]

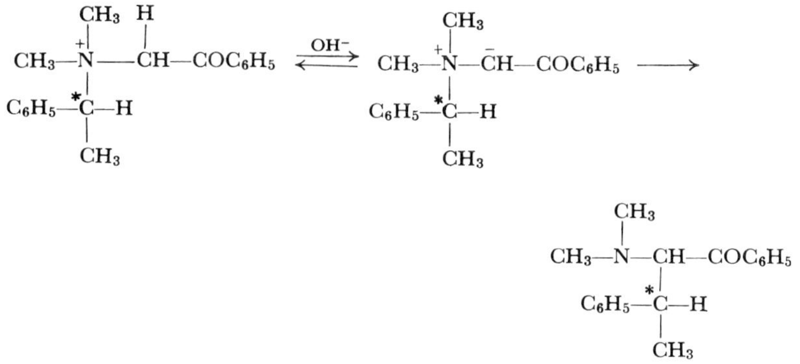

$$CH_3-\overset{+}{N}(CH_3)(CH^*(CH_3)C_6H_5)-CH_2-COC_6H_5 \underset{}{\overset{OH^-}{\rightleftharpoons}} CH_3-\overset{+}{N}(CH_3)(CH^*(CH_3)C_6H_5)-\overset{-}{C}H-COC_6H_5 \longrightarrow CH_3-N(CH_3)-CH(COC_6H_5)-\overset{*}{C}H(C_6H_5)(CH_3)$$

When treated with sodamide in benzene or liquid ammonia, allylbenzyldimethylammonium bromide reacted in competing rearrangements in which the benzyl group underwent 1,2- and 1,4-shifts.[19] In benzene at 80°, the ratio of 1,2- to 1,4-shift was about 1.4. In liquid ammonia at $-33°$, the 1,2-migration predominated by a much larger factor.[19b] When allyltriethylammonium bromide was treated with

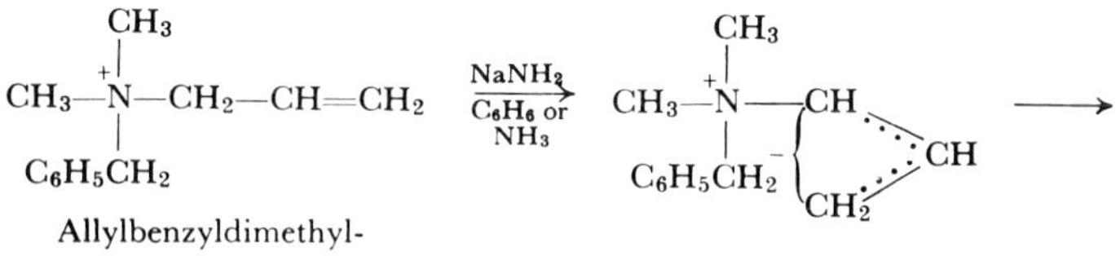

$$CH_3-\overset{+}{N}(CH_3)(CH_2C_6H_5)-CH_2-CH=CH_2 \xrightarrow[C_6H_6 \text{ or } NH_3]{NaNH_2} CH_3-\overset{+}{N}(CH_3)(CH_2C_6H_5)-\overset{-}{CH}\cdots CH \cdots CH_2 \longrightarrow$$

Allylbenzyldimethylammonium bromide

$$CH_3-N(CH_3)-CH(CH_2C_6H_5)-CH=CH_2 + CH_3-N(CH_3)-CH=CH-CH_2CH_2C_6H_5$$

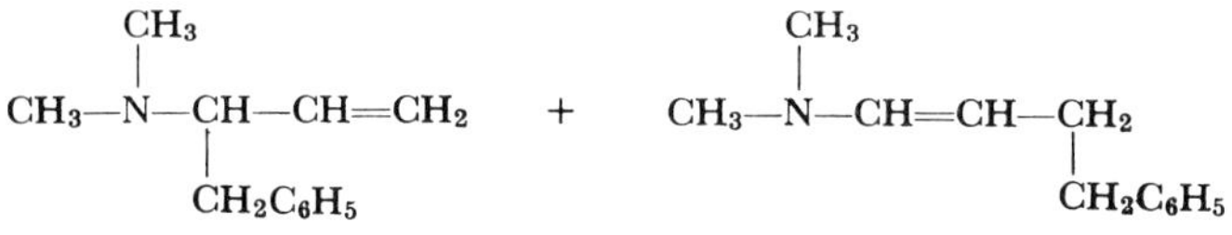

Product of 1,2-rearrangement

Product of 1,4-rearrangement

sodamide in liquid ammonia at $-33°$, 0°, and 65°, the ratios of 1,2- to 1,4-ethyl migration were 1 to 1, 1 to 2, and 1 to 25, respectively.[19d]

[17] A. Campbell, A. H. J. Houston, and J. Kenyon, *J. Chem. Soc.*, p. 93 (1947).

[18] J. H. Brewster and M. W. Kline, *J. Am. Chem. Soc.*, **74**, 5179 (1952).

[19] (a) T. Thomson and T. S. Stevens, *J. Chem. Soc.*, p. 1932 (1932); (b) E. F. Jenny and J. Druey, *Angew. Chem.* (*Intern. Ed. Engl.*), **1**, 155 (1962); (c) see also H. Hellmann, *Osterr. Chemiker. Z.*, **62**, 315 (1961); (d) H. Hellman and G. M. Scheytt, *Ann.*, **654**, 39 (1962).

Jenny and Druey[19b] found that treatment of optically active compound I with sodamide gave products II and III. The asymmetric carbon in the product of the 1,2-shift (II) had retained its configuration to the extent of over 90%, and that in the product of the 1,4-shift (III), about 80% in benzene and 72% in liquid ammonia. These results point to an intimate ion pair mechanism for the two rearrangements.[19b] Rotation of the anionic and cationic parts of this ion pair with respect to

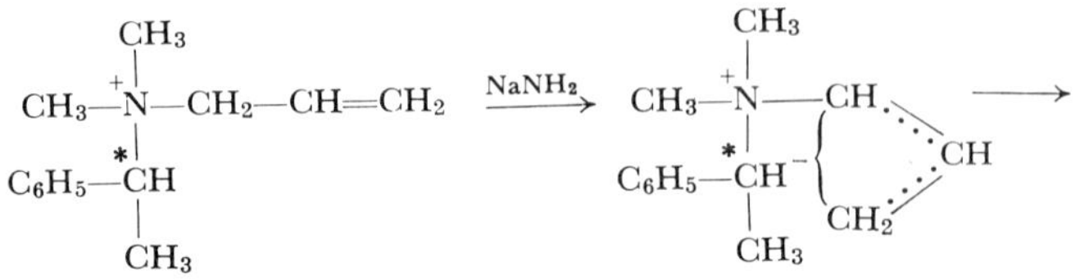

$$(CH_3)_2\overset{+}{N}(CH_2{-}CH{=}CH_2)(\overset{*}{C}H(CH_3)C_6H_5) \xrightarrow{NaNH_2} (CH_3)_2\overset{+}{N}{-}CH{\cdots}CH{\cdots}CH_2 \;\; \overset{*}{C}H^{-}(CH_3)C_6H_5 \longrightarrow$$

I

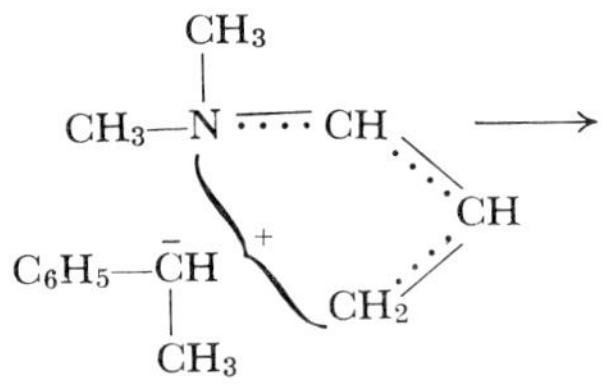

$$\text{Ion pair:}\; (CH_3)_2N{\cdots}CH{\cdots}CH{\cdots}CH_2^{+} \;\; C_6H_5{-}\bar{C}H{-}CH_3 \longrightarrow$$

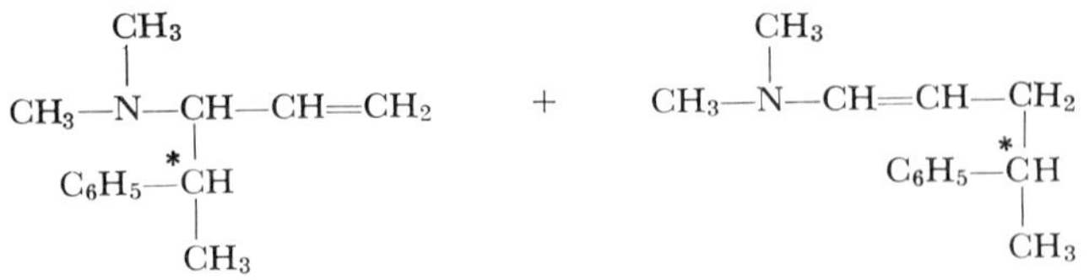

Ion pair

$$(CH_3)_2N{-}CH(CH{=}CH_2){-}\overset{*}{C}H(CH_3)C_6H_5 \quad + \quad (CH_3)_2N{-}CH{=}CH{-}CH_2{-}\overset{*}{C}H(CH_3)C_6H_5$$

II

Product of 1,2-rearrangement, 90% retention of configuration

III

Product of 1,4-rearrangement, 70–80% retention of configuration

one another occurred at comparable rates to that of ion pair collapse. These observations resemble those made in connection with rotations of cations and anions within ion pairs obtained by proton abstraction from carbon acids (see Chapters III and V).

The 1,4-rearrangement formulated above is related to the Sommelet-Hauser rearrangement,[20] in which benzyltrimethylammonium salts are

[20] (a) M. Sommelet, *Compt. Rend. Acad. Sci.*, **205**, 56 (1937); (b) C. R. Hauser and S. W. Kantor, *J. Am. Chem. Soc.*, **73**, 1437, 4122 (1951); (c) G. Wittig, L. Löhman, and W. Happe, *Ann.*, **557**, 205 (1947).

treated with sodamide in liquid ammonia to give *o*-methylbenzyldimethylamine. When the *o*-positions of the benzyl quaternary ammonium salt are already substituted with methyl groups, the rearranged

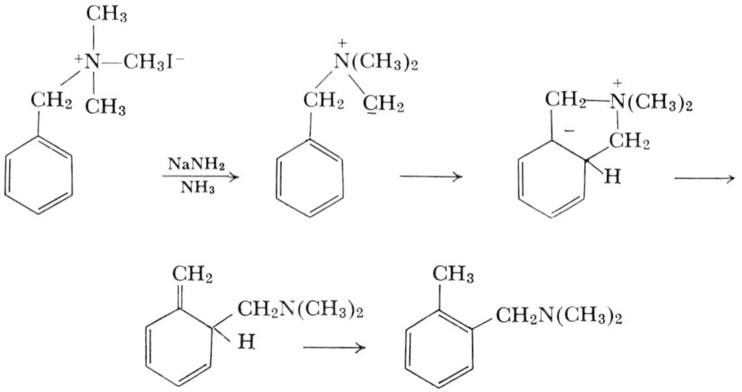

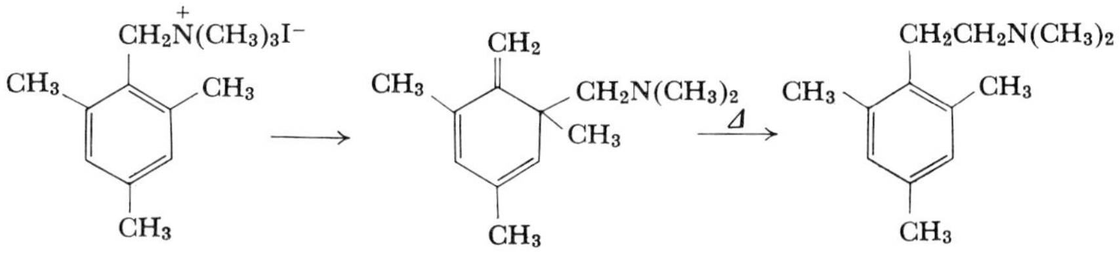

$$\text{C}_6\text{H}_5\text{CH}_2\overset{+}{\text{N}}(\text{CH}_3)_3\text{I}^- \xrightarrow[\text{NH}_3]{\text{NaNH}_2} \text{C}_6\text{H}_5\text{CH}_2\overset{+}{\text{N}}(\text{CH}_3)_2\underline{\text{C}}\text{H}_2 \longrightarrow \longrightarrow \longrightarrow o\text{-CH}_3\text{C}_6\text{H}_4\text{CH}_2\text{N}(\text{CH}_3)_2$$

IV

V

product cannot aromatize, and compounds such as IV result.[21a] These compounds thermally rearrange to re-establish the aromatic ring for

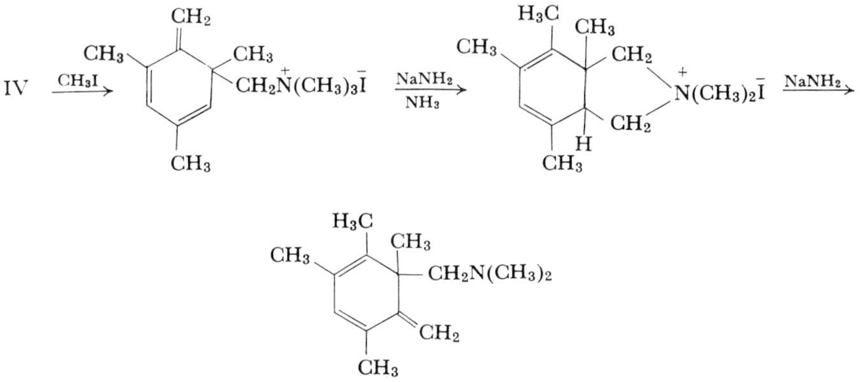

$$\text{IV} \xrightarrow{\text{CH}_3\text{I}} \xrightarrow[\text{NH}_3]{\text{NaNH}_2} \xrightarrow{\text{NaNH}_2} \text{VI}$$

VI

example, IV gives V.[21] Compound IV, when converted to its quaternary ammonium salt, is converted by sodamide in liquid ammonia to VI.[21]

Appropriate sulfonium salts also undergo the Sommelet-Hauser rearrangement when treated with base.[21b,c]

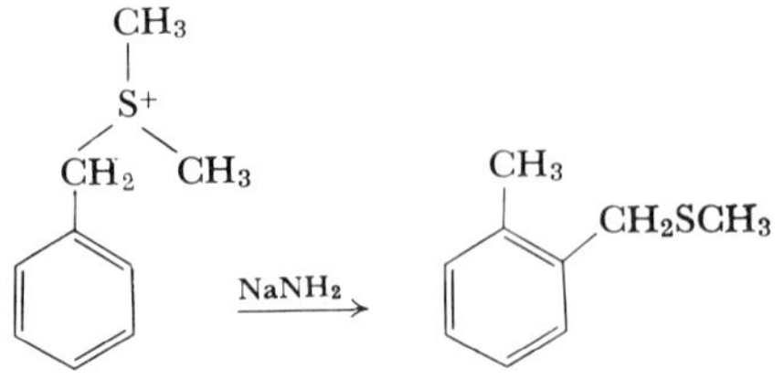

CH_3
S^+
CH_2 CH_3
$\xrightarrow{NaNH_2}$
CH_3
CH_2SCH_3

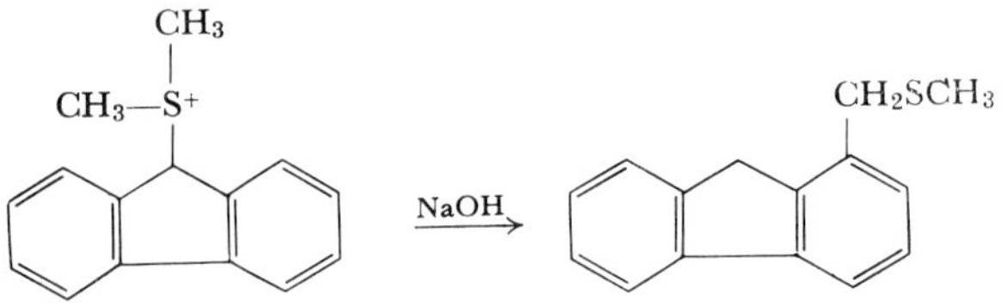

CH_3
$CH_3—S^+$
$\xrightarrow{NaOH}$
CH_2SCH_3

An attempt to capture the methylene ylide intermediate of the Sommelet-Hauser rearrangement with benzophenone as electrophile resulted in reaction only at the benzyl position.[22]

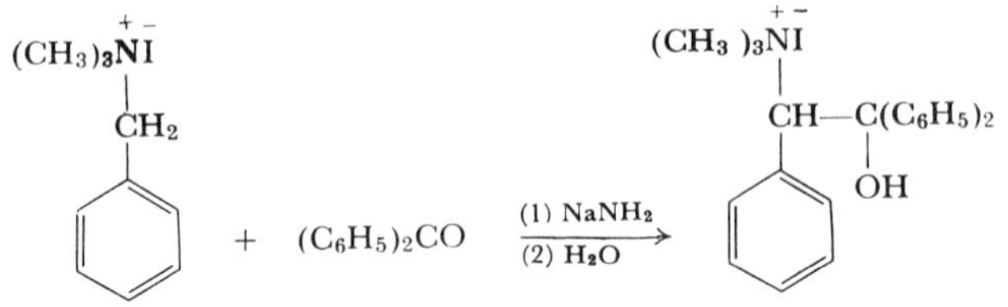

$(CH_3)_3\overset{+\,-}{N}I$
CH_2
$+ \quad (C_6H_5)_2CO \quad \xrightarrow[(2)\ H_2O]{(1)\ NaNH_2}$
$(CH_3)_3\overset{+\,-}{N}I$
$CH—C(C_6H_5)_2$
OH

Studies of substituent effects on migratory aptitude in the rearrangement[23] indicate that carbanion stabilizing substituents such as cyclopropyl, vinyl,[23b] and phenyl[23a] tend to direct the carbon to which they are attached into the *o*-position of the attached benzyl group faster than do hydrogen and methyl as substituents. Moreover, compound VII

[21] (a) C. R. Hauser and D. N. Van Eenam, *J. Am. Chem. Soc.*, **79**, 5512, 6080, 6274, 6277 (1957); (b) C. R. Hauser, S. W. Kantor, and W. R. Brasen, *ibid.*, **75**, 2660 (1953); (c) A. W. Johnson and R. B. LaCount, *ibid.*, **83**, 417 (1961).

[22] W. H. Puterbaugh and C. R. Hauser, *J. Am. Chem. Soc.*, **86**, 1105, 1108, 1394 (1964).

[23] (a) W. Q. Beard, Jr. and C. R. Hauser, *J. Org. Chem.*, **25**, 334 (1960); (b) C. L. Bumgardner, *J. Am. Chem. Soc.*, **85**, 73 (1963); (c) W. Q. Beard, Jr. and C. R. Hauser, *J. Org. Chem.*, **26**, 371 (1961).

rearranges to VIII as the major product, which suggests that proton removal can be the more important step in directing the rearrangement.[23c]

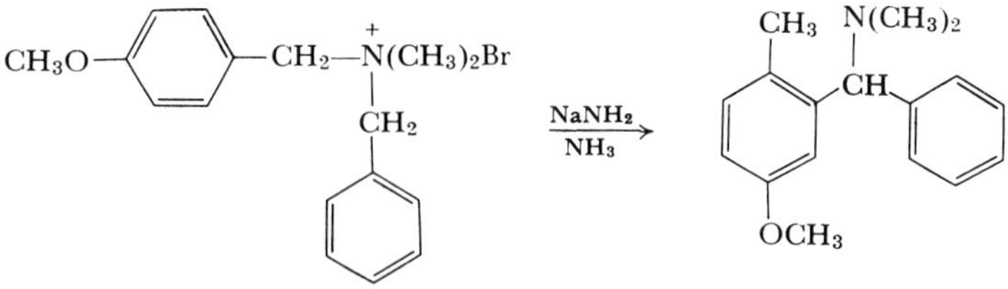

VII $\xrightarrow[NH_3]{NaNH_2}$ VIII

An attractive mechanism that correlates most of the facts about the Sommelet-Hauser rearrangement, and also correlates it with the Stevens rearrangement, involves an ion pair intermediate. The anion of

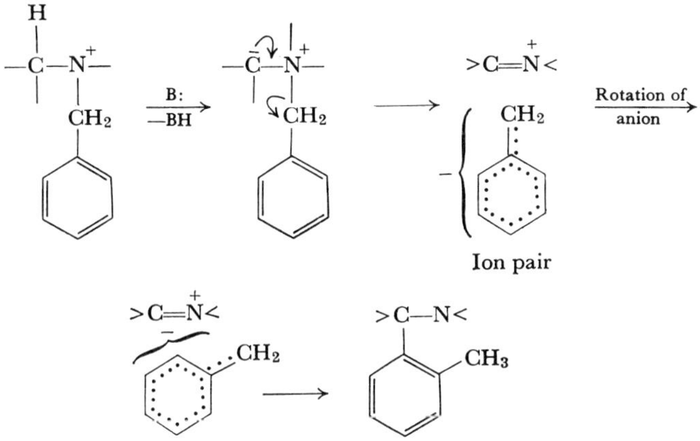

Ion pair

this ion pair undergoes partial rotation before collapse to the covalent state occurs. With such a mechanism, production of VIII from VII would involve the ion pair A.

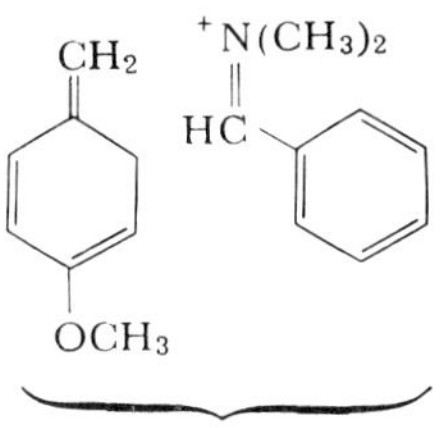

A

Wittig Rearrangement

In the Wittig rearrangement,[24] a benzyl or allyl[20b] ether is treated with a strong base such as phenyllithium or sodamide, and a 1,2-rearrangement occurs to give an alkoxide, which when acidified provides an alcohol.[24] Studies of the migratory aptitudes of groups of various 9-fluorenyl ethers in tetrahydrofuran–phenyllithium[24c] led to the following sequence: allyl, benzyl > methyl, ethyl, *p*-nitrophenyl > phenyl.[24d]

$$C_6H_5CH_2\text{—}O(CH_3) \xrightarrow{C_6H_5Li} C_6H_5\overset{\overset{+}{Li}}{\bar{C}}H\text{—}O(CH_3) \longrightarrow C_6H_5CH(CH_3)\text{—}OLi \xrightarrow{H_2O} C_6H_5CH(CH_3)\text{—}OH$$

$$CH_2{=}CH\text{—}CH_2\text{—}O(CH_2\text{—}CH{=}CH_2) \xrightarrow{NaNH_2} CH{=}CH\text{—}\overset{\overset{+}{Na}}{\bar{C}}H\text{—}O(CH_2\text{—}CH{=}CH_2) \xrightarrow{Ether} CH_2{=}CH\text{—}CH(CH_2\text{—}CH{=}CH_2)\text{—}OH$$

$$\text{9-H, 9-OR-fluorene} \xrightarrow[(CH_2)_4O]{C_6H_5Li} \text{9-R, 9-OH-fluorene}$$

In a different study of migratory aptitude, Hauser and Kantor[20b] metalated a series of benzyl ethers with potassium amide in liquid ammonia, isolated the carbon salts, and caused them to rearrange as suspensions in boiling ether. These authors obtained the following order of migratory aptitude: benzyl > *sec*-butyl > neopentyl, phenyl. In a third study of the rearrangement of several desyl ethers with ethanolic potassium hydroxide, Curtin and co-workers[25] suggested the following order: benzhydryl > benzyl, *p*-nitrophenyl > phenyl.

Stevens and co-workers[26] reported that both 9-fluorenyl crotyl ether

[24] (a) G. Wittig and L. Löhman, *Ann.*, **550**, 260 (1942); (b) G. Wittig and W. Happe, *ibid.*, **557**, 205 (1947); (c) G. Wittig, H. Döser, and L. Lorenz, *ibid.*, **562**, 192 (1949); (d) G. Wittig, *Angew. Chem.*, **66**, 10 (1954).

[25] (a) D. Y. Curtin and S. Leskowitz, *J. Am. Chem. Soc.*, **73**, 2633 (1951); (b) D. Y. Curtin and W. R. Proops, *ibid.*, **76**, 494 (1954).

[26] J. Cast, T. S. Stevens, and J. Holmes, *J. Chem. Soc.*, p. 3521 (1960).

and 9-fluorenyl α-methylallyl ether gave 9-crotyl-9-fluorenol as the major product. The first is a 1,2- and the second a 1,4-rearrangement.

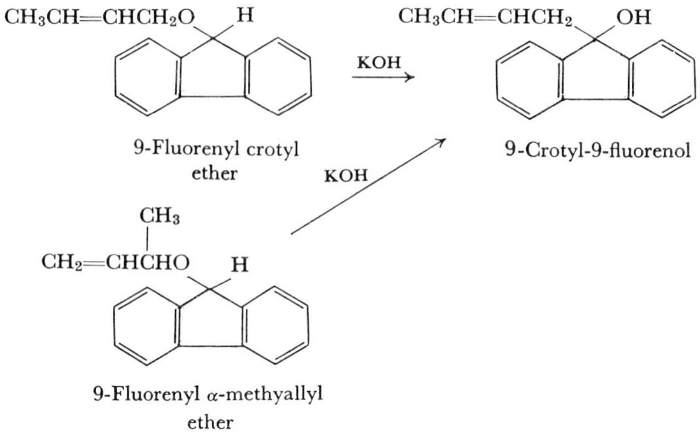

$CH_3CH{=}CHCH_2O$ H → (KOH) → $CH_3CH{=}CHCH_2$ OH

9-Fluorenyl crotyl ether

9-Crotyl-9-fluorenol

CH_3

$CH_2{=}CHCHO$ H → (KOH) → 9-Crotyl-9-fluorenol

9-Fluorenyl α-methyallyl ether

Rearrangement of cyclopropylcarbinyl benzyl ether with methyllithium in tetrahydrofuran gave as product cyclopropylcarbinylphenylcarbinol. Only 6–10% ring opening of the cyclopropyl group occurred during this rearrangement.[27a]

When heated with *N*-methylaniline–potassium *N*-methylanilide, optically active ether IX underwent cleavage to give 2-phenylbutane

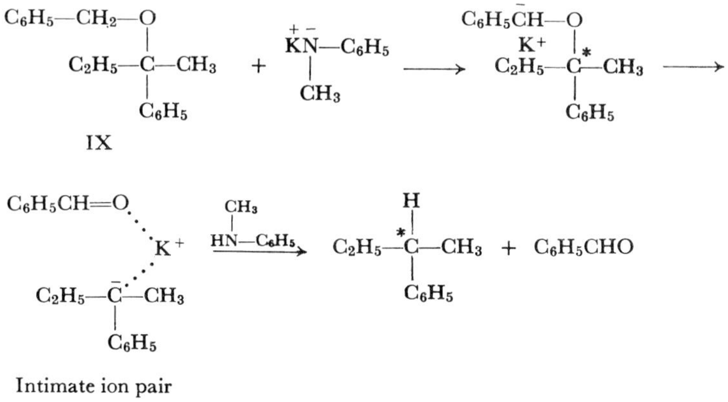

$$C_6H_5{-}CH_2{-}O{-}C^{*}(C_2H_5)(CH_3)(C_6H_5) \;(\text{IX}) + \overset{+\,-}{K}N(CH_3){-}C_6H_5 \longrightarrow C_6H_5\bar{C}H{-}O{-}C^{*}(C_2H_5)(CH_3)(C_6H_5)\;K^+ \longrightarrow$$

$$C_6H_5CH{=}O \cdots K^+ \cdots C_2H_5{-}\bar{C}(CH_3)(C_6H_5) \xrightarrow{HN(CH_3){-}C_6H_5} C_2H_5{-}C^{*}H(CH_3)(C_6H_5) + C_6H_5CHO$$

Intimate ion pair

with at least 29% net retention of configuration.[28] This stereochemical result was part of a pattern of results obtained in cleavage reactions in

[27] (a) P. T. Lansbury and V. A. Pattison, *J. Am. Chem. Soc.*, **84**, 4295 (1962); (b) P. T. Lansbury and V. A. Pattison, *J. Org. Chem.*, **27**, 1933 (1962).

[28] D. J. Cram, C. A. Kingsbury, and A. Langemann, *J. Am. Chem. Soc.*, **81**, 5785 (1959).

which ion pairs were formed, the carbanions of which reacted with proton donors of the medium (see Chapter IV). It was suggested that such cleavage reactions and the Wittig rearrangement had such ion pairs as intermediates. When protons served as electrophiles, cleavage occurred. When the carbonyl group served as the electrophile, the product of the Wittig rearrangement was produced.

Strong support for this ion pair elimination-addition mechanism was obtained by Schöllkopf and co-workers.[29] Cleavage of optically active ether X with butyllithium in a variety of solvents and temperatures gave alcohol XI which was 67–82% racemic. To the extent that XI was optically active, rearrangement had occurred with retention of configuration. Control experiments demonstrated that the observed racemization occurred only during the rearrangement.

$$\underset{\mathbf{X}}{C_6H_5{-}CH_2{-}O{-}\overset{*}{C}H(CH_3)(C_2H_5)} \xrightarrow{BuLi} C_6H_5{-}\overset{\overset{+}{Li}}{\bar{C}H}{-}O{-}\overset{*}{C}H(CH_3)(C_2H_5) \longrightarrow \underset{\text{Partially racemizes}}{C_6H_5{-}CH{=}\overset{+}{O}{-}Li \cdots \bar{C}H(CH_3)(C_2H_5)}$$

$$\longrightarrow C_6H_5{-}CH(OLi){-}\overset{*}{C}H(CH_3)(C_2H_5) \xrightarrow{H_2O} \underset{\mathbf{XI}}{C_6H_5{-}CH(OH){-}\overset{*}{C}H(CH_3)(C_2H_5)}$$

Partially racemic, partially of retained configuration

When ethers XII and XIII were rearranged as a mixture in tetrahydrofuran–ether at −56°, the amounts of cross-bred products produced indicated that only 7% of the reactions occurred by intermolecular

$$\underset{\mathbf{XII}}{p\text{-}DC_6H_4CH_2{-}O{-}CH(CH_3)(C_2H_5)} \qquad \underset{\mathbf{XIII}}{C_6H_5CH_2{-}O{-}CH(CH_3){-}H}$$

[29] (a) U. Schöllkopf and W. Fabian, *Ann.*, **642**, 1 (1961); (b) U. Schöllkopf and D. Walter, *Angew. Chem.*, **73**, 545 (1961); (c) U. Schöllkopf, *Angew. Chem.* (*Intern. Ed. Engl.*), **1**, 126 (1962); (d) U. Schöllkopf, *ibid.*, **2**, 161 (1963); (e) U. Schöllkopf, M. Patsch, and H. Schäfer, *Tetrahedron Letters*, **36**, 2515 (1964).

paths. Under the same conditions, optically active ether X was found to rearrange with about 80% racemization and 20% retention of configuration.[29c]

Lansbury and Pattison[27b] obtained independent evidence that the Wittig rearrangement proceeds by a dissociation-recombination mechanism. Isomerization of benzyl ether to benzylphenylcarbinol with methyllithium gave as a second product methylphenylcarbinol. As solvent was varied from one-to-one ether–tetrahydrofuran to one-to-one tetrahydrofuran–dimethoxyethane, the amount of methylphenylcarbinol produced increased from 3 to 25%. Clearly benzaldehyde is generated as an intermediate in this reaction.

The related Meisenheimer rearrangement of amine oxides to hydroxylamine derivatives with heat was found to occur with about 60–80% racemization.[29d,e] Electron dispersing groups substituted in the aryl group enhanced the reaction rate, whereas electron providing groups inhibited the rate. Again, an elimination-addition mechanism was postulated. At first heterolytic cleavage was formulated,[29d] but

$$(CH_3)_2\overset{+}{N}-\overset{-}{O},\ C_6H_5-\overset{*}{C}(D)-H \xrightarrow{\Delta} (CH_3)_2\overset{+}{N}=O,\ C_6H_5-\overset{-}{C}(D)-H \longrightarrow (CH_3)_2N-O-C(C_6H_5)(D)-H$$

Ion pair 80% Racemic

later[29e] the reaction was thought to go by homolytic cleavage and recombination of the radicals produced within the solvent cage.

1,2-Aryl Migrations from Carbon to Carbon

Anionic 1,2-phenyl migrations have been demonstrated to occur when organometallics of appropriate systems are either prepared, or prepared and then heated. This rearrangement was independently reported at about the same time by Grovenstein[30a] and by Zimmerman and Smentowski.[31a] These workers[30a, 31a] initially found that when 2,2,2-triphenylethyl chloride was treated with sodium metal in dioxane or isooctane–ether, 1,1,2-triphenylethylsodium was produced, which

[30] (a) E. Grovenstein, Jr., *J. Am. Chem. Soc.*, **79**, 4985 (1957); (b) E. Grovenstein, Jr. and L. P. Williams, *ibid.*, **83**, 412, 2537 (1961); (c) E. Grovenstein, Jr. and G. Wentworth, *ibid.*, **85**, 3305 (1963); (d) E. Grovenstein, Jr. and L. C. Rogers, *ibid.*, **86**, 854 (1964).

[31] (a) H. E. Zimmerman and F. J. Smentowski, *J. Am. Chem. Soc.*, **79**, 5455 (1957); (b) H. E. Zimmerman and A. Zweig, *ibid.*, **83**, 1196 (1961).

could be protonated to give 1,1,2-triphenylethane, or carbonated to give 2,2,3-triphenylpropanoic acid.

$$(C_6H_5)_3C{-}CH_2{-}Cl + Na \longrightarrow \underset{\text{}}{\overset{\overset{+}{Na}}{(C_6H_5)_2\bar{C}{-}CH_2{-}C_6H_5}}$$

2,2,2-Triphenylethyl chloride

$$(C_6H_5)_2\bar{C}{-}CH_2{-}C_6H_5 \xrightarrow{ROH} (C_6H_5)_2CH{-}CH_2{-}C_6H_5$$

1,1,2-Triphenylethane

$$(C_6H_5)_2\bar{C}{-}CH_2{-}C_6H_5 \xrightarrow{CO_2} (C_6H_5)_2C(CO_2H){-}CH_2{-}C_6H_5$$

2,2,3-Triphenyl propanoic acid

Subsequently, both groups of workers[30b, 31b] demonstrated that when lithium was substituted for sodium and the reaction or similar reactions were conducted at low temperatures, unrearranged organometallic was formed as demonstrated by its protonation or carbonation to give unrearranged hydrocarbon and acid, respectively. Grovenstein and Williams[30b] worked with 2,2,2-triphenylethyl chloride and Zimmerman and Zweig[31b] with 2,2-diphenylpropyl chloride. The derived organolithium compounds at higher temperatures gave rearranged organolithium compounds (products of 1,2-phenyl migration) as shown by protonation and carbonation experiments. Between $-30°$ and $-65°$, 2,2,2-triphenylethyllithium in tetrahydrofuran was shown to be stable, but rearrangement occurred readily at 0°.[30b] In ether at 0°, 2,2-diphenylpropyllithium could be formed, and rearrangement occurred at about 35°.[31b]

$$(C_6H_5)_3C{-}CH_2{-}Cl \xrightarrow{Li,\ (CH_2)_4O} (C_6H_5)_3C{-}CH_2{-}Li$$

2,2,2-Triphenylethyl chloride → 2,2,2-Triphenylethyllithium

$$(C_6H_5)_3C{-}CH_2{-}Li \xrightarrow{BH} (C_6H_5)_3C{-}CH_3$$

$$(C_6H_5)_3C{-}CH_2{-}Li \xrightarrow{CO_2} (C_6H_5)_3C{-}CH_2{-}CO_2H$$

$$\downarrow 0°$$

$$\overset{Li^+}{(C_6H_5)_2\bar{C}{-}CH_2{-}C_6H_5} \xrightarrow{BH} (C_6H_5)_2CH{-}CH_2{-}C_6H_5$$

1,1,2-Triphenylethyllithium

$$\xrightarrow{CO_2} (C_6H_5)_2C(CO_2H){-}CH_2{-}C_6H_5$$

$$(C_6H_5)_2C(CH_3)-CH_2-Cl \xrightarrow[0^\circ]{\text{Li, Ether}} (C_6H_5)_2C(CH_3)-CH_2-Li \xrightarrow{BH} (C_6H_5)_2C(CH_3)-CH_3$$

2,2-Diphenylpropyl chloride; 2,2-Diphenylpropyllithium

$$(C_6H_5)_2C(CH_3)-CH_2-Li \xrightarrow{CO_2} (C_6H_5)_2C(CH_3)-CH_2-CO_2H$$

$$\downarrow 35^\circ$$

$$C_6H_5-\overset{CH_3}{\underset{Li^+}{\bar{C}}}-CH_2-C_6H_5 \xrightarrow{BH} C_6H_5-CH(CH_3)-CH_2-C_6H_5$$

$$C_6H_5-\overset{CH_3}{\underset{Li^+}{\bar{C}}}-CH_2-C_6H_5 \xrightarrow{CO_2} C_6H_5-C(CH_3)(CO_2H)-CH_2-C_6H_5$$

The ease of phenyl migration in these systems varied widely with the metal used. With 2,2,2-triphenylethyl chloride as starting material, attempts at formation of the unrearranged sodium or potassium organometallics resulted only in the rearranged derivatives.[30b] With 2,2-diphenylpropyl chloride as starting material, the dialkylmagnesium compound was prepared (via the mercury derivative), but could not be induced to rearrange. On the other hand, an attempt to form unrearranged alkylpotassium (via the mercury derivative) gave only rearranged alkylpotassium.[31b] Only in the case of the lithium derivative could unrearranged compounds first be prepared and then be rearranged. Thus the tendency to rearrange is $RK \sim RNa > RLi > R_2Mg$, which is in the decreasing order of the ionic character of the carbon–metal bond. This correlation supports a carbanionic mechansim for the rearrangement.

In an internal competition experiment, phenyl was shown to migrate preferentially to the *p*-tolyl group by a factor of about 11 when 2-phenyl-2-*p*-tolylpropyllithium was heated in ether.[31b] This fact, coupled with the observation that in radical rearrangements of similar systems phenyl and *p*-tolyl have similar migratory aptitudes,[32] indicates that the rearrangement is not radical but anionic in character. Phenyl as a migrating group is better able to stabilize negative charge than *p*-tolyl.

The intramolecular character of these 1,2-phenyl migrations was demonstrated by the following experiment.[30c] In tetrahydrofuran,

[32] (a) M. S. Kharasch, A. C. Poshkus, A. Fono, and W. Nudenberg, *J. Org. Chem.*, **16**, 1458 (1951); (b) W. H. Urry and N. Nicolaides, *J. Am. Chem. Soc.*, **74**, 5163 (1952).

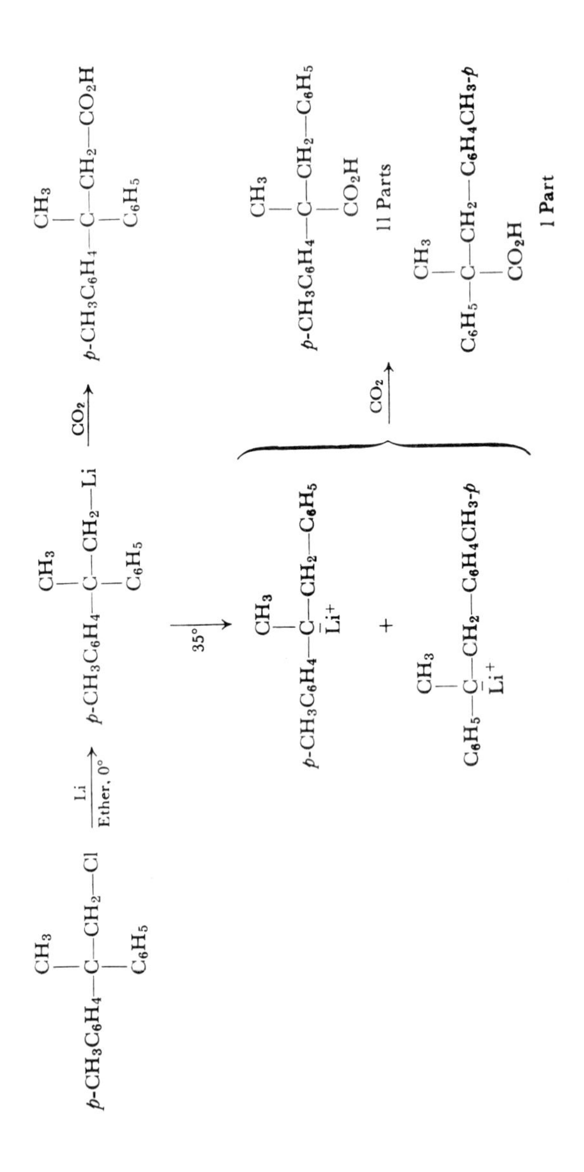
p-CH3C6H4—C—CH2—Cl
CH3
C6H5
Li
Ether, 0°
p-CH3C6H4—C—CH2—Li
CO2
p-CH3C6H4—C—CH2—CO2H
35°
p-CH3C6H4—C—CH2—C6H5
Li+
+
C6H5—C—CH2—C6H4CH3-p
CO2
CO2H
11 Parts
1 Part

2,2,2-triphenylethyllithium was rearranged in the presence of labeled phenyllithium, and the products were carbonated. The 2,2,3-triphenylpropanoic acid produced contained less than 0.05% of the label.

These experiments provide strong support for the 1,2-rearrangements occurring by an anionic mechanism. Two general schemes can be envisioned. In the first aryl as a neighboring group either dislodges lithium or migrates to the free anionic center to form an "ethylene phenanion" as either a discrete intermediate or a transition state. In the carbonium ion counterpart of such a scheme, an ethylene phenonium ion as a discrete intermediate has been demonstrated.[33] Completion of the rearrangement produces the more stable benzyl organometallic compound. In the second general mechanism, an intramolecular elimination-addition mechanism is envisioned, the components being held together as ligands of the metal ion and by ion pairing. These possibilities have not been distinguished experimentally.

1st Mechanism:

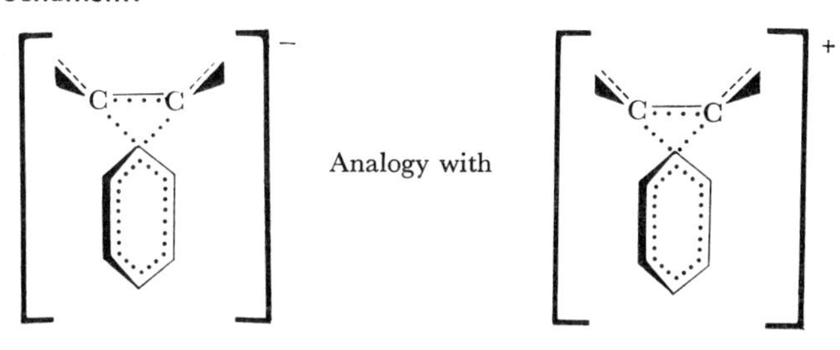

Phenanion as either intermediate or transition state

Phenonium ion, demonstrated to be a discrete intermediate[33]

2nd Mechanism:

$$-\overset{|}{\underset{|}{C}}-\overset{|}{\underset{|}{C}}- \;(C_6H_5,\ M) \longrightarrow \;>C{=}C< \;\vdots\; M^+ \cdots \bar{C}_6H_5 \longrightarrow -\overset{|}{\underset{|}{C}}-\overset{|}{\underset{|}{C}}- \;(M,\ C_6H_5)$$

Ion pair with olefin as ligand of the cation

The fact that in an appropriate system rearrangement can occur by an *intermolecular* elimination-addition mechanism has been demonstrated by Grovenstein and Wentworth.[30c] At −65° in tetrahydrofuran, treatment of 2,2,3-triphenylpropyl chloride with lithium gave unrearranged

[33] (a) D. J. Cram, *J. Am. Chem. Soc.*, **71**, 3863 (1949); (b) D. J. Cram, *ibid.*, **86**, 3767 (1964).

organolithium compound, 2,2,3-triphenylpropyllithium, as shown by carbonation experiments. When warmed to 0° in the presence of labeled benzyllithium, 1,1,3-triphenylpropyllithium was produced as shown by carbonation experiments. The final acids, 2,2,4-triphenylbutanoic and phenylacetic acid, contained labels in about that proportion required for a completely intermolecular rearrangement of the benzyl and lithium residues.

$$C_6H_5CH_2\text{—}\underset{\underset{C_6H_5}{|}}{\overset{\overset{C_6H_5}{|}}{C}}\text{—}CH_2\text{—}Li + C_6\overset{*}{H_5}CH_2Li \xrightarrow[0^\circ]{(CH_2)_4O} \{(C_6H_5)_2C{=}CH_2 + C_6\overset{*}{H_5}CH_2Li\} \longrightarrow$$

2,2,3-Triphenylpropyl-lithium

$$\{(C_6H_5)_2\overset{\overset{+}{Li}}{\bar{C}}\text{—}CH_2\text{—}\overset{*}{CH_2}C_6H_5 + C_6\overset{*}{H_5}CH_2Li\} \xrightarrow{CO_2}$$

$$(C_6H_5)_2\underset{\underset{CO_2H}{|}}{C}\text{—}CH_2\text{—}\overset{*}{CH_2}C_6H_5 + C_6\overset{*}{H_5}CH_2\text{—}CO_2H)$$

2,2,4-Triphenylbutanoic acid

A number of interesting questions remain to be answered in connection with the rearrangement of the 2,2,2-triphenylethyllithium system. (1) Does phenyl participate in the breaking of the carbon–lithium bond, or does a primary anion intervene as a discrete intermediate? (2) To what extent does migration occur for steric reasons, and to what extent for charge delocalization reasons? (3) What is the stereochemical course of the rearrangement? (4) What is the effect of substituents at C_α and C_β on the driving force for rearrangement? (5) How does the rearrangement respond to solvent effects? (6) Can rearrangement be observed when carbanions are generated by other means? These are the kinds of problems that have been studied in the carbonium ion counterpart of this 1,2-rearrangement of aryl.[34]

Benzilic Acid and Related Rearrangements

Although the benzilic acid rearrangement does not involve carbanions, it is nonetheless an anionic rearrangement and is treated here. In most of the examples of the rearrangement,[35] aryl serves as the

[34] D. J. Cram, *in* "Steric Effects in Organic Chemistry" (M. Newman, ed.), Wiley, New York, 1956, Chapter 5.

[35] S. Selman and J. F. Eastham, *Quart. Rev. (London)*, **14**, 221 (1961).

migrating group in a 1,2-migration, although there are notable exceptions to this generalization. Examples of the reaction are formulated.[36]

$$C_6H_5{-}C({=}O){-}C({=}O){-}C_6H_5 \xrightarrow[(2)\ H_3O^+]{(1)\ KOH} (C_6H_5)_2C(OH){-}CO_2H$$

$$\text{Phenanthrenequinone} \xrightarrow[(2)\ H_3O^+]{(1)\ KOH} \text{9-hydroxyfluorene-9-carboxylic acid (HO, } CO_2H)$$

$$CH_2CO_2H{-}C({=}O){-}C({=}O){-}CH_2CO_2H \xrightarrow{(1)\ KOH} HO{-}C(CH_2CO_2H)_2{-}CO_2H$$

$$\text{Alloxan} \xrightarrow[(2)\ H_3O^+]{(1)\ KOH} \text{Alloxanic acid}$$

The mechanism[37] which most satisfactorily correlates and explains a large body of data involves a rapidly reversible addition of hydroxide ion to one of the carbonyl groups, followed by a rate-limiting migration of aryl to the relatively electron-deficient carbon atom (Equation (7)). The rapid proton transfers which produce the final product are those that usually occur when alcohols or carboxylic acids are dissolved in hydroxylic solvents.

The reaction is first order in hydroxide ion and in substrate.[38] The carbonyl oxygens of benzil undergo base-catalyzed exchange with O^{18} enriched water at a rate much faster than rearrangement occurs.[39] The

[36] (a) J. von Liebig, *Ann.*, **25**, 27 (1838); (b) R. Nietzki, *Ber.*, **23**, 3136 (1890); (c) O. Walloch, *Ann.*, **437**, 148 (1924); (d) G. Scheuing, *Ber.*, **56**, 252 (1923).

[37] C. K. Ingold, *Ann. Rep.*, **25**, 124 (1928).

[38] (a) F. H. Westheimer, *J. Am. Chem. Soc.*, **58**, 2209 (1936); (b) F. H. Westheimer, *J. Org. Chem.*, **1**, 1339 (1936).

[39] I. Roberts and H. C. Urey, *J. Am. Chem. Soc.*, **60**, 880 (1938).

rearrangement of benzil to sodium benzilate occurs almost twice as fast in 2:1 dioxane–deuterium oxide as in 2:1 dioxane–water at 50°.[40] This

$$(7)\quad C_6H_5\text{—}\overset{\overset{\displaystyle O}{\|}}{C}\text{—}\overset{\overset{\displaystyle O}{\|}}{C}\text{—}C_6H_5 + OH^- \underset{\text{Fast}}{\overset{K}{\rightleftharpoons}} C_6H_5\text{—}\overset{\overset{\displaystyle O}{\|}}{C}\text{—}\underset{\underset{\displaystyle C_6H_5}{|}}{\overset{\overset{\displaystyle \bar{O}}{|}}{C}}\text{—}OH \xrightarrow[\text{Slow}]{k}$$

$$C_6H_5\text{—}\underset{\underset{\displaystyle C_6H_5}{|}}{\overset{\overset{\displaystyle \bar{O}}{|}}{C}}\text{—}CO_2H \rightleftharpoons C_6H_5\text{—}\underset{\underset{\displaystyle C_6H_5}{|}}{\overset{\overset{\displaystyle OH}{|}}{C}}\text{—}CO_2^-$$

difference is attributed to the greater basicity of deuteroxide in deuterium oxide as compared with hydroxide in water. Doering and Urban[41] observed that benzil reacts with sodium methoxide or potassium *tert*-butoxide to yield the corresponding benzilic acid esters, but that the reaction with sodium ethoxide is seriously complicated by hydride ion transfers from ethoxide to benzil.

Electron-withdrawing substituents in the *m*- or *p*-position of benzil enhance, and electron-providing substituents inhibit the rate of rearrangement of benzil.[42] Both the ground and transition states are affected by these substituents, particularly because of the equilibrium stage prior to the rate-determining step (see Equation (7)). The observed rate constant for rearrangement ($k_{obs.}$) is related to the equilibrium constant (K) and the rate constant for the rate-controlling step (k) by Equation (8). If the rate-limiting transition state resembles in structure the adduct of benzil and hydroxide ion, then electron-withdrawing substituents should stabilize that transition state since it carries

$$(8)\qquad k_{obs.} = Kk$$

a negative charge.[43] The fact that *o*-substituents retard the rate of rearrangement as compared to their *p*-counterparts[42] reflects steric compression in the transition state for rearrangement, and a consequent increase in activation energy.

Through use of isotopic labels, the migrating group has been identified

[40] J. Hine and H. W. Haworth, *J. Am. Chem. Soc.*, **80**, 2274 (1958).

[41] W. von E. Doering and R. S. Urban, *J. Am. Chem. Soc.*, **78**, 5938 (1956).

[42] (a) J. H. Blanksma and W. H. Zaayer, *Rec. Trav. Chim.*, **67**, 883 (1938); (b) E. Pfeil, G. Geissler, W. Jacqueman, and F. Lomker, *Ber.*, **89**, 1210 (1956).

[43] (a) G. Hammond, *J. Am. Chem. Soc.*, **77**, 334 (1955); (b) J. F. Eastham, R. G. Nations, and C. J. Collins, *J. Org. Chem.*, **23**, 1764 (1958).

in unsymmetrically substituted benzils[44] (see Equation (9)). As expected on the basis of the above substituent effects on rate, the aryl group substituted with the more electron-withdrawing group underwent migration at the faster rate. A plot of the logarithm of the migration ratios of mono-substituted *m*- and *p*-substituted benzils against the respective Hammett substituent constants (σ) gave a straight line whose slope yielded a reaction constant of $\rho = 1.43$.[35] This value compares with $\rho = 1.45$ for the reaction of 2,4-dinitrophenyl aryl ethers with potassium methoxide to give aryl methyl ethers[45] (see Equation (10)).

(9)

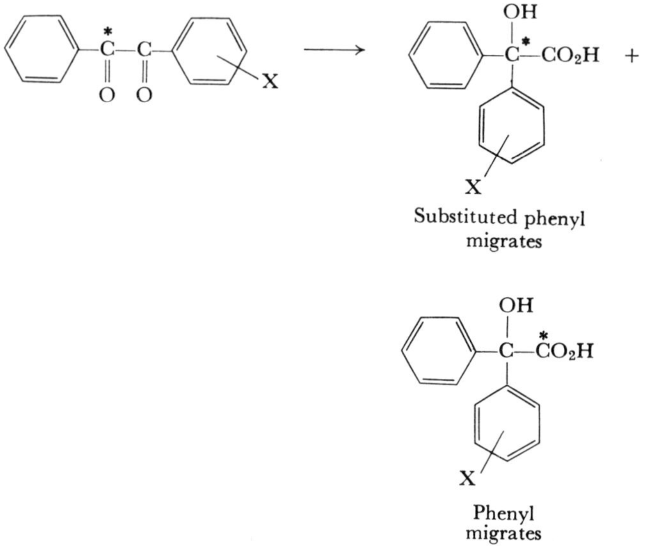

Substituted phenyl migrates

Phenyl migrates

(10)

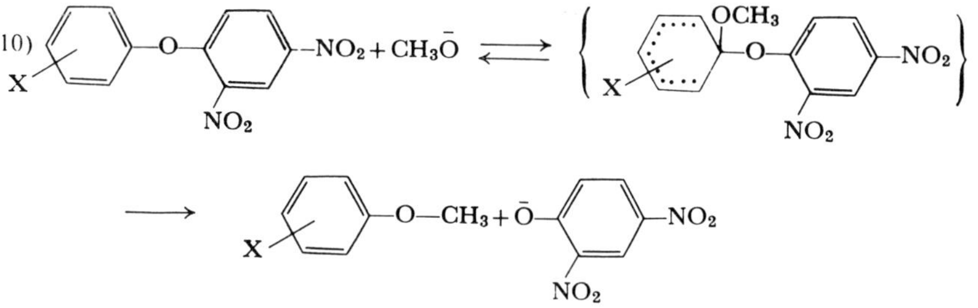

[44] (a) J. D. Roberts, D. R. Smith, and C. C. Lee, *J. Am. Chem. Soc.*, **73**, 619 (1951); (b) C. J. Collins and O. K. Neville, *ibid.*, **73**, 2471 (1951); (c) M. T. Clark, E. G. Hendley, and O. K. Neville, *ibid.*, **77**, 3280 (1955); (d) D. G. Ott and G. G. Smith, *ibid.*, **77**, 2325 (1955).

[45] Y. Ogata and M. Okano, *J. Am. Chem. Soc.*, **71**, 3212 (1949).

Rearrangements related to the benzilic acid rearrangement frequently occur when aryl Grignard reagents are added to benzil or substituted benzils.[35] In general, rearrangement is likely to occur when the organometallic reagent is highly hindered, and a less compressed product can be generated by rearrangement of a less hindering aryl group.

$$\mathrm{Ar{-}\overset{O}{\overset{\|}{C}}{-}\overset{O}{\overset{\|}{C}}{-}Ar' + Ar''MgX \xrightarrow[(2)\ H_3O^+]{(1)\ Ether} Ar{-}\underset{Ar''}{\underset{|}{\overset{OH}{\overset{|}{C}}}}{-}\overset{O}{\overset{\|}{C}}{-}Ar' + Ar{-}\overset{O}{\overset{\|}{C}}{-}\underset{Ar''}{\underset{|}{\overset{OH}{\overset{|}{C}}}}{-}Ar + Ar{-}\underset{Ar'}{\underset{|}{\overset{OH}{\overset{|}{C}}}}{-}\overset{O}{\overset{\|}{C}}{-}Ar''}$$

Rearranged product

Similarly, α-arylbenzoins sometimes undergo rearrangements when treated with base.[25a, 35, 46] For example, Curtin and Leskowitz[25a] observed that XIV gave XV and its cleavage products when treated with alkali. Many other examples of this and similar rearrangements have been reviewed by Selman and Eastham.[35]

$$\underset{\text{XIV}}{\mathrm{C_6H_5{-}\underset{CH(C_6H_5)_2}{\underset{|}{\overset{OH}{\overset{|}{C}}}}{-}\overset{O}{\overset{\|}{C}}{-}C_6H_5}} \xrightarrow{\text{Base}}$$

$$\underset{\text{XV}}{\mathrm{(C_6H_5)_2CH\underset{O}{\underset{\|}{C}}\overset{OH}{\overset{|}{C}}(C_6H_5)_2}} + \mathrm{(C_6H_5)_2CH{-}CO_2H + (C_6H_5)_2CHOH}$$

Several questions arise in the benzilic acid and related rearrangements concerning the exact character of the rate-determining step. For

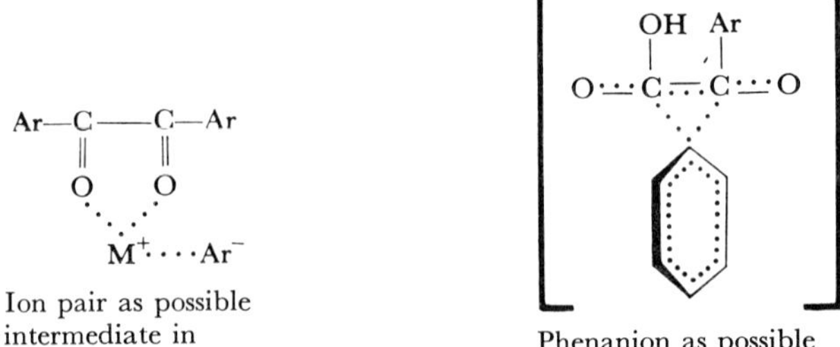

Ion pair as possible intermediate in tertiary ketol rearrangement

Phenanion as possible intermediate in benzilic acid rearrangement.

[46] J. F. Eastham, J. E. Huffaker, V. F. Raaen, and C. J. Collins, *J. Am. Chem. Soc.*, **78,** 4323 (1956).

example, at least in those cases in which the reaction is carried out in non-proton donating media, the possibility exists that ion pair intermediates intervene. Another possibility is that phenanions exist as discrete intermediates in some of these 1,2-aryl rearrangements.

REARRANGEMENTS WITH 1,3-ELIMINATION REACTION STAGES

A family of rearrangements is known that involves the closing of a three-membered ring by base-catalyzed reaction to form an unstable and in some cases nonisolable intermediate. This species in a second step decomposes to give rearranged material. In simplified form, these rearrangements correspond to the sequence of Equation (11). They find

(11) 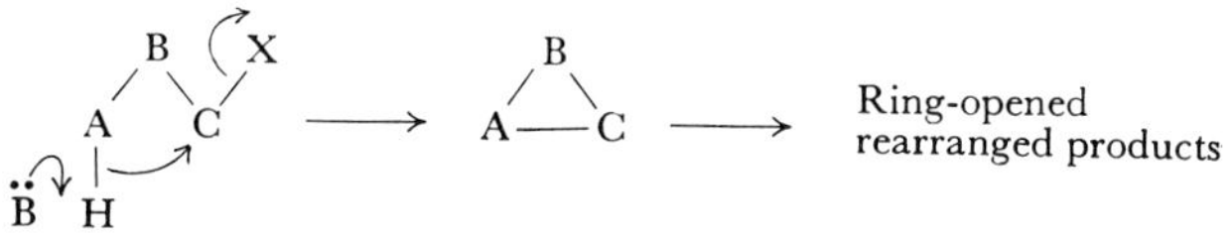 B:⁻ + H–A(–B–C–X) ⟶ A–C (three-membered ring with B) ⟶ Ring-opened rearranged products

their driving force for the ring-closing stage in the high nucleophilicity of carbanions on the one hand, and in the proximity of a site subject to nucleophilic substitution on the other. The first step of these rearrangements is very similar to the time honored method of preparing cyclopropane compounds by 1,3-elimination reactions. An example of the latter process involves the conversion of XVI to XVII.[47]

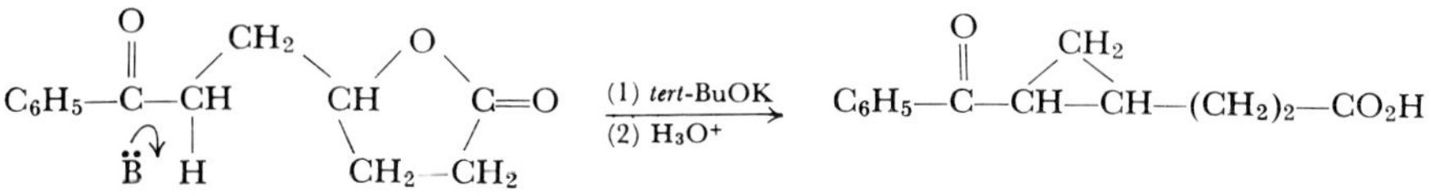

$$C_6H_5{-}\overset{O}{\overset{\|}{C}}{-}CH(H)CH_2{-}\text{(γ-butyrolactone)} \xrightarrow[(2)\ H_3O^+]{(1)\ tert\text{-BuOK}} C_6H_5{-}\overset{O}{\overset{\|}{C}}{-}\underbrace{CH{-}CH}_{CH_2}{-}(CH_2)_2{-}CO_2H$$

XVI XVII

As examples of this type of transformation, the Favorskii, Neber, Ramberg-Bäckland, and related rearrangements are discussed in this section.

Favorskii Rearrangement

Although two reviews of the Favorskii rearrangement are available,[48] the essentials of the reaction are covered here since many new results are

[47] P. Yates and C. D. Anderson, *J. Am. Chem. Soc.*, **85**, 2937 (1963).

[48] (a) R. Jacquier, *Bull. Soc. Chim. France*, [5] **17**, D35 (1950); (b) A. S. Kende, "Organic Reactions," Wiley, New York, 1960, Vol. 11, p. 261.

available, and since it serves as a prototype for some of the less investigated rearrangements that subsequently are treated.

In the Favorskii rearrangement, an α-haloketone when treated with nucleophilic bases ($\bar{O}H$, $\bar{O}R$, or $\ddot{N}HR_2$) gives rearranged carboxylic acids or their derivatives. Appropriate dihaloketones usually give unsaturated acids or their derivatives.

$$CH_3\bar{O}\ H{-}CH_2{-}CO{-}C(CH_3)_2Br \xrightarrow{-Br^-} \left[\text{cyclo-}(CH_2{-}C(CH_3)_2{-}C{=}O) \right] \xrightarrow{CH_3\bar{O}} CH_3O_2C{-}C(CH_3)_3$$

$$H\bar{O}\ H{-}CH_2{-}CO{-}C(Cl)_2{-}CH_3 \longrightarrow \left[\text{cyclo-}(CH_2{-}C(Cl)(CH_3){-}C{=}O) \right] \xrightarrow{OH^-} HO_2C{-}C(CH_3){=}CH_2$$

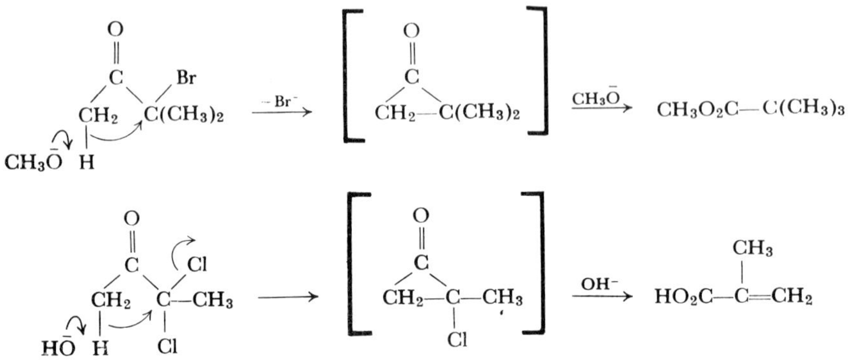

The existence of a three-membered ring intermediate is indicated by the following experiments. When treated with sodium hydroxide, both 1-chloro-3-phenyl-2-propanone and 1-chloro-1-phenyl-2-propanone give 3-phenylpropionic acid. This common product from two different starting materials suggests a common intermediate, phenylcyclopropanone, which is formed in principle by a 1,3-elimination reaction from either starting material.[49]

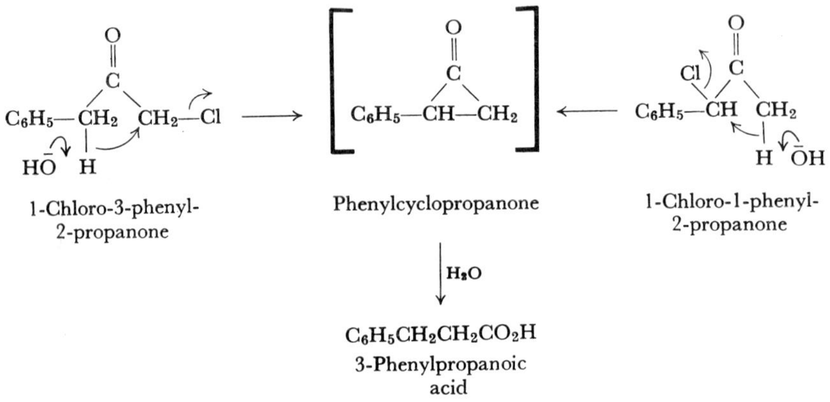

$$C_6H_5{-}CH_2{-}CO{-}CH_2{-}Cl \ (H\bar{O}) \longrightarrow \left[C_6H_5{-}\text{cyclo-}(CH{-}CH_2{-}C{=}O) \right] \longleftarrow C_6H_5{-}CHCl{-}CO{-}CH_3 \ (\bar{O}H)$$

1-Chloro-3-phenyl-2-propanone; Phenylcyclopropanone; 1-Chloro-1-phenyl-2-propanone

$$\downarrow H_2O$$

$C_6H_5CH_2CH_2CO_2H$

3-Phenylpropanoic acid

[49] W. D. McPhee and E. Klingsberg, *J. Am. Chem. Soc.*, **66**, 1132 (1944).

An elegant demonstration of a symmetrizing stage in the mechanism of the rearrangement was reported by Loftfield.[50] The rearrangement of 2-chlorocyclohexanone in ethanol–sodium ethoxide was shown to be first order in substrate and in alkoxide ion. Treatment of 2-chlorocyclohexanone-1,2-C^{14} (isotope equally distributed between carbon atoms 1 and 2) with less than one equivalent of sodium isoamoxide in isoamyl alcohol gave isoamyl cyclopentanecarboxylate and recovered chloroketone. The recovered material had the same isotope distribution as the starting material, and the radiocarbon label in the ester product was distributed 50% on the carboxyl carbon, 25% on the α-carbon of the ring, and 25% on the two ring β-carbon atoms.

Clearly no halogen migration preceded rearrangement. Otherwise chloroketone with the C^{14} scrambled between the 2,6-positions would have been recovered. Furthermore, a mechanism is required which has an intermediate in which the α and α′ carbons of the cyclohexanone molecule become equivalent. The cyclopropanone intermediate satisfies this requirement, and its formation typifies a large family of base-catalyzed 1,3-elimination reactions which lead to three-membered rings. The opening of the postulated cyclopropanone intermediate to give acids and their derivatives is consistent with what is known about the behavior of cyclopropanone derivatives.[51] Cleavage of this ketone,

$NaOC_2H_5$ → HOC_2H_5 → $CO_2C_2H_5$ (50%) + $CO_2C_2H_5$ (50%)

as with other ketones with base, should occur in such a way as to provide the more stable transient carbanion (see Chapter IV). The products of the Favorskii rearrangement are consistent with this generalization, as is clear from the previous examples given. Thus, the rearrangement probably proceeds normally by a cyclopropanone mechanism.

Even more direct evidence of a cyclopropanone intermediate in the Favorskii rearrangement is found in what amounts to an interception of such a species, and its conversion to an isolable product without destruction of the three-membered ring. Breslow and co-workers[52]

[50] (a) R. B. Loftfield, *J. Am. Chem. Soc.*, **72**, 632 (1950); (b) R. B. Loftfield, *ibid.*, **73**, 4707 (1951); (c) R. B. Loftfield and L. Schaad, *ibid.*, **76**, 35 (1954).

[51] (a) P. Lipp, J. Buchkremer, and H. Seeles, *Ann.*, **499**, 1 (1932); (b) A. S. Kende, Ph.D. Thesis, Harvard University, Cambridge, Massachusetts, 1956.

[52] R. Breslow, J. Posner, and A. Krebs, *J. Am. Chem. Soc.*, **85**, 234 (1963).

treated either one pure diastereomer or a mixture of diastereomers of α,α'-dibromodibenzyl ketone with triethylamine in methylene chloride

a is more carbanion-stabilizing than b

at 25°, and obtained diphenylcyclopropenone as product. Similar reactions were carried out with aliphatic and cyclic dibromoketones.

α,α-Dibromodibenzyl-ketone

Diphenylcyclopropenone

$$C_4H_9\text{—CH(Br)—CO—CH(Br)—}C_4H_9 \xrightarrow[CH_2Cl_2]{(C_2H_5)_3N} C_4H_9\text{—C}{=}\text{C—}C_4H_9 \text{ (cyclopropenone)}$$

Other studies have a bearing on the detailed mechanisms of the ring-closing and ring-opening stages of the Favorskii rearrangement. Stork and Borowitz[53] prepared and established the stereochemical structures

[53] G. Stork and I. Borowitz, *J. Am. Chem. Soc.*, **82**, 4307 (1960).

of the two diastereomers of XVIII (a and b). When treated with dry sodium benzyloxide in ether, diastereomer XVIIIa gave rearrangement product XIXa with little or no contamination with XIXb. Under the same conditions, diastereomer XVIIIb gave XIXb with little or no contamination with XIXa. These stereochemical results indicate that

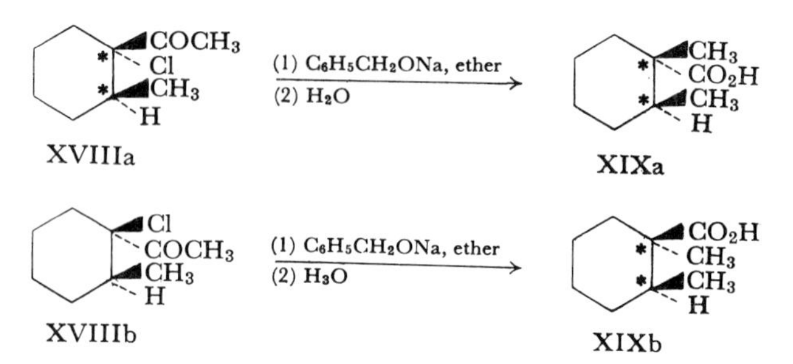

the asymmetric carbon atom attached to chlorine underwent inversion during the formation of the cyclopropanone intermediate.

House and Gilmore[54a] examined the rearrangement of XVIIIa in 1,2-dimethoxyethane–sodium methoxide, and obtained a mixture of 95% methyl ester of acid XIXa and 5% methyl ester of acid XIXb. These results resemble those of Stork and Borowitz, the major reaction course involving overall inversion at the asymmetric carbon atom attached to chlorine. However, in methanol as solvent and sodium methoxide as base, XVIIIa gave a 52% yield of the methyl ester of XIXb (product of retention at $\overset{*}{C}$—Cl) and a 41% yield of the methyl ester of XIXa. Studies of other systems also suggested that the degree of stereospecificity of the Favorskii rearrangement depends on the solvent polarity,[54b] although at least some of the lack of stereospecificity may be associated with isomerization of α-haloketone prior to its rearrangement.[55]

Other investigators[56] have entertained the possibility that the Favorskii rearrangement and some of its competing reactions involve not the simple cyclopropanone intermediates drawn above, but an intermediate in which charge[56a, b, c] or bonds[56d] or both are highly

[54] (a) H. O. House and W. F. Gilmore, *J. Am. Chem. Soc.*, **83**, 3980 (1961); (b) H. O. House and W. F. Gilmore, *ibid.*, **83**, 3972 (1961).

[55] N. L. Wendler, R. P. Graber, and G. G. Hazen, *Tetrahedron*, **3**, 144 (1958).

[56] (a) J. G. Aston and J. D. Newkirk, *J. Am. Chem. Soc.*, **73**, 3900 (1951); (b) A. A. Sacks and J. G. Aston, *ibid.*, **73**, 3902 (1951); (c) J. G. Burr and M. J. S. Dewar, *J. Chem. Soc.*, p. 1201 (1954); (d) A. W. Fort, *J. Am. Chem. Soc.*, **84**, 2620, 2625 (1962).

delocalized, as in the planar resonance hybrid, XX. Since such an intermediate possesses a plane of symmetry, unless asymmetrically solvated, it could not be involved in a stereospecific reaction. Resonance forms d, e, and f involve π- rather than σ-bonding between C_α—O or C_α—$C_{\alpha'}$.

Complete clarification of the intimate details of the Favorskii rearrangement must await further work. Aside from the question of the stereochemistry of the ring closure at $\overset{*}{C}$—Cl are the following points of ambiguity. (1) Does a carbanion intervene as a true intermediate in the 1,3-elimination reaction stage, or is the proton removed in the same transition state that the carbon–chlorine bond is broken? If the latter possibility applies, what is the stereochemistry at C—H? (2) What is the stereochemistry at C_α in the ring-opening stage of the cyclopropenone

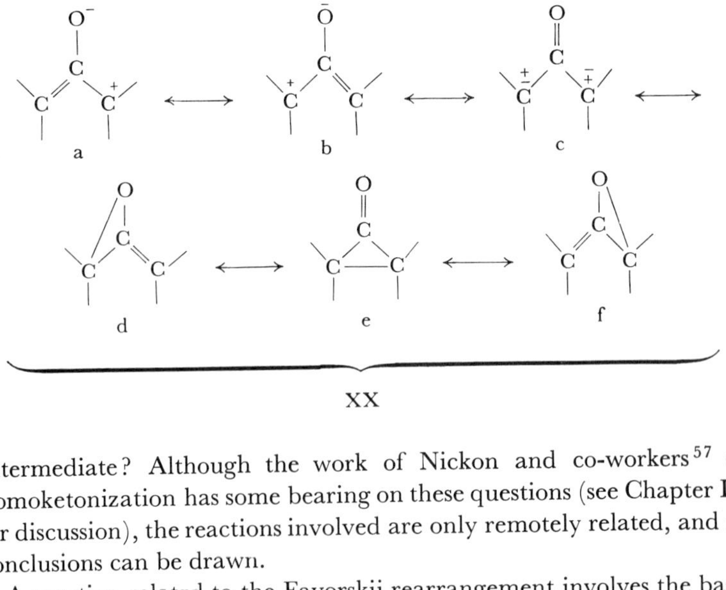

XX

intermediate? Although the work of Nickon and co-workers[57] on homoketonization has some bearing on these questions (see Chapter III for discussion), the reactions involved are only remotely related, and no conclusions can be drawn.

A reaction related to the Favorskii rearrangement involves the base-catalyzed ring closure of XXI to give XXII, and the facile ring opening of XXII.[58] When phenolic bromide XXI was treated with potassium *tert*-butoxide–*tert*-butyl alcohol, spectroscopic evidence for formation of XXII was obtained. The spirodienone was obtained by passing an

[57] (a) A. Nickon, J. H. Hammons, J. L. Lambert, and R. O. Williams, *J. Am. Chem. Soc.*, **85**, 3713 (1963); (b) A. Nickon and J. L. Lambert, *ibid.*, **84**, 4604 (1962).

[58] R. Baird and S. Winstein, *J. Am. Chem. Soc.*, **85**, 567 (1963).

ether solution of XXI through alumina on which was adsorbed potassium hydroxide. Careful evaporation of the ether solution gave XXII. The substance reacted readily with alkoxide or hydroxide ions to give the ring-opened ether or alcohol (XXIII).

XXI ⇌(Base) → XXII →(ROH) XXIII

R = H or CH_3

Neber Rearrangement

In the Neber rearrangement, an oxime tosylate is treated with base to produce (after treatment of the reaction mixture with water) an α-aminoketone.[59] The subject has been thoroughly reviewed by O'Brien.[60]

Successful applications of the reaction have been observed only when the system contained the structural features of formula XXIV.[61a] When two distinguishable α-methylene groups are available, the reaction proceeds in the direction which results in the substitution of the amino group for one of the more acidic hydrogens, irrespective of whether the oxime tosylate is *cis* or *trans* to those hydrogens.[59, 61, 62]

$$R{-}CH_2{-}C(=NOTs){-}R' \xrightarrow[(2)\ H_3O^+Cl^-]{(1)\ NaOC_2H_5,\ HOC_2H_5} R{-}CH(\overset{+}{N}H_3\,\overset{-}{Cl}){-}C(=O){-}R'$$

XXIV

Two types of intermediates have been isolated in the Neber reaction, XXV[61a] and XXVI[59c] on the one hand, and XXVII[61b] on the other.

[59] (a) P. W. Neber and A. Friedolsheim, *Ann.*, **449**, 109 (1926); (b) P. W. Neber and A. Uber, *ibid.*, **467**, 52 (1928); (c) P. W. Neber and A. Burgard, *ibid.*, **493**, 281 (1932); (d) P. W. Neber and G. Huh, *ibid.*, **515**, 283 (1935); (e) P. W. Neber, A. Burgard, and W. Thier, *ibid.*, **526**, 277 (1936).

[60] C. O'Brien, *Chem. Rev.*, **64**, 81 (1964).

[61] (a) M. J. Hatch and D. J. Cram, *J. Am. Chem. Soc.*, **75**, 38 (1953); (b) D. J. Cram and M. J. Hatch, *ibid.*, **75**, 33 (1953).

[62] (a) H. O. House and W. F. Berkowitz, *J. Org. Chem.*, **28**, 307 (1963); (b) H. O. House and W. F. Berkowitz, *ibid.*, **28**, 2271 (1963).

Intermediates XXV and XXVI are adducts of XXVII with appropriate proton donors, XXVII being obtained by treatment of XXVI with sodium carbonate.[61b]

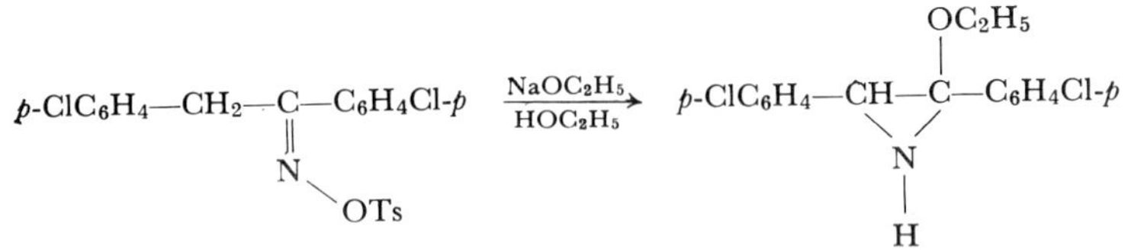

$$p\text{-ClC}_6\text{H}_4\text{—CH}_2\text{—C(=N—OTs)—C}_6\text{H}_4\text{Cl-}p \xrightarrow[\text{HOC}_2\text{H}_5]{\text{NaOC}_2\text{H}_5} p\text{-ClC}_6\text{H}_4\text{—CH—C(OC}_2\text{H}_5\text{)—C}_6\text{H}_4\text{Cl-}p \text{ (aziridine, N—H)}$$

XXV

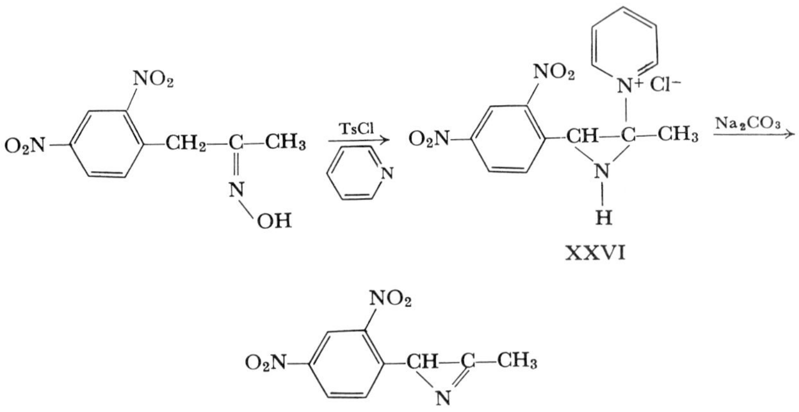

$$\text{O}_2\text{N—C}_6\text{H}_3(\text{NO}_2)\text{—CH}_2\text{—C(=N—OH)—CH}_3 \xrightarrow[\text{pyridine}]{\text{TsCl}} \text{O}_2\text{N—C}_6\text{H}_3(\text{NO}_2)\text{—CH—C(N}^+\text{C}_5\text{H}_5\ \text{Cl}^-\text{)—CH}_3 \text{ (aziridine, N—H)} \xrightarrow{\text{Na}_2\text{CO}_3}$$

XXVI

$$\text{O}_2\text{N—C}_6\text{H}_3(\text{NO}_2)\text{—CH—C—CH}_3 \text{ (azirine, N)}$$

XXVII

The only results that bear on the stereochemistry of the reaction at the carbanionic site involve the production of equatorially oriented amino group in the Neber reaction of XXVIII.[60] This result does not

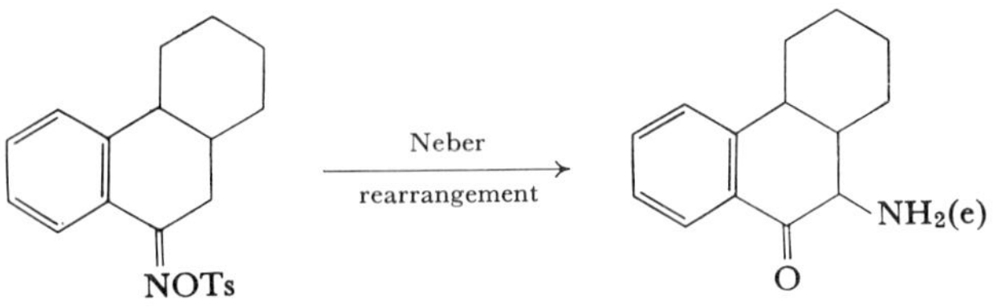

$$\text{XXVIII (=NOTs)} \xrightarrow[\text{rearrangement}]{\text{Neber}} \text{(C=O, NH}_2\text{(e))}$$

XXVIII

reveal whether or not a carbanion intervenes as a discrete intermediate in the reaction or what its stereo-chemical fate may be. The overall stereochemistry of electrophilic substitution at the α-carbon in the Neber reaction is as yet an unsolved problem.

The ring-closing stage of the Neber rearrangement seems to combine the features of a base-induced 1,3-elimination reaction, superimposed on which is a 1,2-addition reaction. The stereochemically indiscriminate character of the reaction suggests that the azirine ring system of XXVII is not formed directly, but that a vinyl nitrene is the first intermediate, which in a second stage gives an azirine. This, in turn, produces an aziridine by an addition reaction.

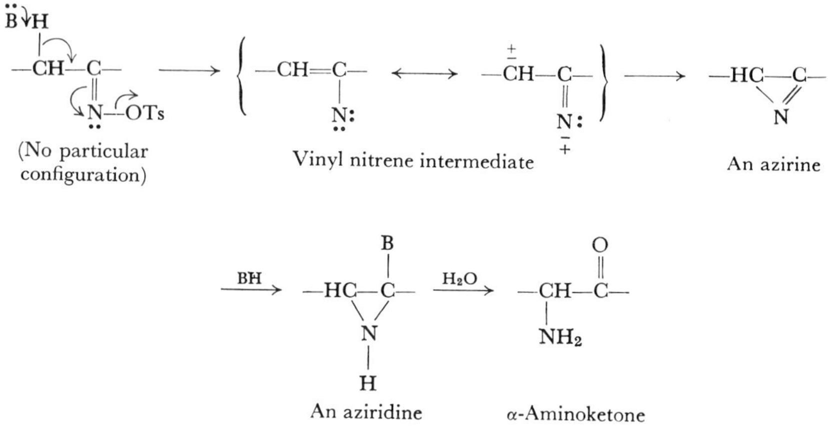

B:H —CH—C(=N—OTs)— ⟶ { —CH=C(—N:)— ⟷ —$\overset{+}{\underset{-}{C}}$H—C(=$\overset{-}{\underset{+}{N}}$:)— } ⟶ —HC—C— (ring with N)

(No particular configuration) Vinyl nitrene intermediate An azirine

$\xrightarrow{BH}$ —HC—C(B)— (ring N—H) $\xrightarrow{H_2O}$ —CH(NH$_2$)—C(=O)—

An aziridine α-Aminoketone

Additional support for this mechanism is found in the observation[63] that pyrolysis of α-azidostyrene produced 2-phenylazirine, presumably through a vinylnitrene as intermediate. Although this intermediate was

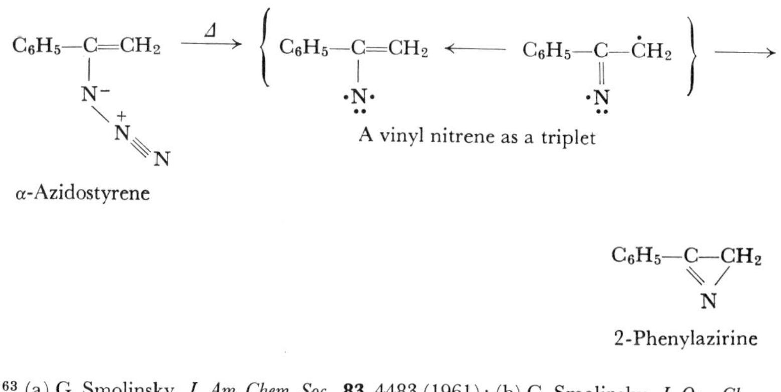

C_6H_5—C(=CH_2)—N^-—$\overset{+}{N}$≡N $\xrightarrow{\Delta}$ { C_6H_5—C(=CH_2)—·N̈· ⟵ C_6H_5—C(=·N̈)—ĊH$_2$ } ⟶

α-Azidostyrene A vinyl nitrene as a triplet

C_6H_5—C—CH_2 (ring with N)

2-Phenylazirine

[63] (a) G. Smolinsky, *J. Am. Chem. Soc.*, **83**, 4483 (1961); (b) G. Smolinsky, *J. Org. Chem.*, **27**, 3557 (1962).

formulated as existing in the triplet state, no evidence bearing on this point was available.

A number of reactions related to the Neber rearrangement have been observed[64] that probably involve azirine intermediates. Campbell *et al.*[64a] treated an oxime (XXIX) with a Grignard reagent and obtained aziridine XXX. Baumgarten *et al.*[64b,c] found that *N,N*-dichloro-*sec*-

$$\underset{\text{XXIX}}{C_6H_5-\underset{\underset{NOH}{\|}}{C}-CH_2-CH_3} \xrightarrow{C_6H_5MgBr} \left[C_6H_5-C\overset{N}{=}CH-CH_3\right] \xrightarrow{C_6H_5MgBr}$$

$$\left[C_6H_5-C(C_6H_5)-CH-CH_3 \text{ (aziridine, N–MgBr)}\right] \xrightarrow{H_2O} \underset{\text{XXX}}{C_6H_5-C(C_6H_5)-CH-CH_3 \text{ (aziridine, N–H)}}$$

alkylamines (XXXI) and sodium methoxide gave α-amino ketones (XXXII). Smith,[64d] Parcell,[64e] Morrow,[64f] and their respective co-workers have studied the base-catalyzed rearrangement of dimethyl-

$$\underset{\text{XXXI}}{R-\underset{\underset{NCl_2}{|}}{CH}-CH_3} \xrightarrow{NaOCH_3} R-C\overset{N}{=}CH_2 \xrightarrow{CH_3OH}$$

$$R-\overset{\overset{OCH_3}{|}}{C}-CH_2 \text{ (aziridine, N–H)} \xrightarrow{H_2O} \underset{\text{XXXII}}{R-\underset{\underset{O}{\|}}{C}-CH_2-NH_2}$$

hydrazone methiodides to α-amino ketones. In the case of compound XXXIII, both types of Neber intermediates have been isolated.[64e]

[64] (a) K. N. Campbell, B. K. Campbell, J. F. McKenna, and E. P. Chaput, *J. Org. Chem.*, **8**, 103 (1943); (b) H. E. Baumgarten and F. A. Bower, *J. Am. Chem. Soc.*, **76**, 4561 (1954); (c) H. E. Baumgarten, J. E. Dirks, J. M. Peterson, and D. C. Wolf, *ibid.*, **82**, 4422 (1960); (d) P. A. S. Smith and E. E. Most, *J. Org. Chem.*, **22**, 358 (1957); (e) R. F. Parcell, *Chem. & Ind.* (*London*), p. 1396 (1963); (f) D. F. Morrow and M. E. Butler, *J. Heterocyclic Chem.*, **1**, 53 (1964).

$$C_6H_5\text{—}C(=N\text{—}\overset{+}{N}(CH_3)_3\ \bar{I})\text{—}CH(CH_3)_2 \xrightarrow{(CH_3)_2CH\bar{O}\overset{+}{K}} \underset{\text{XXIII}}{C_6H_5\text{—}C\text{—}C(CH_3)_2\ (\text{azirine, } =N\text{—})} \xrightarrow{(CH_3)_2CHOH}$$

$$C_6H_5\text{—}C(OCH(CH_3)_2)\text{—}C(CH_3)_2\ (\text{aziridine, N—H}) \xrightarrow[HCl]{H_2O} \underset{\text{XXXIV}}{C_6H_5\text{—}C(=O)\text{—}C(CH_3)_2NH_2}$$

Ramberg-Bäckland Reaction

An interesting reaction that involves a base-catalyzed 1,3-elimination reaction step consists of conversion of α-bromosulfones to alkenes.[65]

$$CH_3CH_2SO_2CH(Br)CH_2CH_3 + 3KOH \longrightarrow CH_3CH{=}CHCH_2CH_3 + KBr + K_2SO_3 + 2H_2O$$

The mechanism of this reaction has been studied by Bordwell and co-workers.[66] These investigators found[66a] that the rate of chloride ion release from chloromethyl sulfone and like compounds in dilute sodium hydroxide–40% dioxane–water solution at 50° was first order in sulfone and first order in hydroxide ion. The authors proposed the following mechanism for the reaction with α-chloroethyl ethyl sulfone.

$$\underset{\alpha\text{-Chloroethyl ethyl sulfone}}{CH_3\text{—}CH(H)\text{—}SO_2\text{—}CH(Cl)\text{—}CH_3} + \bar{O}H \underset{}{\overset{\text{Fast}}{\rightleftharpoons}} CH_3\text{—}\bar{C}H\text{—}SO_2\text{—}CH(Cl)\text{—}CH_3 + H_2O \xrightarrow{\text{Slow}}$$

$$\underset{\text{An episulfone}}{CH_3\text{—}CH\text{—}CH\text{—}CH_3\ (\text{bridged by } SO_2)} \xrightarrow{\text{Fast}} CH_3\text{—}CH{=}CH\text{—}CH_3$$

79% *cis*-2-Butene
21% *trans*-2-Butene

[65] L. Ramberg and B. Bäckland, *Arkiv Kemi Mineral Geol.*, **13A**, No. 27 (1940).

[66] (a) F. G. Bordwell and G. D. Cooper, *J. Am. Chem. Soc.*, **73**, 5187 (1951); (b) N. P. Neurieter and F. G. Bordwell, *ibid.*, **85**, 1209 (1963); (c) see also L. A. Paquette, *ibid.*, **86**, 4085 (1964).

Although the episulfone was not isolated from the base-catalyzed reaction, it was prepared by the reaction of diazoethane and sulfur dioxide.[66b] The *cis*-isomer was isolated in pure form. When heated, it gave pure *cis*-2-butene. A mixture of 78% *trans*–22% *cis*-episulfone gave, when heated, a mixture of 78% *trans*-2-butene and 22% *cis*-2-butene. When decomposed in deuterium oxide–sodium deuteroxide, *cis*-episulfone gave 100% *cis*-2-butene (non-deuterated), whereas in *tert*-butyl alcohol-*O*-*d*–potassium *tert*-butoxide, the 2-butene produced was over 90% deuterated in the 2,3-positions, and was 19% *cis* and 81% *trans*. Thus the *cis*-episulfone underwent exchange and epimerization in *tert*-butyl alcohol-*O*-*d*–potassium *tert*-butoxide faster than it gave alkene, the reverse being true in deuterium oxide–sodium deuteroxide. Finally, the 2-butene produced from α-chloroethyl ethyl sulfone in deuterium oxide–sodium deuteroxide was 98% deuterated in the 2,3-position, and was 78% *cis*.

$$2CH_3CHN_2 + SO_2 \xrightarrow{\text{Ether}} \text{cis-episulfone} + \text{trans-episulfone}$$

H, CH$_3$ — C—C — H, CH$_3$ (ring with SO_2) + H, CH$_3$ — C—C — CH$_3$, H (ring with SO_2)

cis-Episulfone   *trans*-Episulfone

↓ Δ   ↓ Δ

H, CH$_3$ C=C H, CH$_3$ H, CH$_3$ C=C CH$_3$, H

The results led to the following conclusions.[66b] (1) The α-chlorosulfone undergoes reversible carbanion formation faster than ring closure occurs. (2) The Ramberg-Bäckland reaction involves an episulfone intermediate. (3) This intermediate decomposes stereospecifically to alkene. (4) The *trans*-episulfone is more stable than the *cis*. (5) In aliphatic systems, the thermodynamically less stable *cis*-episulfone is formed preferentially in kinetically controlled processes.

The last conclusion is particularly interesting, and recalls the fact that isomerization of terminal to internal alkenes with potassium *tert*-butoxide in dimethyl sulfoxide produces the less stable *cis*-olefins as the predominant isomers under conditions of kinetic control.[67] Bordwell and co-workers[66b] suggested that at the distances involved in the transition state for ring closure of the episulfone, attractive (London) forces between the two methyl groups outweigh the more usually observed

[67] A. Schriesheim and C. A. Rowe, Jr., *Tetrahedron Letters*, **10**, 405 (1962).

repulsive forces. An alternative explanation is that less steric inhibition of solvation is present in the transition state for formation of the internally more compressed *cis*-isomer than in that of the *trans*-isomer, and that steric effects in solvation can be more important in this type of reaction than internal steric effects (see Chapter V for discussion).

$$HN(OH^{-}\cdot H)-SO_2-CH_2-Cl \rightleftarrows HN^{-}-SO_2-CH_2-Cl \longrightarrow HN-SO_2-CH_2 \text{ (ring)} \xrightarrow{OH^-}$$

$$SO_3^{=} + H_2O + CH_2{=}N{-}H \longrightarrow CH_2O + NH_3$$

A transformation related to the Ramberg-Bäckland reaction involves the base-catalyzed conversion of chloromethanesulfonamide to formaldehyde, chloride ion, and sulfite ion.[68] This reaction probably goes by a similar mechanism.

α-Lactam Formation and Reaction

A number of base-catalyzed rearrangements that involve 1,3-elimination stages to give α-lactams will be mentioned here. Baumgarten and co-workers[69] succeeded in isolating α-lactam XXXV by treating either XXXVI or XXXVII with potassium *tert*-butoxide.[69a,c] When optically active XXXVII was employed, the lactam produced was optically active. When treated with various nucleophiles (HNu), the α-lactam underwent ring opening to give amide derivative, XXXVIII.

$$\underset{\text{XXXVI}}{C_6H_5{-}CH(H\cdots\bar{O}Bu\text{-}tert){-}C(=O){-}N(Cl){-}C(CH_3)_3} \longrightarrow \underset{\text{XXXV}}{C_6H_5{-}CH{-}C(=O){-}N{-}C(CH_3)_3 \text{ (ring)}} \longleftarrow \underset{\text{XXXVII}}{C_6H_5{-}CH(Cl){-}C(=O){-}N(H\cdots\bar{O}Bu\text{-}tert){-}C(CH_3)_3}$$

$$\text{XXXV} \xrightarrow{HNu} \underset{\text{XXXVIII}}{C_6H_5{-}CH(Nu){-}C(=O){-}NH{-}C(CH_3)_3}$$

[68] T. B. Johnson and I. B. Douglass, *J. Am. Chem. Soc.*, **63**, 1571 (1941).

[69] (a) H. E. Baumgarten, *J. Am. Chem. Soc.*, **84**, 4975 (1962); (b) H. E. Baumgarten, R. L. Zey, and U. Krolls, *ibid.*, **83**, 4469 (1961); (c) H. E. Baumgarten, J. F. Fuerholzer, R. D. Clark, and R. D. Thompson, *ibid.*, **85**, 3303 (1963).

Sheehan and Lengyel[70a] formed α-lactam XXXIX and observed similar reactions with nucleophiles. In addition, when warmed in ether, XXXIX underwent both ring opening and a cleavage reaction.

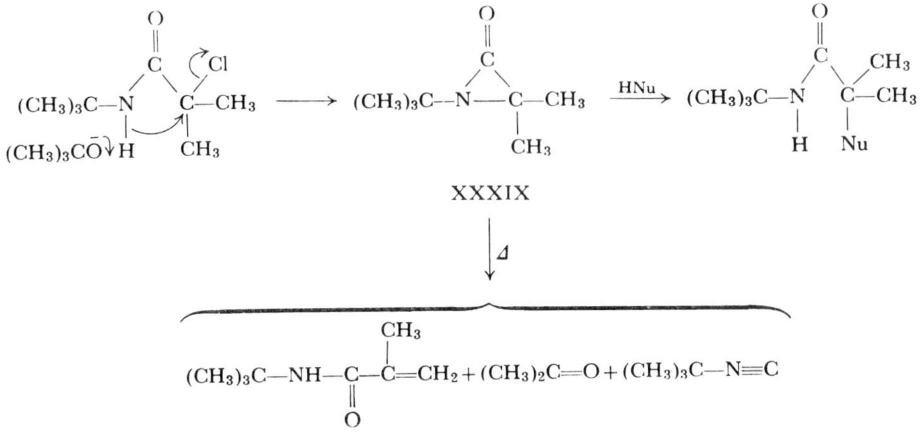

$$(CH_3)_3C{-}N(H){-}C(=O){-}C(Cl)(CH_3)_2 \xrightarrow{(CH_3)_3C\bar{O}} (CH_3)_3C{-}N{-}C(=O){-}C(CH_3)_2 \ \text{(XXXIX)} \xrightarrow{HNu} (CH_3)_3C{-}N(H){-}C(=O){-}C(CH_3)_2{-}Nu$$

$$\text{XXXIX} \xrightarrow{\Delta} (CH_3)_3C{-}NH{-}C(=O){-}C(CH_3){=}CH_2 + (CH_3)_2C{=}O + (CH_3)_3C{-}N{\equiv}C$$

A better example of the cleavage reaction involved production of cyclohexanone and *tert*-butyl isocyanide from amide XL.[70b] An isomer of the α-lactam, oxirane XLI, was postulated as an intermediate in this transformation.

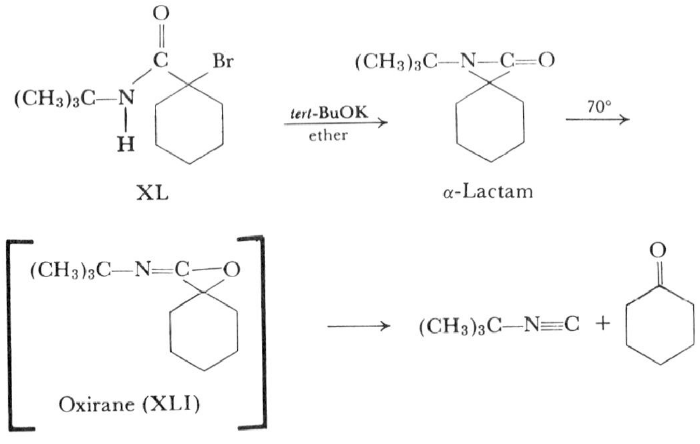

$$\text{XL} \xrightarrow[\text{ether}]{tert\text{-BuOK}} \alpha\text{-Lactam } \xrightarrow{70^\circ} \left[(CH_3)_3C{-}N{=}C{-}O \ \text{Oxirane (XLI)}\right] \longrightarrow (CH_3)_3C{-}N{\equiv}C + \text{cyclohexanone}$$

XL

α-Lactam

Oxirane (XLI)

[70] (a) J. C. Sheehan and I. Lengyel, *J. Am. Chem. Soc.*, **86**, 1356 (1964); (b) J. C. Sheehan and I. Lengyel, *ibid.*, **86**, 746 (1964).

Author Index

Numbers in parentheses are footnote numbers and are inserted to enable the reader to locate a reference when the authors' names do not appear in the text.

A

Adams, F. H., 86
Agranoff, B. W., 184
Albin, J., 50
Albrecht, H., 86
Alikhanov, P. P., 25
Allinger, J., 32, 86, 139, 143(2, 8), 144, 146(11), 149(11), 151(2, 11), 154(11), 165(8)
Allinger, N. L., 133
Andersen, K. K., 58
Anderson, C. D., 243
Andreades, S., 56, 69, 70(19)
Anet, F. A. L., 81, 82
Applequist, D. E., 5, 6(7a), 19(7), 48, 124
Arcus, C. L., 170
Aston, J. G., 247
Aylward, F., 161

B

Bach, J. L., 63
Bäckland, B., 253
Bafus, D. A., 127
Baird, R., 248
Baker, J. W., 201
Baker, V. B., 24
Balfe, M. P., 170
Bank, S., 176, 180(4a)
Barash, L., 125, 126(42d)
Barie, W. P., 159, 162(18)
Barnes, R. K., 50
Bartlett, P. D., 53
Baskov, Yu. V., 202
Bates, R. B., 176, 201, 205(27e)
Baumgarten, H. E., 252, 255
Bayer, H. O., 193
Beard, W. Q., Jr., 228, 229(23c)
Beckett, C. W., 201
Belelskaya, I. P., 117
Bell, R. P., 9, 13(8b), 20
Benkeser, R. A., 63
Benson, R. E., 67
Bent, H. A., 50, 56(6b)
Bergson, G., 176, 183
Bergstrom, F. W., 138, 142(1)
Berkowitz, W. F., 249
Birch, A. J., 166
Blackledge, J., 67
Blanksma, J. H., 240
Bohnert, E., 86, 116
Boozer, C. E., 170
Bordwell, F. G., 73, 131, 253, 254
Borisov, A. E., 131, 132(51)
Borowitz, I., 246
Bottini, A. T., 52, 81, 82(11), 113
Boutan, P. J., 73
Bower, F. A., 252
Bradshaw, J. S., 160, 161(23), 212, 215(2)
Brasen, W. R., 228
Braude, E. A., 130
Brauman, J. I., 5, 15, 18(5), 42(5), 48, 56(4b), 66(4b), 146, 153(14)
Breitbeil, F. W., 115
Breslow, R., 79, 80, 215, 245
Brewster, J. H., 193, 200, 225
Broaddus, C. D., 202
Brody, R. G., 172(36c), 173
Brown, H. C., 200
Brown, M. L., 24
Brown, T. L., 127
Buchkremer, J., 245
Bugg, C., 52
Bumgardner, C. L., 213, 228
Burgard, A., 249
Burr, J. G., 247
Burwell, R. L., Jr., 5, 37
Butler, M. E., 252
Butler, T. J., 159, 162(18)

C

Caldwell, R. A., 18, 19(18), 56, 57(20), 59(20)
Campbell, A., 225
Campbell, B. K., 252
Campbell, K. N., 252
Carnighan, R. H., 176, 201, 205(27e)
Carpino, L. A., 164
Carr, M. D., 194, 195(20g)
Carter, P., 164
Cast, J., 230
Catchpole, A. G., 201, 204
Chaput, E. P., 252
Charman, H. B., 60, 118, 119(36b, c), 121(36), 122(e), 123(a, b)
Chaykovsky, M., 40, 41, 44, 77, 80(52)
Chen, A., 13, 14(12), 15(12), 48, 51, 59(3), 60(3)
Cilento, G., 74
Clark, J. R. P., 194, 195(20g)
Clark, M. T., 241
Clark, R. D., 255
Clement, W. H., 219
Closs, G. L., 50
Closs, L. E., 50
Cohn, L. A., 58, 68(23c), 73(23c)
Coles, J. A., 130
Collins, C. J., 240, 241, 242
Conant, J. B., 1, 201
Coombs, R. D., 24
Coon, M. J., 184
Cooper, G. D., 73, 253
Cope, A. C., 66
Corey, E. J., 40, 41, 44, 77, 78(54), 80(52), 105, 108, 161, 199
Coulson, C. A., 23, 48, 50
Craig, D. P., 74
Cram, D. J., 28, 29(35), 32, 33, 34(39), 37, 38, 45, 56, 57, 59, 65, 82, 83, 86, 88, 93, 96, 97, 98(7, 11), 101, 103(6, 7, 11), 105(11), 106, 108, 109, 110(23), 113(22), 133, 139, 140, 141, 142, 143(2, 3, 4, 5, 6, 7, 8), 144(4), 145(3), 146(5, 10,11), 148(10), 149(3, 4, 5, 11), 151(2, 4, 11, 12, 16), 154(11), 156(16), 157 (16), 158(16), 160, 161(23), 165 (3, 8), 168, 169(6, 30), 170, 172, 173, 176, 177, 179, 186, 187(12a), 189, 190(15), 193(15), 195(19), 207(5, 12), 208(19), 209, 212, 215(2), 231, 237, 238, 249, 250 (61b)
Creighton, E. M., 223
Cremonini, B., 73
Cruickshank, D. W. J., 74
Crump, J. W., 131, 132(53)
Cullis, M. J., 51
Curtin, D. Y., 123, 130, 131, 132(53), 134, 230, 242

D

Dankner, D., 188
Danner, P. S., 41, 42
Dauben, H. J., Jr., 15, 55
Davies, A., 165
Davis, G. T., 58
Dayton, J. C., 21
DeBoer, E., 67
DePuy, C. H., 115
Desiderato, R., 52
Dessy, R. E., 5, 7, 13, 14(12), 15, 19(7), 48, 51, 59(3), 60(3)
Dewar, M. J. S., 247
Dewey, R. S., 161
DeWolfe, R. H., 193
Dickerhoof, D. W., 127
Dillon, R. L., 9, 11(9), 12(9), 20
Dirks, J. E., 252
Dittmer, B., 105
Doering, W. von E., 55, 56, 71, 72(17), 76, 83, 176, 181(4b), 240
Döser, H., 230
Douglas, I. B., 255
Dreiding, A. S., 131
Druey, J., 225, 226(19b)
Dunn, J. L., 223
Dykhno, N. M., 20, 24(21c, d), 27, 28(32)

E

Eastham, J. F., 127, 238, 240, 241(35), 242
Eggerer, H., 184

Eigen, M., 13, 14, 93
Ela, S. W., 209
Eliel, E. L., 157
Elizarova, A. N., 176, 182 (4e)
Emerson, M. T., 100
Escales, R., 159, 162 (20)
Evans, A. A., 170

F

Fabian, W., 232
Faivush, M., 20, 24 (21d)
Fava, A., 173
Felletschin, G., 223
Fernelius, W. C., 138, 142 (1)
Ferstandig, L. L., 29
Fischer, H. P., 101, 189
Flanagan, P. W. K., 51
Flournoy, J. M., 21
Fono, A., 235
Fookson, A., 213
Fort, A. W., 247
Fox, I. R., 58
Freedman, H. H., 54
Freeman, J. P., 213
Friedolsheim, A., 249
Fuerholzer, J. F., 255

G

Gale, L. H., 118
Gall, J. S., 172 (36a), 173
Garner, C. S., 117, 118, (35a, b, c)
Garratt, P. J., 67
Gaspar, P. P., 55, 176, 181 (4b)
Gassman, P. G., 33, 146, 147
Gaston, L. K., 141, 143 (5), 146 (5), 149 (5)
Geissler, G., 240
Gibson, J. W., 127
Gilbert, J. M., 40, 41, 42, 77, 146, 153 (14)
Gilmore, W. F., 247
Goering, H. L., 105, 116, 172 (36c), 173, 185
Goggin, D., 63
Golovnya, R. V., 63
Gordon, A. B., 223
Gordy, W., 47, 62
Gosser, L., 45, 88, 93, 96, 98, (7, 11), 103 (6, 7, 11), 105 (11), 169, 172, 173, 176, 189, 190 (15), 193 (15)
Graber, R. P., 247
Graham, W. H., 212
Granger, M. R., 18, 19 (18), 56, 57 (20), 59 (20)
Gross, O., 159
Grovenstein, E., Jr., 233, 234, 235 (30b, c), 237
Grunwald, E., 9, 100
Guthrie, R. D., 186, 187 (12a), 207 (12)

H

Haag, W. O., 193, 194, 195 (20a)
Haberfield, P., 34, 93, 97, 142, 143 (6, 7), 169 (6)
Hall, G. H., 60, 61 (26c), 62 (26c)
Hamilton, C. L., 215, 216 (8), 217 (8)
Hammond, G., 240
Hammons, J. H., 5, 18 (5), 42 (5), 48, 56 (4b), 66 (4b), 115, 116 (30), 146, 153 (14), 248
Happe, W., 226, 230
Harnsberger, H. F., 159, 162 (18)
Harris, E. E., 130
Hatch, M. J., 249, 250 (61b)
Hauck, F., 32, 140, 143 (3), 144, 145 (3), 146 (10, 11), 148 (10), 149 (3, 11), 151 (11, 12), 154 (11), 165 (3)
Hauser, C. R., 226, 227 (21a), 228, 229 (23c), 230 (20b)
Hausser, J. W., 134
Haworth, H. W., 240
Haynes, P., 176, 183
Hazen, G. G., 247
Hellmann, H., 225
Hendley, E. G., 241
Henning, U., 184
Henriques, F. C., Jr., 47, 113
Herb, S. F., 194, 195 (20h)
Hermann, R. B., 133
Higginson, W. C. E., 14, 19, 48
Hill, A. S., 213
Hill, E. A., 221, 222 (12b)
Hine, J., 41, 42, 62, 69, 240

Hine, M., 41, 42, 62
Hinze, J., 129
Hochstein, F. A., 66
Hoffmann, A. K., 56, 71, 72
Hofmann, J. E., 27, 28, 29(34), 30(34), 31(34), 33, 194
Hogen-Esch, T., 2
Hoijtink, G. J., 67
Hojo, M., 172(36a), 173
Holmes, J., 230
Holtz, J., 55
House, H. O., 247, 249
Houston, A. H. J., 225
Howard, F. L., 213
Howden, M. E., 128, 129(44), 219
Hsu, S. K., 85, 185, 186(10), 187(10), 188(10)
Huffaker, J. E., 242
Hugh, W. E., 205
Hughes, E. D., 118, 119(36b, c), 120, 121, 122(36e, f), 123(36a, b), 170, 185, 201, 204
Huh, G., 249
Hunig, S., 161
Hunter, D. H., 133, 193, 208(19)
Hurd, D. T., 129
Hush, N. S., 67
Huyser, E. S., 215

I

Iakushin, F. S., 28, 29(35)
Illiceto, A., 173
Impastato, F. J., 125, 126(d, f, g)
Ingold, C. K., 85, 118, 119(36b, c), 120(36d), 121, 122(36e), 123(36a, b), 170, 185, 186, 187, 188(10), 200, 201, 204, 205(32), 239
Ingraham, L. L., 23
Isaewa, A. A., 28, 29(35)
Isajewa, G. G., 27, 28(32)
Ives, D. J. G., 85
Izrailevich, E. A., 20, 24(21c), 25, 26, 27(31), 28(32), 60, 61(26b), 63

J

Jacobs, T. L., 188
Jacqueman, W., 240
Jacquier, R., 243
Jäger, H., 106, 141, 143(5), 146(5), 149(5)
Jaffé, H. H., 58, 74, 129
Jakuhin, F. S., 27, 28(32)
Janiak, P. St., 106
Jenny, E. F., 225, 226(19b)
Jensen, F. R., 118, 119, 123(37a, c)
Johnson, A. W., 228
Johnson, H. W., Jr., 130, 131
Johnson, J. R., 116
Johnson, T. B., 255
Johnstone, R. A. W., 223
Jones, W. M., 58, 68(23c), 73(23c)
Jordan, T., 74, 77(49e), 83(49e)

K

Kaiser, E. T., 105
Kalb, L., 159
Kantor, S. W., 226, 228, 230(20b)
Kaplan, F., 128, 129(44)
Karabatsos, P. J., 100
Karpov, T. P., 120
Katz, T. J., 66, 67
Kawahara, F. S., 184
Kazanski, B. A., 20, 22(21e)
Kende, A. S., 243, 245
Kenner, J., 162(25), 164
Kenyon, J., 116, 165, 170, 225
Kharasch, M. S., 235
Kilpatrick, J. E., 48, 201
Kimball, G. E., 74
Kincaid, J. F., 47, 113
Kingsbury, C. A., 34, 37, 93, 97, 165, 168, 169(30), 179, 231
Kistiakowsky, G. B., 201
Klager, K., 66
Kline, M. W., 225
Klingsberg, E., 244
Knight, E. C., 164
Knox, G. R., 33, 34(39)
Koch, H. F., 27, 28(33), 29(33)
Koch, H. P., 74
Koehl, W. J., 123, 131(40), 132(40)
Kon, C. A. R., 205
Konig, H., 77, 78(54), 108

Kopecky, K. R., 32, 109, 139, 140, 143(2, 3), 144, 145(3), 146(10, 11), 148(10), 149(3, 11), 151(2, 11, 16), 154(11), 156(16), 157(16), 158(16), 165(3)
Kornblum, N., 86
Kovalenko, T. T., 64
Krebs, A., 245
Kreevoy, M., 60
Krolls, U., 255
Kromhout, R. A., 100
Kuhn, R., 86

L

LaCount, R. B., 228
LaLancette, E. A., 67
Lambert, J. L., 65, 114, 115, 116(30), 248
Lampher, E. J., 24, 193
Landgrebe, J. A., 118, 119(37a, c, f), 123(37a, c)
Landis, P. S., 131
Lang, F. T., 157
Langemann, A., 32, 86, 109, 139, 140, 143(2, 3), 144, 145(3), 149(3, 11), 151(2, 11, 12), 154(11), 156(16), 157(16), 158(16), 165(3), 231
Langford, C. H., 5, 37
Langworthy, W. C., 15, 28, 29(35)
Lansbury, P. T., 219, 231, 233
Lawler, R. G., 25, 26(30), 27
Ledwith, A., 42
Lee, C. C., 241
Lee, W. G., 133
Leffler, J. E., 9
Lengyel, I., 256
Leskowitz, S., 230, 242
Letsinger, R. L., 24, 85, 116
Levy, J. F., 172(36c), 173
Levy, L. K., 76, 83
Lewis, A., 50
Lewis, E. S., 170
Lewis, I. C., 58, 61 (23a)
Lewis, R. N., 129
Lipp, P., 245
Lipscomb, W. N., 74, 77(49e), 83(49e)
Little, E. L., 24
Löhman, L., 226, 230
Loftfield, R. B., 245
Lohr, L. L., Jr., 74, 77(49e), 83(49e)
Lomker, F., 240
Long, F. A., 37
Lorand, J. P., 59
Lorenz, L., 230
Lovett, W. E., 72, 73, 202
Lowry, T. H., 77, 78(54), 108
Lukina, M., 20, 22(21e)
Lwowski, W., 109, 151, 156(16), 157(16), 158(16)
Lynen, F., 184
Lyness, W. I., 202, 204(30c)
Lyons, A. L., 24

M

McCarron, F. H., 116
McCarty, C. G., 134
Maccoll, A., 74
McElhill, E. A., 68
McEwen, W. K., 2, 3, 14, 15, 19, 42, 48, 56(4a), 70(4a)
McFadyen, J. S., 159, 162(19)
McFarlane, N., 42
McKenna, J. F., 252
McLean, S., 176, 183
MacNicol, M., 223
McPhee, W. D., 244
Maioli, L., 73
Malhotra, K., 206
Malyanov, V. A., 120
Mangold, R., 223
Manochkina, P. N., 25
Mardaleishvili, E. L., 117
Marsh, F. D., 24
Martin, K. J., 213
Mateos, J. L., 32, 144, 149(11), 151(11), 154(11)
Matteson, D. S., 211
Mayo, D. W., 63
Mazur, R. H., 212, 215, 216(8)
Mears, T. W., 213
Meinwald, J., 50

Meyers, C. Y., 73
Miller, S. I., 133
Mironov, V. A., 176, 182(4e)
Mock, W. L., 161
Moffitt, W. E., 23, 48, 50, 74
Mohacsi, E., 79, 80
Morrow, D. F., 252
Morton, A. A., 2, 14, 24
Most, E. E., 252
Motes, J. M., 113, 114(28)
Muller, H. R., 161
Muller, R. J., 27, 28(34), 29(34), 30(34), 31(34), 198, 199(23)
Muzzucato, U., 173

N

Nandi, K. N., 170
Naslund, L., 196, 197(21b), 198(21)
Nations, R. G., 240
Neber, P. W., 249
Nesmeyanov, A. N., 63, 117, 131, 132(51)
Nesmeyanova, O. A., 63
Neurieter, N. P., 253, 254(66b)
Neville, O. K., 241
Newkirk, J. D., 247
Nichols, P. L., 194, 195(20h)
Nichols, R. E., 28
Nickon, A., 65, 114, 115, 116, 160, 213, 248
Nicolaides, N., 235
Nielsen, W. D., 28, 29(35), 32, 105, 144, 146(10, 11), 148(10), 149(11), 151(11), 154(11), 176
Nietzki, R., 239
Nordlander, J. E., 215, 216(8), 219(8b)
Nudenberg, W., 235
Nyholm, R. S., 74

O

Oae, S., 72, 83, 84(44)
O'Brien, C., 249, 250(60)
O'Brien, D. F., 5, 6(7a), 19(7), 48
O'Connor, D. E., 202, 204(30c)
O'Donnell, J. P., 38, 176
Ogata, Y., 241
Ohno, A., 72, 83, 84(44)
Okano, M., 241
Okuzumi, Y., 13, 14(12), 15(12), 48, 51, 59(3), 60(3)
Onsager, L., 12
Orgel, L. E., 74
Ossorio, R. P., 185
Osyany, J. M., 81
Ott, D. G., 241

P

Paquette, L. A., 253
Parcell, R. F., 252
Parker, A. J., 146
Partos, R. D., 83, 106, 113(22)
Pasto, D. J., 161, 211
Patel, D. J., 215, 216(8), 217(8c)
Patsch, M., 232, 233(29e)
Pattison, V. A., 219, 231, 233
Patton, J. T., 86
Paul, M. A., 37
Pauling, L., 201
Pearson, R. G., 9, 11(9), 12(9), 20
Penner, S. E., 24
Perevalova, E. G., 63
Peterson, A. H., 124
Peterson, J. M., 252
Pfeil, E., 240
Phillips, H., 170
Piccolini, R., 60, 61(26c), 62(26c)
Pine, S. H., 83, 106
Pines, H., 193, 194, 195(20a)
Pitt, C. G., 125, 126(42e)
Pittman, V. P., 170
Pitzer, K. S., 201
Poddubraya, S. S., 117
Pomerantz, P., 213
Porter, C. W., 116
Poshkus, A. C., 235
Posner, J., 245
Pratt, R. J., 131
Price, C. C., 194, 195(20c, e), 202(20c, e)
Price, E., 58
Price, G. G., 40, 41
Prichard, R. H., 116
Proops, W. R., 230
Prosen, E. J., 201

Prosser, T. J., 194, 195(20b), 202(20b)
Pudjaatmaka, A. H., 5, 18(5), 42(5), 48, 56(4b), 66(4b), 146, 153(14)
Purlee, E. L., 100
Puterbaugh, W. H., 228

R

Raaen, V. F., 242
Ramberg, L., 253
Ramsden, H. E., 24
Ranneva, Yu. I., 27, 28(32), 29(35), 64
Rapaport, H., 70
Rausch, M., 63
Redman, L. M., 24
Rees, T. C., 221
Renk, E., 212
Reppe, W., 66
Reutov, O. A., 117, 120, 129
Richey, H. G., Jr., 221, 222(12b)
Rickborn, B., 33, 34(39), 37, 38, 97, 105, 141, 143(4), 144(4), 149(4), 151(4), 168, 169(30), 176, 179
Riemenschneider, R. W., 194, 195(20h)
Rifi, M. R., 15, 55
Rikhter, M. I., 20, 22(21e)
Rilling, H. C., 184
Ringold, H. J., 206
Ritchie, C. D., 44, 77
Roberts, G., 116
Roberts, I., 239
Roberts, J. D., 52, 60, 61(26c), 62(26c), 68, 81, 82(11), 113, 128, 212, 215, 216, 217(8c), 219(8b), 241
Rochow, E. G., 129
Rogers, L. C., 233
Rosenberger, M., 67
Rossetto, U., 173
Rossini, F. D., 201
Roth, W. R., 176, 183
Rothberg, S., 213
Rowe, C. A., Jr., 33, 176, 180(4a), 194, 195(20f), 196, 197(21b), 198(21), 199(23), 200(21a), 201(21a), 254
Ruchardt, C., 215, 216(8), 219(8b)
Russell, G. A., 95

S

Sacks, A. A., 247
Sahyun, M. R. V., 33
Salinger, R. M., 5, 7, 19(7), 48
Salome, L., 165
Sandel, V. R., 54
Sass, R. L., 52
Saunders, M., 47
Savistova, M., 161
Schaad, L., 245
Schäfer, H., 232, 233(29e)
Schaffernicht, W., 146
Schechter, H., 51
Scheuing, G., 239
Scheytt, G. M., 225
Schlafer, H. L., 146
Schlenk, W., 55
Schlichting, O., 66
Schöllkopf, U., 232, 233(29c, d, e)
Schriesheim, A., 27, 28, 29(34, 35), 30(34), 31(34), 33(41), 34, 176, 180(4a), 194, 195(20f), 196, 197(21b), 198(21), 199(23), 200(21a), 201(21a), 254
Schroll, G., 63
Schwartz, A. M., 116
Schwarzenbach, G., 20
Scott, D. A., 28, 29(35), 101, 105, 176, 189
Seeles, H., 245
Seibert, W., 159, 162(18)
Selman, S., 238, 241(35), 242
Servis, K. L., 215, 216(8)
Seyferth, D., 133
Shafer, P. R., 212, 215, 216(8), 219 (8b)
Shapiro, I. O., 27, 28(32), 29(35)
Sharman, S. H., 193
Shatenshtein, A. I., 13, 14, 20, 22, 23(21b), 24(21c, d), 25, 26, 27(31), 28, 29(35), 60, 61(26), 62, 63, 64
Shaw, P. D., 63
Shechter, H., 200
Sheehan, J. C., 256
Sheppard, W. A., 68, 73
Shoppee, C. W., 116

Shriner, R. L., 86
Sidler, J. D., 219
Silver, M. S., 215, 216(8), 219(8b)
Singer, L. A., 57
Sinz, A., 160, 213
Sleezer, P. D., 122, 193
Smentowski, F. J., 233
Smid, J., 2
Smith, D. R., 241
Smith, G. G., 241
Smith, H. W., 74, 77(49e), 83(49e)
Smith, P. A. S., 252
Smith, S., 172(36a), 173
Smith, W. V., 47
Smolinsky, G., 70, 251
Snyder, W. H., 194, 195(20c, e), 202(20c, e)
Sobolev, E. V., 176, 182(4e)
Sommelet, M., 226
Spitzer, R., 48
Stanford, S. C., 62
Staples, C. E., 176, 201, 205(27e)
Stearns, R. S., 41, 42
Stefaniak, L., 202
Steiner, E. C., 40, 41, 42, 77, 130, 131, 146, 153(14)
Stevens, T. S., 159, 162(19), 164, 223, 225, 230
Stewart, R., 38, 176
Stone, H., 51
Stork, G., 246
Streitwieser, A., Jr., 5, 15, 18(5), 19(18), 25, 26(30), 27, 28, 29 (33, 35), 42(5), 48, 56(4b), 57(20), 59(20), 65, 66(4b), 146, 153(14), 170
Sutton, L. E., 74
Swithenbank, C., 50
Symant, H. H., 159, 162(18)

T

Taft, R. W., Jr., 58, 61(23a), 198
Tagaki, W., 72, 83, 84(44)
Talalay, P., 184
Taliaferro, J. D., 215
Tarbell, D. S., 72, 116, 202
Thaler, W., 193
Thier, W., 161, 249
Thompson, R. D., 255
Thomson, T., 223, 225
Thorpe, F. G., 118, 119(36c), 120(36d), 121(36)
Thyagarajan, B. S., 76
Timmons, R. J., 161
Toepel, T., 66
Towns, D. L., 105
Trambarillo, R. F., 47
Traylor, T. G., 81, 117, 118(35a–c), 121, 124
Traynham, F. G., 51
Trepka, R. D., 82, 106

U

Uber, A., 249
Uglova, E. U., 120
Unmack, A., 3
Urban, R. S., 240
Urbanski, T., 202
Urey, H. C., 239
Urry, W. H., 235
Uyeda, R. T., 33, 56, 176, 177, 207(5)

V

van der Meij, P. H., 67
Van Eenam, D. N., 228
Van Sickle, D. E., 27, 28(33), 29(33, 35)
van Tamelen, E. E., 161
Vasil'eva, L. N., 20, 23(21b), 24(21c, d), 27, 28(32)
Vaughan, L. G., 133
Vidal, F., 86, 116
Vinard, D. R., 60
Volger, H. C., 118, 120(36d), 121(36), 122 (36e, f)
Volkenau, N. A., 131, 132(51)
von Liebig, J., 239

W

Walborsky, H. M., 113, 114(28), 125, 126(42d, e, f, g), 127(42c, h)
Waldbillig, J. O., 211
Wallace, T. J., 33
Walling, C., 193

Wallis, E. S., 86
Walloch, O., 239
Walsh, A. D., 23, 48, 50(5c)
Walter, D., 232
Webb, R. L., 68
Wedegaertner, D. K., 118, 119(37a, c), 123(37a, c)
Weidler, A. M., 176, 183
Weijland, W. P., 67
Weiner, M. A., 127
Weiss, M., 116
Wendler, N. L., 247
Wentworth, G., 233, 235(30c), 237
Wenz, A., 116
West, R., 127
Westheimer, F. H., 239
Wetterling, M. H., 55
Wheland, G. W., 1, 41, 42
Whipple, L. D., 118, 119(37a, c), 123(37a, c)
Whitesides, G. M., 128, 129(44)
Whiting, M. C., 40, 41, 194, 195(20g)
Wiberg, K. B., 50
Wilen, S. H., 157
Wilken, P. H., 157
Wilks, G. C., 85
Willey, F., 103, 189
Williams, L. P., 233, 234, 235(30b)
Williams, R. O., 115, 116(30), 248
Wilmarth, W. K., 21
Wilson, C. L., 85, 185, 186(10), 187(10), 188(10)
Wingrove, A. S., 96, 108, 110(23)
Winstein, S., 117, 118(35a, b, c), 121, 122, 124, 172(36a), 173, 185, 193, 248
Witanowski, M., 128, 202
Wittig, G., 55, 86, 116, 223, 226, 230
Wolf, D. C., 252
Wollschitt, H., 2
Wooding, N. S., 14, 19, 48
Woodman, J. B., 86
Woods, G. F., 53
Wright, G. F., 117

Y

Yakovleva, E. A., 20, 22(21e)
Yamada, F., 47
Yates, P., 243
Young, A. E., 125, 126(42d, g), 127(h)
Young, J. H., 86
Young, W. G., 122, 185, 193
Youssef, A. A., 113, 114(28)
Yu, M., 20, 22(21e)
Yurygina, E. N., 25

Z

Zaayer, W. H., 240
Zalar, F. V., 33, 146, 147
Zeiss, H. H., 165, 170
Zey, R. L., 255
Ziegler, K., 2, 116
Zimmerman, H. E., 76, 233, 234(31b), 235(31b)
Zuyagintseva, E. N., 62
Zweig, A., 233, 234(31b), 235(31b)

Subject Index

A

Acetamide, 12
Acetic acid, 12, 41
Acetoacetic ester, 10
Acetone, 10, 12
Acetonitrile, 12
Acetonitrile anion, 52
Acetophenone, 4, 13, 14, 41
1-Acetoxynortricyclene, 114
1-Acetoxynortricyclene cleavage, 114
Acetylacetone, 10, 12
Acetyl anion, 52
N-Acetylaziridine, 82
Acetylene, 19
  acidity, 49
  pK_a, 48
  s-character effects, 49
Acetylene-allene rearrangement, 101
  base catalyst effects in, 189
  intramolecularity in, 189, 190
  mechanism of, 190
  solvent effects in, 189, 190
Acid-base reactions, 138
Acidity, 25
  aromatic hydrogens, 25
  bicyclic β-diketone, 53
  bridgehead sulfones, 76
  cyclic ketones, 51
  ferrocene, 62
  nitrocyclanes, 51
  s-orbital effects, 23
  trisulfonylmethanes, 76
Acidity scale
  in dimethyl sulfoxide, 42
  McEwen, 2, 4, 19, 48
  method of establishing, 2
  MSAD scale, 19, 48
  solvent dependence, 44
  Steiner, 43
  Streitwieser, 4, 19
  thermodynamic, 1
Activation energy for
  inversion of dialkylmagnesium compounds, 128
  organometallic rearrangement, 216
  ring-chain isomerizations, 217
Activation energy-reaction coordinate profiles, 206
Additivity of methyl substituent effects, 29
Aggregation of carbon salts, 5
Alcohol acidities
  in alcohol, 41
  in dimethyl sulfoxide, 41
  in ether, 41
Alcohol cleavage reaction, 33
1-Alkene isomerization, 197
Alkylazamine, 164
Alkyl chlorosulfite, 171
Alkyl diimide, 159, 160, 161, 163, 212, 213, 215
Alkyl group effects on olefin isomerization, 197
Alkyl substituent effects, 30
Alkylvinyllithium compounds, 132
Allene, 54
Allenic anions, 175
Allylbenzene, 213
Allylbenzyldimethylammonium bromide, 225
Allylcarbinyl bromide, 216, 218
Allylcarbinylmagnesium bromide, 217, 219
Allyl hexyl ether, 73
Allyl hexyl sulfide, 73
Allylic anion configuration, 54
Allylic anions, 175
  comparison with allylic cations, 193
  comparison with allylic radicals, 193
  geometric stability, 193
Allylic anions, *cis* and *trans*, 194
Allylic hydrogens, 23
Allylic organomercurials, 123

Allylic rearrangement intramolecularity, 177
Allylic rearrangements, 24
  base character effects on, 181
  intramolecularity in, 181
  of alkenes, 33
  solvent effects on, 181
Allyl sulfides, 202
Allyl sulfones, 202
Allyl sulfoxides, 202
Allyltriethylammonium bromide, 225
Ambident carbanion, 95
Amine inversion rate, 47
α-Amino ketones, 252, 253
Ammonia, 25
  hybridization, 47
  in dimethylsulfoxide, 92
  isotopic exchange solvent, 20
  pK_a, 14, 20
tert-Ammonium carbanide ion pair, 183
Ammonium-cyanocarbanide ion pair, 105
Ammonium ion rotation within ion pair, 89
Aniline, 4, 32, 41
Anion and cation rotation in ion pair, 105
Anionic and radical cleavage reactions, 156
Anionic catalysis in SE_2 reaction, 121
Anion rotation within ion pair, 179
Anisole, 61, 64
1-Anthracyl-*t*, 26
9-Anthracyl-*t*, 26
Applequist acidity scale, 7, 8
Arenesulfonhydrazide, 159
Aromatic hydrogens, acidity, 25
Aromatization effects, 65
Aryl anion as neighboring group, 249
Aryl as neighboring group, 237
Arylhydrazine oxidation, 160
Aryl hydrogen-deuterium exchange rates, 61
Aryl hydrogens, kinetic acidity, 27
1,2-Aryl migrations, 233
Arylvinyllithium compounds, 132
Asymmetrically solvated
  anion, 179
  carbanion, 97, 154, 163
  ion pair, 166
Asymmetric carbanions, 108
Asymmetric hydrogen bonding, 96
1,3-Asymmetric induction, 183
1,5-Asymmetric induction, 193
Asymmetric ion pair, 154, 171
Asymmetric solvation, 96
Aza-allylic anion, 187
Azamine, 164
α-Azidostyrene, 251
Aziridine inversion rates, 81, 83
  conjugative effects on, 113
  hydrogen bonding effects on, 113
Azirine, 252
Azirines in Neber rearrangement, 250

B

Back reaction, 28
Back reaction of carbanions, 176
Back reactions, 204
Base catalysis, nitrogen as leaving group, 161
Base-catalyzed proton transfers, 176
Benzene, 19, 25, 29, 32, 144
Benzene-*t*, 18
Benzil, 239, 240
Benzilic acid, 238
Benzilic acid rearrangement
  mechanism, 239
  rate expression, 240
  scope and limitations, 239
  substituent effects, 240
Benzhydrol, 156
Benzofluorenes, 6
Benzoic acid, 41
Benzophenone, 156, 157
Benzophenone as electrophile, 228
Benzophenone ketyl, 157, 158
Benzoylacetone, 13
Benzoyl chloride as electrophile, 220
Benzyl anions, 95
Benzylcyclopropane, 220
Benzyl ether, 233
Benzylic anions, 175
Benzyllithium, 54, 55

Benzyl methyl ether, 230
2-Benzyloxy-2-phenylbutane, 165
Benzylphenylcarbinol, 233
Benzylsodium, 55
α-Benzylstyrene, 193, 194, 208
Benzyltrimethylammonium iodide, 227
Bibenzyl, 164
Bicyclic β-diketone acidity, 53
Bicyclobutanide anion, 216, 217
Bicyclobutonium cations, 216
Bicyclobutonium ion, 217
[3.1.0]Bicyclohexane-3-carboxylic acid, 147
Bimolecular electrophilic substitution at saturated carbon, 138
Bimolecular mechanism of prototropy, 185, 186
Biphenyl, 25
Biphenylyl anion, resonance in, 26
p-Biphenylyldiphenylmethane, 5, 43, 44
Bis-1,4-dimethylpentylmercury, 130
Bond hybridization of cyclopropane, 48
Boric acid, 13
Bridged allylic chloride, 185
Bridged ions, 217
Bromination of
  carbanions, 9
  cycloalkanones, 198
  organolithium compounds, 124, 125
  organometallics, 119
Bromine as oxidizing agent, 164
Bromobenzene, 33
2-Bromobutane, 119
1-Bromo-3,5-hexadiene, 218
cis-α-Bromostilbene, 130
trans-α-Bromostilbene, 130
Brønsted equation, 9
Brønsted linear relationships, 9
2-Butanone, 151
1-Butene, 122, 199, 213, 217, 218
cis-2-Butene, 253, 254
trans-2-Butene, 253
cis-2-Butenyllithium, 131
trans-2-Butenyllithium, 131
n-Butyl alcohol, 41
sec-Butyl alcohol, 144
tert-Butyl alcohol, 4, 32, 41, 106, 144
tert-Butyl alcohol–potassium *tert*-butoxide, 37, 92
tert-Butylbenzene, 29, 30, 63
tert-Butyl isocyanide, 256
Butyllithium, 63, 67, 86, 131, 133
sec-Butylmercuric bromide, 119, 120
tert-Butyl methylphenylacetate, 93
p-*tert*-Butyltoluene, 30, 31

C

Camphenilone, 65, 114
4-Camphyllithium, 124
4-Camphylmercuric iodide, 117
Carbanion and carbonium ion comparison, 170
Carbanion-carbene resonance, 25
Carbanion configuration, 95
  conjugatively stabilized, 52
  sulfonylcarbanions, 80
Carbanion hybridization, 47, 70
Carbanion rearrangements
  1,3-elimination stages, 211
  metal ions as leaving groups, 214
  neighboring groups in 1,2-shifts, 211
  ring-chain, 211
Carbanion rotation within ion pair, 103
Carbanion stabilization effects
  aromatization, 48
  s-character, 48-52
  conjugative, 48, 52-55
  homoconjugative, 48, 63-65
  inductive, 48, 55-63
  mechanism of, 47
  solvent, 32
Carbanion structure, 47
Carbanions as
  conjugate bases, 1
  intermediates in cleavage reaction, 153
  intermediates in methyleneazomethine rearrangement, 186
  intermediates in ring-chain rearrangements, 211
Carbon acidity scales
  MSAD scale, 19
  Steiner scale, 43, 146
  Streitwieser scale, 146

Carbon acids, 1
  contrasted with other acids, 3
  definition, 1
Carbon as leaving group in SE_1 reaction, 138
Carbonation of
  carbon salts, 3
  cyclooctatetraene dianion, 66
  cyclopentadiene, 13
  rearranged organometallics, 216
Carbonium ion and carbanion comparison, 170
Carbon-metal bond character, 235
Carbon salts, 2
Carbon tetrafluoride, 69
Case I conformation of sulfones, 74, 77, 78
Case II conformation of sulfones, 74 77, 78
Cation effects on carbon salt basicity, 41
Cation of base effects on SE_1 reaction, 145
Cation rotation within ion pair, 179
Cesium cylohexylamide, 5, 18
Cesium fluorenide, spectra of, 2
s-Character effects, 48, 70
  acetylene, 49
  acidity relationships, 49
  allenic anions, 54
  cyclopropane, 49
  ethylene, 49
  on exchange of strained systems, 50
  spin-spin coupling constants, 50
  vs. conjugative effects, 52
Charge distribution in carbanions, 11
Charge distribution in phenyl-substituted methyllithium compounds, 55
Charge type effects, 32
4-Chlorocamphane, 124
2-Chlorocyclohexanone, 245
2-Chlorocyclohexanone-1,2-C^{14}, 245
α-Chloroethyl ethyl sulfone, 253
5-Chloro-1-hexene, 221, 222
Chloromethanesulfonamide, 255
cis-p-Chlorophenyl-1,2-diphenylethylene, 130
trans-p-Chlorophenyl-1,2-diphenylethylene, 130
cis-2-*p*-Chlorophenyl-1,2-diphenylvinyllithium, 132
trans-2-*p*-Chlorophenyl-1,2-diphenylvinyllithium, 131, 132
1-Chloro-1-phenyl-2-propanone, 244
1-Chloro-3-phenyl-2-propanone, 244
Cholestenone, 184
Cinnamylamine, 213
Claisen reaction, 138
Cleavage of
  benzyl alcohols, 165
  benzyl ethers, 165
  nortricyclanone, 147
  2-phenyl-2-butyl diimide, 162
Cleavage reaction
  comparison with hydrogen-deuterium exchange, 155
  deuterated solvents, 149
  electrophile source, 149, 151
  in dimethyl sulfoxide, 146
  internal electrophiles, 152
  isotope effects, 150
  metal cation effects on, 32
  solvent effects on, 32
  stereochemistry effects of mixed solvents, 147
  substituent effects on stereochemistry, 150
  temperature effects, 152
Coalescence temperature, 81, 82
Collapse ratio
  allylic anions, 204, 205, 206, 209
  aza-allylic anion, 207, 208
  cyclohexadiene system, 206
  1,2-diphenylpropenyl system, 208
  2-phenylbutenyl system, 207
Collapse ratios and equilibria, 208
Color during radical cleavage, 156, 158
Common ion rate depression, 38
Comparison of
  carbanions and carbonium ions, 170
  carbanions, carbonium ions, and radicals, 175
  leaving groups in SE_1 reaction, 155, 168, 169

1,3-rearrangements of different charge type, 184
SE_1 and SN_1 reactions stereochemical course, 170, 171
Competition between retention and isoracemization mechanism, 104
Concentration of electrophiles in SE_1 reactions, 144
Conducted tour mechanism, 101, 102, 105, 189, 190, 193
Conductivity of carbon salts, 2
Configurational relationships
2-cyano-2-phenylbutanoic acid and 2-phenylbutyronitrile, 143
2-dimethylamino-2-phenylpropanoic acid and 1-dimethylamino-phenylethane, 143
2-methoxy-2-phenylpropanoic acid and 1-phenylmethoxyethane, 143
2-methyl-2-phenylbutanoic acid and 2-phenylbutane, 143
2-phenyl-2-butanol and 2-phenyl-butane, 165
2-phenylpropanoic acid and 1-deuterophenylethane, 143
Configurational requirement for *d*-orbital effects, 74
Configurational stability
bis-3,3-dimethylbutylmercury, 129
cyclopropylorganometallics, 128
3,3-dimethylbutyllithium, 129
3,3-dimethylbutylmagnesium bromide, 128
3,3-dimethylbutylmagnesium chloride, 128
2,2-dimethylcyclopropylmagnesium bromide, 128
function of ionic character of bond, 129
of open-chain organometallics, 128
cis-propenyllithium, 132
trans-propenyllithium, 132
solvent effects on, 128
1,3,3-trimethylcyclobutylmagnesium bromide, 128
tris-3,3-dimethylbutylaluminum, 129
cis-vinyllithium compounds, 130
trans-vinyllithium compounds, 130
Configuration of carbanions
benzyl anion, 95
stabilized by conjugative effect, 52
stabilized by inductive effect, 56
sulfonylcarbanions, 75, 76, 79
Configuration of leaving group, 148, 149
Configurations of propenyllithium compounds, 133
Conformational effects on olefin equilibria, 201
Conformational requirement for *d*-orbital effects, 74
Conformation of carbanions stabilized by second row elements, 112
Conformation of sulfonyl carbanions, 110
Conjugative effects, 52
blended with *d*-orbital effects, 80
carbonyl group, 52
competition with inductive effect, 59
cyano group, 52
cyclopropane ring, 212
nitro group, 52
olefin equilibria, 201
unsaturated sulfide equilibria, 203
unsaturated sulfoxide equilibria, 203
vs. *s*-character effects, 52
σ^* Constants for olefin isomerization, 198
Contact ion pairs, spectra of, 2
Cope elimination reaction, 33
Correlation between
configuration and rotation, 143
thermodynamic and kinetic acidity, 8
Correlation of
kinetic and thermodynamic acidities, 11
rates and equilibria, 13
Cross breed experiments, 232
in Stevens rearrangement, 223
9-Crotyl-9-fluorenol, 231
Crotylmercuric bromide, 122
Crystal structure,
ammonium tricyanomethide, 52
bisdimethylamino sulfone, 74
pyridinium dicyanomethylide, 53
Cumene, 2, 4, 19, 22, 27, 30

Cumene-α-t, 18
3-Cyano-2,3-dimethyl-4-phenyl-2-butanol, 142
4-Cyano-3,5-dimethylphenol, 77
1-Cyano-2,2-diphenylcyclopropane, 114
Cyano group, role in conducted tour mechanism, 105
2-Cyano-2-methyl-3-phenylpropanoic acid, 142
p-Cyanophenol, 77
2-Cyano-2-phenylbutanoic acid, 142, 143
Cyclic ketones, 51
Cyclic sulfones, 78, 80
Cyclic transition state, 118
Cyclobutadiene dianion, 65, 66
Cyclobutane, 19
Cyclobutane-t, 18
Cyclobutanone, 51
Cyclobutylcarbinyl chloride, 221
Cyclobutylmethylmetallics, 222
Cyclobutyl phenyl ketone, 51
Cycloheptanone, 51
Cycloheptatriene, 15, 17, 19, 65
Cycloheptatriene, pK_a, 66
Cycloheptatrienyl anion, 182
1,3-Cyclohexadiene, 206
1,4-Cyclohexadiene, 206
Cyclohexadiene system, 205
Cyclohexane, 19, 21, 32
Cyclohexane-t, 18
Cyclohexanone, 51, 256
Cyclohexene, 24
Cyclohexylamine, 5, 15, 26, 27, 29
N-Cyclohexylaziridine, 81
2-Cyclohexyl-2-butyl anion, 158
2-Cyclohexyl-2-butyl radical, 158
2-Cyclohexyl-1,1-diphenyl-2-methyl-1-butanol, 158
Cyclohexylmethyl chloride, 222
Cyclohexyl phenyl ketone, 51
Cyclononatetraene, 65
Cyclononatetraenide anion, 66
Cyclooctatetraene, 66, 67
Cyclooctatetraene dianion, 65, 66
Cyclopentadiene, 13, 14, 15, 19, 41, 67
Cyclopentadiene isomerizations, 182
Cyclopentadiene, pK_a, 66
Cyclopentadienecarboxylic acid, 13
Cyclopentadienide anion, 65, 66
Cyclopentane, 21
Cyclopentanone, 51
Cyclopentylmethyl chloride, 222
Cyclopentyl phenyl ketone, 51
Cyclopropane, 49
  acidity, 49
  s-character effects, 49
  conjugation, 212
Cyclopropane-t, 18
Cyclopropanones as rearrangement intermediates, 244, 245
Cyclopropenones, 246
Cyclopropyl anions, 147
Cyclopropyl anions, symmetry properties, 113
Cyclopropylcarbinylamine, 213
Cyclopropylcarbinyl benzyl ether, 231
Cyclopropylcarbinyl bromide, 215, 217, 218
Cyclopropylcarbinyllithium, 219
Cyclopropylcarbinyllithium-α,α-d_2, 220
Cyclopropylcarbinylmagnesium bromide, 217
Cyclopropylcarbinylphenylcarbinol, 220, 231
Cyclopropylcarbinyl radical, 215
Cyclopropylcarbinyl tosylate, 216
Cyclopropylcarboxaldehyde, 212
Cyclopropyldiphenylmethyl methyl ether, 220
Cyclopropyllithium, 127
Cyclopropylmethylcarbinylamine, 213
Cyclopropyl phenyl ketone, 51
Cyclopropyl phenyl sulfone, 76
p-Cymene, 30, 31

D

Decarboxylation reactions, 142
  comparison with hydrogen-deuterium exchange, 170
  cyclic sulfone acid, 110
Delocalization energies, 217
Dessy acidity scale, 8
Deuterated ammonia, 14, 21, 22, 23
Deuterated tripropylammonium iodide, 100

Deuterated triphenylmethane, 44
Deuteration of triphenylmethylsodium, 45
Deuterium, 21
1-Deuterocyclopentadiene, 182
2-Deuterocyclopentadiene, 182
5-Deuterocyclopentadiene, 182
1-Deuterophenylethane, 141
Diallyl ether, 230
Diazoethane, 254
1,1-Dibenzylhydrazine, 164
1,1-Dibenzyl-2-*p*-tolylsulfonylhydrazine, 164
α,α-Dibromodibenzyl ketone, 246
cis-Dibromoethylene, 133
trans-Dibromoethylene, 133
Di-*tert*-butylketone, 65
Di-*sec*-butylmercury, 119, 123
Di-4-camphylmercury, 117
Dichloroacetone, 10
Dicyanomethane, 10, 12
1,1-Dideuterocycloheptatriene, 181
Dielectric constant
  effect on stereochemistry, 144, 162
  of dimethyl sulfoxide, 146
  of medium, 44, 92
Diethylene glycol, 144
N,N-Diethyl-2-methyl-3-phenylpropionamide, 93
Diethyl 2-octylphosphonate, 107
Diffusion control of rate of proton transfer, 12, 14, 44
Difluoramine, 213
Dihydropentalene, 67
Dilithium pentalenide, 67
2-Dimethylacetamido-9-methylfluorene, 96, 104
2-Dimethylacetamido-9-methyl-7-nitrofluorene, 93, 98, 99, 104, 191
N,N-Dimethylallylamine, 195
α,α-Dimethylallyl chloride, 185
γ,γ-Dimethylallyl chloride, 185
3-Dimethylamino-2,3-dimethyl-3-phenyl-2-butanol, 146
3-Dimethylamino-2-methyl-3-phenyl-2-butanol, 141
1-Dimethylaminophenylethane, 141
2-Dimethylamino-2-phenylpropanoic acid, 143
Dimethylaniline, 64
3,3-Dimethylbutyllithium, 129
3,3-Dimethylbutylmagnesium bromide, 128
3,3-Dimethylbutylmagnesium chloride, 128
2,2-Dimethylcyclopropylmagnesium bromide, 128
3,4-Dimethyl-3,4-diphenylhexane, 156
Dimethyl formamide, 32, 51, 60
2,5-Dimethylhexanoic acid, 130
Dimethylhydrazone methiodide rearrangement, 252
1,4-Dimethylpentylmagnesium bromide, 130
3,5-Dimethyl-4-methylsulfonylphenol, 77
3,4-Dimethyl-4-phenyl-3-hexanol, 139, 144, 151
2,3-Dimethyl-3-phenyl-2-pentanol, 139, 146
N,N-Dimethylpropenylamine, 195
Dimethyl sulfone, 12
Dimethylsulfonylmethane, 12, 78
Dimethylsulfonylmethane anion, 79
Dimethyl sulfoxide, 27, 29, 32, 33, 34, 37, 38, 40, 42, 43, 103, 146, 161, 180, 181, 201
  acidity scale in, 42
  alcohol equivalence points in, 41
  as scavenger, 36
  pK_a, 44
  quenching in, 45
  rate enhancements in, 35
  solvent for acidity scales, 42, 43
Dimethyl sulfoxide-d_6, 221
Dimethyl sulfoxide–methanol, 106
Dimethyl sulfoxide–potassium *tert*-butoxide, 197, 198
Dimsylpotassium, 40, 42
Dimsylsodium, 42
2,4-Dinitroaniline, 43
Dinitromethane, 10, 12
2,4-Dinitrophenyl aryl ethers, 241
Dioxane, 32
γ,γ-Diphenylallylcarbinyl bromide, 219

Diphenylamine, 4, 41
cis-1,3-Diphenyl-1-butene, 209
cis-2,4-Diphenyl-2-butene, 209
trans-1,3-Diphenyl-1-butene, 209
trans-2,4-Diphenyl-2-butene, 209
Diphenylcyclopropenone, 246
Diphenylketone as leaving group, 156
Diphenylmethane, 4, 5, 6, 14, 43, 44
Diphenylmethane-α-*t*, 18
2,3-Diphenyl-3-methoxy-2-butanol, 146
Diphenylmethylamine, 64
2,3-Diphenyl-2-methyl-1-butanol, 151
1,2-Diphenyl-2-methyl-1-butanone, 148
2,2-Diphenylmethylcyclopropyl bromide, 125, 127
2,2-Diphenylmethylcyclopropyllithium, 127
2,2-Diphenylmethylcyclopropyl-magnesium, 127
Diphenylmethyllithium, 54, 55
2,3-Diphenyl-3-methyl-2-pentanol, 145, 146, 147, 150, 151
Diphenyl 2-octyl phosphine oxide, 107
Diphenylphosphinoxyaziridine, 82
2-Diphenylphosphinoxyoctane, 107
1,2-Diphenyl-1-propanone, 93
2,2-Diphenylpropyl chloride, 234, 235
2,2-Diphenylpropyllithium, 235
Diphenylsulonylcarbethoxymethane, 80
Diphenylsulfonylmethane, 80
cis-1,2-Diphenylvinyllithium, 131
trans-1,2-Diphenylvinyllithium, 131
Dipropylamine, 89
Dipropyl ketone, 51
Double bond-no-bond resonance, 69
Double ion pair, 167

E

Eclipsing effects, 254
Eclipsing effects in alkene isomerization, 199
Electrophile in SE reaction, 137
Electrophile source in cleavage reaction, 149, 151
Electrophilic substitution
  bridgehead organomercurial, 118
  closed transition state, 121
  cyclic transition state, 119
  open transition state, 119, 121
  oxygen as leaving group, 166
  at saturated carbon, 131, 137
  with *p*-tolyl disulfide, 131
Electrostatic effects, 55
Electrostatic inhibition of
  inversion effects, 83, 134
  inversion of carbanions, 111
  rotation of carbanions, 111
Electrostatic requirements for asymmetric carbanions, 111
Elimination-addition mechanism, 233, 237
1,2-Elimination reactions, 33
1,3-Elimination reactions, 243, 255
Elimination reactions of 2-phenyl-2-butylhydrazine, 162
1,1-Elimination vs. 1,2-elimination of hydrazine tosylate, 164
Enol tautomers, 93
Entropy effects in olefin isomerization, 200
Enzymatic intramolecular proton shifts, 184
Episulfones, 253, 254
cis-Episulfone, 254
trans-Episulfone, 254
Episulfone stereochemistry, 254
Equality of racemization, exchange and bromination rates of a ketone, 85
Equilibria
  alkyllithiums and alkylmercury compounds, 7
  alkyllithiums and iodides, 7
  between olefins, 193, 194, 200
  comparison of Applequist and Dessy scale, 8
  cyclohexadiene system, 206
  inductive effects in olefins, 200
  in methyleneazomethine systems, 187
  lithium alkyls and iodides, 5
  lithium salts and iodides, 6
  ring-chain Grignard reagents, 217
  ring-chain organometallics, 216, 221
  substituent effects on alkenes, 202
  substituent effects on olefins, 200
  unsaturated sulfides, 203

unsaturated sulfoxides, 203
unsaturated sulfones, 203
Equilibria and collapse ratios, 208
Equilibrium constants for hydrocarbon group-metal redistribution reactions, 8
Equivalence points in titration of acids with dimsylmetallics, 41
Esters from benzilic acid rearrangement, 240
Ethane, 19
Ethane acidity, 49
Ethanol, 41, 62, 144
Ether, 32
Ethylbenzene, 27, 30, 63
2-Ethyl-1-butene, 199
Ethylcyclobutane, 22
Ethyl cyclopentenylmalonate, 205
Ethylcyclopropane, 22, 212
Ethylene, 19, 23
acidity, 49
s-character effects, 49
Ethylenediamine, 144
Ethylene glycol, 32, 34, 62, 144
Ethylene glycol-potassium ethylene glycoxide, 37
m-Ethyltoluene, 30, 31
o-Ethyltoluene, 30
p-Ethyltoluene, 30, 31

F

Favorskii rearrangement, 243, 244
mechanism, 245
stereochemistry, 247
Field effects, 55
Fluoradene, 16, 19
Fluoradene, pK_a, 70
Fluorene, 4, 5, 14, 15, 17, 19, 43
Fluorene, pK_a, 66, 70
9-Fluorenyl
crotyl ether, 230
ethers, 230
α-methylallyl ether, 231
Formaldehyde, 255
Formamide, 62
Formanilide, 41
Functional groups that do not induce carbanion asymmetry, 108
Functional groups that induce carbanion asymmetry, 108

G

Geometric stability of allylic anions, 193
Glycerol, 40, 41, 144
Glycol, 62
Grignard reagent
optically active, 127
oxygenation, 219
rearrangements, 216, 217
ring-chain multiple rearrangement, 218

H

Hammett σ p-plot, 29
Hammett substituent constants, 58
benzilic acid rearrangement, 241
σ^-, σ_I^-, σ_m^-, σ_p^-, σ_R^- constants, 58
σ^- for methylsulfinyl, 73
σ^- for methylsulfonyl, 73
trifluoromethyl σ's, 68
n-Heptane, 21
Heterolylic vs. homolytic cleavage, 157, 158
Hexamethylbenzene, 30, 31
1-Hexene, 24, 222
Hexyl propenyl sulfide, 73
H_- in sulfolane-water mixtures, 37
Homoconjugated carbanions, 114
Homoconjugative effects, 63
Homoenolate anion, 65, 114
Homolytic cleavage, 157, 233
Homolytic cleavage, nitrogen leaving group, 161
Homolytic racemization, 110
H_- plot against log k, 39
Hückel rule, 65, 217
Hughes-Ingold tautomer rule, 204, 206
Hybridization
nitrogen in bisdimethylamino sulfone, 74
sulfonylcarbanions, 75, 76, 79, 80
Hydrocarbon acidity, 4, 6
Hydrogen bonding, 29, 40
asymmetric, 96
during 1,5-proton shift, 192
effect on anion activity, 42
effect on aziridine inversion rates, 113

effect on reaction rates, 36
in dimethyl sulfoxide, 37
sites, 181
Hydrogen deuterium exchange, 9
acetals, 72
acetylene, 59
activation parameters, 71
allylic rearrangements, 177, 193
anisole, 64
azomethylene system, 187
benzylcyclobutane, 221
tert-butylbenzene, 63
tert-butyl methylphenylacetate, 93
camphenilone, 65, 114
comparison with cleavage reaction, 155
comparison with decarboxylation, 169
conducted tour mechanism, 101
1-cyano-2,2-diphenylcyclopropane, 114
cyclic ketones, 51
cis-dibromoethylene, 133
trans-dibromoethylene, 133
N,N-diethyl-2-methyl-3-phenylpropionamide, 93
diethyl 2-octylphosphonate, 107
2-dimethylacetamido-9-deutero-9-methylfluorene, 88
2-dimethylacetamido-9-methylfluorene, 96
2-dimethylacetamido-9-methyl-7-nitrofluorene, 93, 98, 99
dimethylaniline, 64
N,N-dimethylmethylphenylacetamide, 93
2,2-dimethyl-3-phenylsulfonylbutane, 109
diphenyl 2-octyl phosphine oxide, 107
2-diphenylphosphinoxyoctane, 107
1,2-diphenyl-1-propanone, 93
effect of carbonyl on stereochemistry, 92
effect of dielectric constant of medium, 92
entropy of activation, 72
ethylbenzene, 63
heat of activation, 72
1-hydroperfluoroheptane, 70
2-hydroperfluoropropane, 70
in dimethylformamide, 59
in dimethyl formamide-water, 51
inversion mechanism, 95, 96, 97, 179, 180
methyl 2-octylsulfonate, 107
2-methyl-1-phenyl-1-butanone, 85
N-methyl-*N*-phenyl-2-octylsulfonamide, 107
2-methyl-3-phenylpropionitrile, 114
monohydrofluorocarbons, 69
2-octyl phenyl sulfone, 105, 106, 107, 109
2-octyl sulfonate, 107
orthothioesters, 83, 84
[1.*n*]paracyclophanes, 57
phenylacetylene, 59
2-phenylbutyronitrile, 96, 98, 99
phenyl-*p*-tolylacetic acid, 85
2-phenyl-1,1,1-trifluorobutane, 96
quinaldine, 62
racemization mechanisms, 92
retention mechanisms, 88, 89, 90, 179, 180
ring-chain rearrangements in, 220
solvent effects on stereochemistry, 106
stereochemistry, 179
stereochemistry of 2-octyl sulfone, 106
cis-stilbene, 133
trans-stilbene, 133
substituent effects on aryl, 61
substituted acetylenes, 60
substituted phenylacetylenes, 59, 60
tetramethylammonium salt, 71
tetramethylphosphonium salt, 71
thioacetals, 72
thioketals, 84
toluene, 64
trifluoromethane, 70
tristrifluoromethylmethane, 70
trimethylsulfonium, 71
1,1,1-triphenylethane, 63
2-triphenylmethylpropionitrile, 98, 99
vinyl carbon, 133
vs. allylic rearrangement, 179
Hydrogen
heterocycle, 29
isotope effects, 28

kinetic acidity, 21
1,5-Hydrogen shift, 101, 190
1-Hydroperfluoroheptane, 70
2-Hydroperfluoropropane, 70
Hydroxylamine-*O*-sulfonic acid, 160
Hyperconjugation, 68

I

Indene, 14, 15, 17, 19, 41, 43
Indene, pK_a, 66
Inductive effect, 55
alkyl groups, 31
aryl, 18
bond length, 71
tert-butyl, 31
competition with conjugative effect, 59
dimethylamino, 61
ethyl, 31
fluorine, 56, 61
from β-position, 63
in triphenylmethane, 18
in triptycene, 18
isopropyl, 31
methoxy, 61
methyl, 27, 31, 61
methyl group, additivity of, 30
on bond angles, 56
on hybridization, 56
on olefin equilibria, 202
phenoxy, 61
phenyl, 57, 58, 61
quaternary nitrogen, 56
relation to negative hyperconjugation, 70
separation from *d*-orbital effects, 71, 73
separation from resonance effects, 58
tetramethylammonium, 71
tetramethylphosphonium, 71
trifluoromethyl, 61
trimethylsulfonium, 71
variation with position, 31
Intermolecular 1,2-benzyl rearrangement, 238
Intermolecular elimination-addition mechanism, 237
Internal electrophiles in cleavage reaction, 152
Internal vs. external compression, 196
Internal vs. external electrophile, 151
Intimate ion pairs, 154
Intramolecularity in
acetylene-allene rearrangement, 101
allylic rearrangement, 177, 194
1,2-aryl rearrangements, 236
1,5-hydrogen migration, 101, 191
methyleneazomethine rearrangement, 188
triene to benzene rearrangement, 191
Intramolecular proton shift, 87, 176, 183
acetylene-allene rearrangement, 189
allylic anion as intermediate, 178
base character effects on, 177, 178
enzymatic, 184
hydrogen bonds in intermediate carbanion, 178
isotope effects in, 177, 178
mechanisms for, 177, 178
nitrile racemization, 189
solvent effects in, 177, 178
Inversion mechanisms, 95, 96
1-acetoxynortricyclene cleavage, 115
allylic system, 179
cation of base effects, 145
cleavage reaction, 154
comparison of leaving groups, 168, 169
conditions for in cleavage reaction, 153
criterion for, 95
Favorskii rearrangement, 247
for SE_1 reactions, 172
for SN_1 reactions, 172
hydrogen-deuterium exchange, 97
in cleavage reaction, 144
nitrogen leaving group, 161, 162, 163
pK_a dependence, 97
reductive cleavage of alcohols, 167
reductive cleavage of ethers, 167
solvent effects in, 95, 97
Inversion of configuration, 87
Inversion rates of aziridine, 82
Inversion solvents, 105, 144, 145, 155
Invisible reaction, 90
Ionic bond character, 220, 235

Ionic character of carbon-metal bonds, 129
Ionization rates, relation to pK_a, 11
Ionization vs. dissociation rates, 28
Ion pairs
  tert-ammonium carbanide, 183
  anion and cation rotation in, 105
  anion rotation, 98, 103, 179
  1,2-aryl rearrangement, 237
  as intermediates in cleavage reaction, 154
  benzilic acid rearrangement, 242
  carbanion containing, 153
  carbanion rotation in, 100
  cation rotation in, 89, 90, 179
  elimination-addition mechanism, 232
  from *cis*-stilbene, 134
  from *trans*-stilbene, 134
  intimate, 154
  in triene-benzene rearrangement, 192
  mechanisms, 226
  metallation and protonation reactions, 127
  product separated, 154, 168
  reorganization reaction, 98
  solvent separated, 154
  Sommelet-Hauser rearrangement, 229
  vinyllithium reactions, 132
  Wittig rearrangement, 231
Isoamyl cyclopentanecarboxylate, 245
Isobutane, 21
Isobutene, 23
Isobutyrophenone, 51
cis-trans-Isomerism of allyl anions, 175
Isomerization of
  1-alkenes, 195
  5-deuterocyclopentadiene, 182
  5-methylcyclopentadiene, 183
Isopentane, 21, 22
Isopropyl, 76
Isopropyllithium, 124
1-Isopropyl-3-methylindene, 183
Isoracemization, 87
  competition with retention, 103
  conducted tour mechanism, 101, 102
  definition, 98
  2-dimethylacetamido-9-methyl-7-nitrofluorene, 99
  fluorene system, 102
  mechanism, 98, 100, 189
  nitrofluorene system, 193
  2-phenylbutyronitrile, 98, 99, 102
  2-triphenylmethylproprionitrile, 98, 99
Isotope effects
  back reaction, 29
  hydrogen, 28
  in cleavage reaction, 150
  negative, 28, 29, 176
  solvent, 29
  toluene, 29
Isotope exchange rates, solvent effects on, 34
Isotopic exchange, 38

K

Ketol rearrangements, 242
Ketone cleavage reactions, 33
Ketyl of benzophenone, 156
Kinetic acidity,
  allylic hydrogen, 23
  1-anthracyl-*t*, 26
  9-anthracyl-*t*, 26
  aromatic hydrogens, 25
  aryl hydrogens, 27
  arylmethanes, 17
  benzene, 25
  benzene-*t*, 18
  biphenyl, 25
  tert-butylbenzene, 30
  p-*tert*-butyltoluene, 30, 31
  comparison with thermodynamic acidity, 8
  composite character, 28
  cumene, 22, 28, 30
  cumene-α-*t*, 18
  cyclic ketones, 51
  cyclobutane-*t*, 18
  cyclobutanone, 51
  cyclobutyl phenyl ketone, 51
  cyclohexane, 21
  cyclohexane-*t*, 18
  cyclohexanone, 51
  cyclohexene, 24
  cyclohexyl phenyl ketone, 51
  cyclopentane, 21
  cyclopentanone, 51

cyclopentyl phenyl ketone, 51
cyclopropane hydrogens, 23
cyclopropane-*t*, 18
cyclopropyl phenyl ketone, 51
p-cymene, 30, 31
definition, 1
determination, 9
diphenylmethane-α-*t*, 18
dipropyl ketone, 51
ethylbenzene, 28, 30
ethylcyclobutane, 22
ethylcyclopropane, 22
m-ethyltoluene, 30, 31
o-ethyltoluene, 30
p-ethyltoluene, 31
n-heptane, 21
hexamethylbenzene, 30, 31
1-hexene, 24
hydrogen isotope effects in, 28
α-hydrogens, 28
isobutyrophenone, 51
isopentane, 21, 22
2-methylanthracene, 17
2-methylnaphthalene, 17
3-methylphenanthrene, 17
naphthalene, 25
1-naphthyl-*t*, 26
2-naphthyl-*t*, 26
[1.*n*] paracyclophanes, 57
pentamethylbenzene, 30, 31
9-phenanthryl-*t*, 26
2-phenylbutane-2-*t*, 18
phenylcyclopropane, 22
phenyl-*t*, 26
propene, 24
propylcyclohexene, 24
1-pyrenyl-*t*, 26
2-pyrenyl-*t*, 26
4-pyrenyl-*t*, 26
solvent effects, 27, 28, 39
tautomers, 205
tetramethylammonium hydroxide, 56
1,2,3,4-tetramethylbenzene, 30, 31
1,2,3,5-tetramethylbenzene, 30, 31
1,2,4,5-tetramethylbenzene, 30, 31
toluene, 17, 28, 30, 31
toluene-2-*t*, 18
toluene-3-*t*, 18
toluene-4-*t*, 18
toluene-α-*t*, 18
1,2,3-trimethylbenzene, 30, 31
1,2,4-trimethylbenzene, 30, 31
1,3,5-trimethylbenzene, 30, 31
triphenylmethane-α-*t*, 18
tryptycene-α-*t*, 18
vinylcyclopropane, 22
vinyl hydrogens, 23
weak hydrogen acids, 20
m-xylene, 28, 30, 31
o-xylene, 28, 30
p-xylene, 28, 30, 31
Kinetic and thermodynamic acidity correlations, 9
Kinetic and thermodynamic comparison of carbon, nitrogen and oxygen acids, 13
Kinetic basicity of ambident anions, 205
Kinetic order of isotopic exchange reaction, 34
Kinetic order of racemization reaction, 34
Kinetic vs. thermodynamic control of products in 1-alkene isomerization, 195

L

α-Lactam formation, 255
α-Lactams, 256
Leaving group as electrophile, 151
Leaving group in SE reaction, 137
carbon, 138
comparison, 168
nitrogen, 161
oxygen, 165
Leaving groups
configurational effects, 148
metal ions in rearrangments, 215
nitrogen in rearrangements, 213
shielding effect in inversion mechanisms, 155
steric effects, 148
Lifetime and stereochemical fate of carbanions, 92
Linear free energy plot, 39
Linear free energy relationships,
1-alkene isomerizations, 14, 198
Brφnsted, 9 deviations, 11, 13

Hammett, 58, 61, 73, 198
Liquid ammonia, 27
Lithiation
cis-2-bromo-2-butene, 131
trans-2-bromo-2-butene, 131
organomercury compounds, 123
Lithium carbanide ion pairs, 124
Lithium cyclohexylamide, 4, 15, 18
Lithium cyclononatetraenide, 67
Lithium cyclopentadienide, 67
Lithium fluorenide,
spectra of, 2

M

Maximum rotations, 143
McEwen acidity scale, 2, 4, 14, 19, 56
McFayden-Stevens reaction, 159
Mechanism of acetylene allene rearrangement, 190
Mechanism of 1,2-aryl rearrangements, 237
Mechanism of base-catalyzed rearrangement, 180
Mechanism of benzilic acid rearrangement, 239
Mechanism of cleavage reaction, 154
Mechanism of Favorskii rearrangement, 245
Mechanism of Ramberg-Bäckland rearrangement, 254
Mechanism of SE_1 reaction carbon as leaving group, 144, 147, 148, 153, 154, 156
Mechanism of SE_1 reaction comparison with SN_1 reaction, 171
Mechanism of SE_1 reaction
hydrogen as leaving group, 40, 87, 88, 89, 90, 91, 94, 97 98, 100, 102, 104, 111, 112, 115
nitrogen as leaving group, 163
oxygen as leaving group, 166
Mechanism of Wittig rearrangement, 232
Medium effects, 36
different alcohols as solvents, 37
Medium effects on acidity, 32
Meisenheimer rearrangement, 233
Menthol, 3, 4
Mercuric chloride as electrophile, 118
Mercury-lithium exchange reactions, 132
Metal alkyls, 32
Metalation
at vinyl carbon, 130
4-chlorocamphane, 124
cyclopentadiene, 13
cyclopropyl phenyl sulfone, 76
Metal cations as ligands, 155
Metal cation effects in heterolytic cleavage, 157
Metal cation effects in homolytic cleavage, 157
Metal character effects on 1,2-rearrangements, 235
Metal hydrides, 32
Metal ions as leaving groups, 215
Metals in reduction of ethers, 166
Methane, 17, 19
Methane, pK_a, 15, 20
N-Methanesulfonylaziridine, 81
Methanol, 4, 32, 34, 62, 144
Methanol-potassium methoxide, 37
3-Methoxy-2,3-diphenyl-1-butanone, 141
3-Methoxy-2,3,diphenyl-2-butanol, 141
2-Methoxy-2-phenylbutane, 165, 166, 167
1-Methoxy-2-propanol, 41
2-Methoxy-1,1,2-triphenyl-1-propanol, 158
N-Methylaniline, 140, 144
N-Methylaniline, potassium *N*-methylanilide, 165
Methyl anion hybridization, 47
2-Methylanthracene, 17
9-Methylanthracene, 43
o-Methylbenzyldimethylamine, 227
2-Methylbutanoic acid, 123
1-Methyl-3-*tert*-butylindene, 183, 184
3-Methyl-1-*tert*-butylindene, 183, 184
Methylcyclobutane, 221
1-Methylcyclopentadiene, 183
2-Methylcyclopentadiene, 183
5-Methylcyclopentadiene, 183
Methylcyclopropane, 212, 217, 218
cis-2-Methylcyclopropyl bromide, 124, 125

trans-2-Methylcyclopropyl bromide, 124, 125
cis-2-Methylcyclopropyllithium, 124, 125
trans-2-Methylcyclopropyllithium, 124, 125
2-Methyl-1,2-diphenyl-1-butanol, 140
2-Methyl-1,2-diphenyl-1-butanone, 140
3-Methyl-2,3-diphenyl-2-pentanol, 139
Methyleneazomethine rearrangement, 185, 186
Methyleneazomethine rearrangement
 aza-allylic anions as intermediates, 188
Methyleneazomethine rearrangement
 carbanions as intermediates in, 186
 collapse ratio in, 186, 187, 188
 hydrogen deuterium exchange in, 186, 188
 hydrogen deuterium exchange stereochemistry, 187
 intramolecularity in, 186, 188
 mechanism of, 186, 187
 solvent effects in, 187
Methylenecycloalkane isomerization
 activation parameters, 198
 rates, 198
Methylenecyclobutane, 199
Methylenecycloheptane, 199
Methylenecyclohexane, 198, 199
Methylenecyclooctane, 199
Methylenecyclopentane, 199
5-Methyl-2-hexylmercuric salts, 120
1-Methylindene, 183
3-Methylindene, 183
2-Methylnaphthalene, 17
2-Methyloctanoic acid, 86
Methyl 2-octylsulfonate, 107
2-Methyl-1-pentene, 200
2-Methyl-2-pentene, 200
3-Methyl-1-pentene, 221, 222
4-Methyl-1-pentene, 200
2-Methylpentenes in equilibria, 201
3-Methylphenanthrene, 17
2-Methyl-2-phenylbutanoic acid, 143
2-Methyl-3-phenyl-2-butanol, 141
2-Methyl-1-phenyl-1-butanone, 85
2-Methyl-2-phenylbutyramide, 140
2-Methyl-2-phenylbutyronitrile, 139
Methylphenylcarbinol, 233
cis-1-Methyl-2-phenylcyclopropanol, 115
4-Methyl-4-phenyl-1-hexanone, 140
N-Methyl-N-phenyl-2-octylsulfonamide, 107
3-Methyl-3-phenyl-2-pentanol, 140
2-Methoxy-2-phenylpropanoic acid, 143
2-Methyl-3-phenylpropionitrile, 33, 35, 38, 114, 142
2-Methyl-2-phenylsulfonyloctanoic acid, 109
cis-α-Methylstilbene, 193, 208
trans-α-Methylstilbene, 193, 195, 208
p-Methylsulfinylbenzoic acid, 73
Methylsulfinyl group, 73
p-Methylsulfinylphenol, 73
Methylsulfinyl substituent effect, 73
p-Methylsulfonylbenzoic acid, 73
Methylsulfonyl group, 73
p-Methylsulfonylphenol, 73, 77
Methylsulfonyl substituent effect, 73
2-Methyl-1,1,2-triphenyl-1-butanol, 139, 156
2-Methyl-1,1,2-triphenyl-1-butanol racemization, 110
3-Methyl-1,1,3-triphenylpentane, 140
Michael reaction, 138
Migrating groups
 1,2-aryl rearrangements, 235
 Sommelet-Hauser rearrangement, 228
 Wittig rearrangements, 230
Migratory aptitude, 228
MO calculated delocalization energies, 16
Molecular orbital calculations, 15
Molecular orbital calculations, *d*-orbital effects, 74
Molecular orbital theory, 65
Molecular rearrangements, 211
Monohydrofluorocarbons, 69
Monomolecular electrophilic substitution at saturated carbon, 137
MSAD acidity scale, 17, 48, 56, 60, 180
Multiple rearrangement of Grignard reagents, 218

N

Naphthalene, 25
1-Naphthyl-*t*, 26

2-Naphthyl-*t*, 26
Neber reaction intermediates, 249, 252
Neber rearrangement, 243, 249
Negative hyperconjugation, 48, 68, 69
Negative hyperconjugation, relation to inductive effect, 70
Negative isotope effects, 28, 176
Neopentane, 19
Neutralization of carbon salts, 44
Nickon-Sinz reaction, 213
Nitroacetone, 10
4-Nitroaniline, 43
Nitrocyclanes, neutralization rates, 51
Nitrocyclobutane, 51
Nitrocycloheptane, 51
Nitrocyclohexane, 51
Nitrocyclooctane, 51
Nitrocyclopentane, 51
Nitrocyclopropane, 51
Nitromethane, 10, 12, 13
Nitromethane anion, 52
Nitrogen acids, 3, 13
Nitrogen as leaving group, 161
Nitrogen as leaving group in
  inversion mechanism, 163
  racemization mechanism, 163
  retention mechanism, 163
  ring-chain rearrangement, 212
  SE_1 reaction, 159, 163
Nitrogen trifluoride, 56
Nitro group, role in conducted tour mechanism, 105
3-Nitropentane, 51
No-bond resonance, 68, 69
No-bond resonance structures, 70
Non-proton donating non-polar solvents, 32
Non-proton donating polar solvents, 32
Nortricyclanone, 147
Nuclear magnetic resonance, 54
  aromatic carbon salts, 66, 67
  aziridine inversion rates, 81, 113
  configurational stability of cyclopropylorganometallics, 128
  cyclic Grignard reagent detection, 219
  in Grignard reagent structure study, 215
  in organometallic rearrangement, 215
  lithium pentalenide, 68
  rates of imine inversion, 134
  solvent effects on organometallic configurational stability, 128
  time scale, 129

O

cis-3-Octene, 195, 196
cis-4-Octene, 196
trans-4-Octene, 195, 196
2-Octyl iodide, 86, 116
2-Octyllithium, 86, 116, 123
2-Octyl phenyl sulfone, 105, 107, 108
2-Octyl sulfonate, 107
Olefin equilibria, 200
  conjugative effects in, 202
  functional group effects in, 202
  inductive effects in, 202
Olefin isomerization
  eclipsing effects in, 199
  substituent effects, 197
One-anion catalysis of SE_1 reaction, 122
Open transition state, 119
Optically active organolithium compounds, 123
Optically active organomagnesium compounds, 123, 127
d-Orbital effects,
  aziridine inversion rates, 81, 82, 83
  blended with conjugative effects, 80
  case I conformation, 74
  case II conformation, 74
  configurational requirement, 74, 79
  case I conjugation, 78
  case II conjugation, 78
  conformational requirements, 74, 77, 78
  molecular orbital calculation, 74
  separation from inductive effects, 71, 73
d-Orbital effects in
  unsaturated sulfides, 203
  unsaturated sulfones, 203
  unsaturated sulfoxides, 203
p-Orbital conjugative effects, 23
s-Orbital effect on acidity, 23
d-Orbitals, 48, 71

Organolithium rearrangements, 220
Organomercury compounds, 116, 221
Organomercury compounds, radical substitution reaction, 118
Organomercury exchange reactions, 121
Organometallics
carbonation, 216
isomerization, 216
oxygenation, 216
stereochemistry of substitution, 116
Orthohydrogen, 21
Oxidation of 2-phenyl-2-butylhydrazine, 162
Oxygen acids, 3, 9, 13
Oxygen as leaving group in SE_1 reaction, 165
Oxygenation of Grignard mixture, 219
Oxygenation of rearranged organometallics, 216
Oxygen isotopic exchange, 239
Oxygen-metal bond character effects, 158

P

[1.*n*]Paracyclophanes, 57
Parahydrogen, 21
Pentamethylbenzene, 30, 31
2-Pentanol, 151
1-Pentene, 180, 221
2-Pentene, 180, 213
Perdeutero-1-pentene, 181
Perdeutero-2-pentene, 181
p-Perfluoroisopropylaniline, 69
Perfluoroisopropyl group, 69
Phenanion, 64, 242
9-Phenanthryl-*t*, 26
Phenol, 41
Phenylacetylene, 4, 13, 14, 19
Phenylacetylene, pK_a, 48
2-Phenylazirine, 251
2-Phenylbutane, 139, 142, 144, 147, 151, 157, 160, 161, 165, 166, 167, 179, 214, 231
2-Phenylbutane-2-*d*, 150, 151
2-Phenylbutane-2-*t*, 18
2-Phenyl-2-butanol, 165, 166, 167
4-Phenyl-2-butanone, 115
3-Phenyl-1-butene, 177, 178, 181, 189, 190, 207
2-Phenyl-2-butene, 207
cis-2-Phenyl-2-butene, 177, 178
trans-2-Phenyl-2-butene, 177, 178
2-Phenyl-2-butylamine, 160
2-Phenyl-2-butylhydrazine, 160, 162, 214
2-Phenyl-2-butylhydrazine sulfonamide, 160
2-Phenyl-2-butylpotassium, 151
2-Phenyl-2-butyl radical, 156, 157, 158
2-Phenylbutyronitrile, 96, 98, 99, 104, 142, 189
Phenylcyclopropane, 22
Phenylcyclopropanone, 244
9-Phenylfluorene, 4, 5, 6, 19
Phenyl group inductive effect, 57
Phenyllithium, 55, 76
1-Phenylmethoxyethane, 140, 141, 179
1-Phenylmethoxyethane-1-*d*, 37
Phenylnitromethane, 205
Phenylpotassium, 76
3-Phenylpropanoic acid, 244
2-Phenylpropanoic acid and 1-deuterophenylethane, 143
2-Phenyl-2-propylpotassium, 95
Phenylsodium, 76
α-Phenylstyrene, 140
Phenylsulfenylaziridine, 82
Phenylsulfinylaziridine, 82
Phenylsulfonylaziridine, 82
Phenylsulfenylphenylsulfonylmethane, 80
Phenyl-*t*, 26
Phenyl-*p*-tolylacetic acid, 85
2-Phenyl-1,1,1-trifluorobutane, 96
Phenyltrimethylammonium hydroxide, 37
Phenyl vinyl ketone, 140
pK_a
acetamide, 12
acetic acid, 12, 41
acetoacetic ester, 10
acetone, 10, 12
acetonitrile, 12
acetophenone, 4, 14
acetylacetone, 10

acetylene, 19, 48
alcohols, 62
ammonia, 14, 20
aniline, 4, 41, 48
benzene, 19
benzofluorenes, 6
benzoic acid, 41
benzoylacetone, 13
p-biphenylyldiphenylmethane, 43
boric acid, 13
n-butyl alcohol, 41
tert-butyl alcohol, 4
cumene, 3, 4, 19
4-cyano-3,5-dimethylphenol, 77
p-cyanophenol, 77
cyclic sulfones, 78, 80
cyclobutane, 19
cycloheptatriene, 15, 17, 19, 66
cyclohexane, 19
cyclopentadiene, 13, 14, 19, 41, 66
cyclopentane, 19
cyclopropane, 19
cyclopropene systems, 50
determination, 2
dichloroacetone, 10
dicyanomethane, 10, 12
difference and rate of proton
 transfer, 91
difference in control of
 stereochemistry, 91, 93, 97
2-dimethylacetamido-9-methyl-
 fluorene, 91
2-dimethylacetamido-9-methyl-
 7-nitrofluorene, 93
3,5-dimethyl-4-methylsulfonylphenol,
 77
dimethyl sulfone, 12
dimethylsulfonylmethane, 12, 78
dimethyl sulfoxide, 43, 44, 145
2,4-dinitroaniline, 43
dinitromethane, 10, 12
diphenylamine, 4, 41
diphenylmethane, 4, 6, 14, 43
diphenylsulfonylcarbethoxymethane,
 80
diphenylsulfonylmethane, 80
effects on intramolecularity, 180
electrophiles in cleavage reaction, 145
electrophiles in SE_1 reaction, 155
ethane, 19
ethanol, 41
ethylene, 19
fluoradene, 16, 19, 70
fluorene, 4, 14, 17, 19, 43, 66, 70
formanilide, 41
glycerol, 41
1-hydroperfluoroheptane, 70
2-hydroperfluoropropane, 70
indene, 14, 17, 19, 41, 43, 66
menthol, 3, 4, 20
methane, 15 17, 19, 57
1-methoxy-2-propanol, 41
9-methylanthracene, 43
p-methylsulfonylphenol, 77
neopentane, 19
nitroacetone, 10
4-nitroaniline, 43
nitroethane, 10
nitromethane, 10, 12, 13
phenol, 41
phenols, substituted, 58
phenylacetylene, 4, 14, 19, 48
9-phenylfluorene, 4, 5, 6, 19
planar hydrocarbons, 16
propane, 19
propene, 19
pyrrole, 4
relationships in allylic
 rearrangements, 180
relationships in 1,5-proton
 shift, 192
relation to ionization rates, 11
tetramethylammonium hydroxide, 56
toluene, 4, 17, 19
triacetylmethane, 12
tricyanomethane, 12
tricyclic systems, 50
trifluoromethane, 56, 70
p-trifluoromethylaniline, 68
trimethylammonium ion, 13
1,3,3-trimethylcyclopropene, 50
trinitromethane, 12
triphenylmethane, 4, 6, 14, 19, 41, 43,
 57
1,3,3-triphenylpropene, 19
triptycene, 19, 57

tripropylammonium ion, 91
tristrifluoromethylmethane, 70
water, 41
xanthane, 4, 43
Planar carbanions, 96
Pole-dipole electrostatic repulsions, 76
Potassium amide, 21, 22, 23, 25, 27
Potassium as solvent orienting ion, 168
Potassium *tert*-butoxide, 27, 33, 34, 37
Potassium cyclohexylamide, 26, 27
Potassium cyclononatetraenide, 67
Potassium iodate, 163
Potassium ion and ligands rotation within ion pair, 89, 90
Potassium periodate as oxidizing agent, 164
Product separated ion pair, 92, 154, 168, 182, 192
cis-Products of isomerization, 196
Propane, 19
2-Propanol, 62
Propene, 19, 23
Propene, rearrangement of, 24
cis-Propenyllithium, 131, 132, 133
trans-Propenyllithium, 131, 132, 133
cis-Propenyltrimethyltin, 133
trans-Propenyltrimethyltin, 133
Propylamine, 89
Propylcyclohexene, 24
Propynyl anion configuration, 54
Protonation of salt of cyclopropyl phenyl sulfone, 76
Proton-donating non-polar solvents, 32
Proton-donating polar solvents, 32
Proton recapture, 28
1,5-Proton shift, 190
Proton transfers in unsaturated systems, 175
Proton transfer rates, 48
Proton transfer rates and ΔpK_a, 91
Prototropy, 175
1-Pyrenyl-*t*, 26
2-Pyrenyl-*t*-, 26
4-Pyrenyl-*t*, 26
Pyridine, 32
Pyridinium dicyanomethylide, crystal structure, 53
Pyrrole, 4

Q

Quaternary ammonium salt rearrangement, 228
Quench of carbon salts, 44, 76, 77, 80, 221, 222
Quench of 2-phenyl-2-propylpotassium, 95

R

Racemization, 87
sec-butyllithium, 123
in cleavage reaction, 144
2-methyl-1-phenyl-1-butanone, 85
phenyl-*p*-tolylacetic acid, 85
reversible cleavage, 156
Racemization during neutralization, 45
Racemization mechanisms, 92
carbonyl-containing compounds, 94
cleavage reaction, 146, 154
comparison of leaving groups, 168, 169
effect of solvent on cleavage reaction, 146
for SE_1 reaction, 171
for SN_1 reaction, 171
nitrogen as leaving group, 162, 163
reductive cleavage of alcohols and ethers, 167
Racemization rates,
camphenilone, 65
solvent effects on, 33
Racemization solvents, 105, 144, 146
Radical and anion cleavage reactions, 156
Radical cage reaction, 214
Radical capture by alkoxides, 158
Radical dimerization, 156
Radical substitution reaction of organomercury compounds, 118
Radiolabeled *sec*-butylmercuric bromide, 119
Radiolabeled mercury salts, 120
Ramberg-Bäckland reaction, 253
Rate enhancement in dimethyl sulfoxide, 35, 38
Rates of cleavage reaction, dependence on metal cation, 146

Rates of hydrocarbon exchange with cyclohexylamine, 18
Rates of imine inversion, 134
Rates of mixing, 44
Rates of proton abstraction, 10
Ratio k_e/k_a and stereochemistry, 87
Rearrangements,
- allylic, 24
- amine oxides, 233
- aryl as migrating group, 223
- benzilic acid, 223
- competing reactions, 223
- dimethylhydrazone methiodide rearrangement, 252
- ethers, 231
- Favorskii, 243
- leaving groups, 223
- Neber, 243, 249
- nitrogen leaving group, 213
- organolithium reagents, 220
- Ramberg-Bäckland, 243
- Stevens, 223
- with 1,3-elimination reaction stages, 243
- Wittig, 223, 230

1,4-Rearrangements, 225
- allylic quaternary ammonium, 226
- of ethers, 231
- Sommelet-Hauser, 226

Reduction of crotylmercuric bromide, 123
Reductive cleavage of
- 2-methoxy-2-phenylbutane, 167
- 2-phenyl-2-butanol, 167

Resonance effects
- carbene-carbanion, 25
- in biphenylyl anion, 26
- separation from inductive effects, 58
- trifluoromethylsulfonyl, 58

Retention and isoracemization mechanisms in competition, 103
Retention mechanisms
- 1-acetoxynortricyclene cleavage, 115
- allylic system, 179
- carbonation of organometallics, 123, 127, 130
- cation of base effects, 145
- cleavage reactions, 142, 144, 154
- comparison of leaving groups, 168, 169
- competition with isoracemization, 103
- conditions for in cleavage reaction, 153
- decarboxylation reactions, 109, 142
- electrophile concentration effects, 145
- ether cleavage reaction, 231
- for SE_1 reaction, 171
- for SN_1 reaction, 171
- Grignard formation, 127
- hydrogen-deuterium exchange, 88
- hydrogen-deuterium exchange at vinyl carbon, 133
- in dimethyl sulfoxide, 147, 148
- metallations, 123, 126, 127, 130
- nitrogen leaving group, 161, 162, 163
- 2-octyl phenyl sulfone, 105
- oxygen as leaving group, 166, 231
- protonation of organometallics, 126, 127
- SE_i reaction, 119
- SE_2 reaction, 119
- substitution at vinyl carbon, 130, 131
- substitution of organomercury compounds, 117, 121
- Wittig rearrangement, 233

Retention mechanisms and dielectric constant of medium, 92
Retention of configuration
- cyclopropyl anion, 114
- in cleavage reactions, 140
- metallation and carbonation, 86

Retention solvents, 144, 145, 151
Retention with tripropylammonium alkoxide ion pair as base, 91
Reversible ring-chain rearrangements, 216
Rho value for isotopic exchange, 29
Ring-chain anionic rearrangements, 211
Ring-chain anionic rearrangements
- Grignard reagents, 217

Ring-chain rearrangements, 214
- four-membered rings, 222
- five-membered rings, 222
- labelling experiments, 216
- organolithium reagents, 220

six-membered rings, 222
Ring-opening of cyclopropanone intermediate, 246
Ring strain effects, on hybridization, 52
Ring vs. α-deuteration of cumylpotassium, 95
Rotations within ion pairs, 104
ammonium ions, 172
carbanions, 172
carbonium ions, 172
carboxylate ions, 172

S

Saturated hydrocarbons, 5
Secondary solvent effects, 36
Second-row elements, 71, 105
SE_i reaction, 118, 121, 123, 147, 151
SE_i reaction in dimethyl sulfoxide, 148
Separation of inductive and *d*-orbital effects, 73
SE_1 reaction
carbon as leaving group, 154
comparison of leaving groups, 169
definition, 137, 138
inversion mechanism, 154
nitrogen as leaving group, 159
racemization mechanism, 154
retention mechanism, 154
survey of reactions with carbon leaving group, 138
SE_2 reaction, 118, 121, 122, 123, 138
SE_2' reaction, 123
Shell expansion in sulfur, arylsulfonylphenol substituent effects, 74
Shielding by nitrogen, 163
1,2-Shifts vs. 1,4-shifts, 225
SN_1 and SE_1 reaction comparison, 170
Sodium menthoxide, rotation of, 3
Sodium-potassium alloy, 220, 221
Solvated ion pairs, 163, 166
Solvated metal alkoxide, 171
Solvation energies, 32
Solvation of alkoxide anions, 37
Solvent-base effects, kinetic acidity, 27
Solvent cage, 161, 233
Solvent classification, 32
Solvent effects, 36, 161
acetylene allene rearrangement, 190
aryl 1,2-rearrangement, 234
asymmetric solvation, 155
control of stereochemistry in SE_1 reaction, 155
hydrogen-deuterium exchange, 106
dimethyl sulfoxide rate enhancements, 33
intramolecularity in rearrangements 177, 181
isotopic exchange rates, 35
kinetic acidity, 28, 39
methyleneazomethine rearrangement, 188
nitrogen as leaving group, 161
organometallics configurational stability, 128
racemization rates, 35
stereochemistry of cleavage reaction, 162
stereochemistry of mercury-lithium exchange reactions, 132
steric course of cleavage reaction, 144
Solvent-separated ion pairs, 154
Solvent-separated ion pairs, spectra of, 2
Sommelet-Hauser rearrangement, 226, 228, 229
migrating groups, 228
Sommelet-Hauser substituent effects, 228
Spin-spin coupling constants to determine *s*-character, 50
Starting material racemization, 157
Steady state approximation, 28
Steiner acidity scale, 42, 43
Stereochemical capabilities of carbonium ions and carbanions, 170
Stereochemical course of cleavage reactions, 144
Stereochemical similarity of carbanions and carbonium ions, 173
Stereochemistry
1-acetoxynorticyclene cleavage, 115
alcohol reduction, 166
bromination of organometallics, 119, 124, 125
carbon acid substitution, 85

carbonation, 117, 123, 130
cleavage of alcohols, 165
cleavage of ethers, 165
cleavage reaction, 145, 153
decarboxylation reaction, 109, 110
dielectric constant correlation, 144
effect of leaving group on, 155
episulfone formation, 254
ethers reduction, 166
Favorskii rearrangement, 247
homolytic cleavage reaction, 157
hydrogen-deuterium exchange, 107, 177, 179
internal electrophiles, 152
inversion conditions, 153
metallation, 117, 123, 130
metallation of cyclopropyl halides, 126
metallation of cyclopropyl organometallics, 126
Neber reaction, 250
nitrogen as leaving group, in SE reaction, 160, 161, 162
oxygen as leaving group in SE reaction, 166
protonation of cyclopropyl organometallics, 126
racemization conditions for SE reactions, 153
1,2-rearrangement, 225
1,4-rearrangement, 226
retention conditions for SE reactions, 153
SE reaction and k_e/k_a ratios, 87
SE reaction at unsaturated carbon, 131
solvent effects in SE reaction, 161
Stevens rearrangement, 225
substituent effects in SE reaction, 150
substitution of organomercury compounds, 117
substitution of organometallic compounds, 85, 116
temperature effects in SE reaction, 152
vinyl organometallic compounds, 130
Wittig rearrangement, 232
Stereospecific metallation and carbonation, 86
Stereospecific proton transfer, 183
Stereospecificity in cleavage reaction, configuration of leaving group, 149
Stereospecificity extremes in cleavage reactions, 144
Steric effects,
1-alkene isomerization, 195, 198
angle strain, 200
aziridine inversion rates, 82
cleavage reaction, 139, 150
collapse ratios, 209
eclipsing in olefin isomerization, 199
episulfone formation, 254
ketol rearrangements, 242
leaving group, 148, 149
olefin isomerization, 197
cis-products of olefin isomerization, 196
solvation, 255
Steric inhibition of resonance, 12, 27, 52, 53, 57, 198
Steric inhibition of solvation, 255
Stevens rearrangement, 229
migrating groups, 223
quaternary ammonium ketones, 224
stereochemistry, 225
substituent effects in, 223
sulfonium ketones, 224
cis-Stilbenyl anion, 134
trans-Stilbenyl anion, 134
cis-Stilbenyllithium, 132
Strained systems, acidity of, 50
Strained systems, *s*-character effects in, 50
Streitwieser acidity scale, 5, 6, 70
Substituent effects, 83
accumulative, 12
acetamido, 58
acetyl, 58
acetylene exchange rates, 60
aldehydo, 58
1-alkene isomerization, 197, 198
alkyl on carbanion stability, 197
aziridine inversion rates, 83
benzilic acid rearrangement, 240
bromo, 58
tert-butyl, 30

carboethoxy, 58
carboxy, 58
cyano, 58
diphenylphosphinoxy, 83
ethyl, 30
isopropyl, 30
Meisenheimer rearrangement, 233
methoxy, 58
methyl, 30
methylsulfinyl, 58
methylsulfonyl, 58
nitro, 58
on acidity, 9, 10, 11, 12
on stereochemistry, 150
phenylacetylene exchange rates, 60
phenyl, on ring-chain equilibria, 219
phenylsulfenyl, 83
phenylsulfinyl, 83
phenylsulfonyl, 83
ranking, 12
rates of imine inversion, 134
Sommelet-Hauser rearrangement, 228
Stevens rearrangement, 223
substituted phenylacetylenes, 59
sulfonate, 58
Wittig rearrangement, 230, 233
Substitution of organomercury compounds, 119
Sulfolane, 5, 32, 37
Sulfonamides of arylhydrazine, 159
Sulfonate group, 73
Sulfonium salts, 228
α-Sulfonylcarbanions, 81
configuration, 80
conformations of, 80
Sulfur dioxide, 254
Symmetrical carbanions, 86, 92, 93, 108
Symmetrical sulfonyl carbanions, 110
Symmetrically solvated carbanion, 154, 171

T

Temperature effects
aryl 1,2-rearrangements, 234
Tetrabutylammonium iodide, 98, 100
Tetraethylammonium cyclononatetraenide, 67
Tetrahydrofuran, 32
Tetramethoxymethane, 69
Tetramethylammonium, 71
Tetramethylammonium hydroxide
isotopic exchange, 56
Tetramethylammonium phenoxide, 92
1,2,3,4-Tetramethylbenzene, 30, 31
1,2,3,5-Tetramethylbenzene, 30, 31
1,2,4,5-Tetramethylbenzene, 30, 31
Tetramethylene sulfone, 5, 37
Tetramethylethylene, 23
Tetramethylphosphonium, 71
Tetraphenylethylene glycol, 156, 157
Thermal decomposition of alkyl diimide, 214
Thermal rearrangement, 227
Thermodynamic acidity,
comparison with kinetic acidity, 8
correlation with delocalization energies, 15
definition, 1
Thermodynamic acidity scale, 1
Thermodynamic and kinetic acidity correlations, 9
Titration of dimsylmetallics, 41
Toluene, 4, 15, 17, 19, 27, 29, 30, 31
Toluene-α-*t*, 18
Toluene-2-*t*, 18
Toluene-3-*t*, 18
Toluene-4-*t*, 18
Tosylate of 2-phenyl-2-butylhydrazine, 162
Tosylhydrazide in elimination reaction, 164
Triacetylmethane, 12
Tricyanomethane, 12
Tricyanomethide ion, crystal structure, 53
Triene benzene rearrangement, 190, 191
Triethylamine, 51, 98
Triethylcarbinol, 181
Triethylcarbinol-*O*-*d*, 192
Triethylenediamine, 103, 183
Triethylene glycol, 62
Trifluoroamine, 56
Trifluoromethane, 70
p-Trifluoromethylaniline, 68

Trifluoromethyl group, 68
Trifluoromethylsulfonyl group, 73
Trigonal carbanions stabilized by
 second row elements, 112
Trigonal hybridization of carbon, 52
Triiodide ion as electrophile, 118
Trimethylamine, 56
Trimethylammonium ion, 13
1,2,3-Trimethylbenzene, 30, 31
1,2,4-Trimethylbenzene, 30, 31
1,3,5-Trimethylbenzene, 30, 31
1,3,3-Trimethylcyclobutylmagnesium
 bromide, 128
Trimethylethylene, 23
Trimethylsulfonium, 71
Trimethylsulfonylmethane, 12
Trinitromethane, 12
Triphenylallene, 101, 189
2,2,4-Triphenylbutanoic acid, 238, 239
1,1,1-Triphenylethane, 63, 64
1,1,2-Triphenylethane, 234
1,1,2-Triphenylethane-1-*d*, 64
2,2,2-Triphenylethyl chloride,
 233, 234, 235
1,1,2-Triphenyllithium, 234
2,2,2-Triphenylethyllithium,
 234, 237, 238
1,1,2-Triphenylethylsodium, 233
Triphenylmethane, 4, 5, 6, 14, 19, 41
 43, 44
Triphenylmethane as indicator, 40, 41
Triphenylmethane, pK_a, 57
Triphenylmethane-α-*t*, 18
Triphenylmethyllithium, 54, 55
Triphenylmethylpropionitrile, 98, 99
2,2,3-Triphenylpropanoic acid,
 234, 237
1,3,3-Triphenylpropene, 19
2,2,3-Triphenylpropyl chloride, 237
1,1,3-Triphenylpropyllithium, 238
2,2,3-Triphenylpropyllithium, 238
1,3,3-Triphenylpropyne, 101, 189
Tripropylamine, 32, 90, 91, 103, 192
Tripropylammonium *tert*-butoxide, 90
Tripropylammonium carbanide, 90
Tripropylammonium carbanide ion
 pairs, 98
Tripropylammonium iodide, 98, 103
Triptycene, 19
Triptycene system, 56
Triptycene-α-*t*, 18
Tris-3,3-dimethylbutylaluminum, 129
Tris-trifluoromethylmethane, 70
Two-anion catalysis of SE_i reaction, 122

U

Ultraviolet spectra,
 carbon salts, 2, 5
 cesium salts, 18
 lithium salts, 18
Unsymmetrical carbanions, 105

V

Vinylcyclopropane, 22
Vinyl nitrenes in Neber reaction, 251
Vinyl sulfides, 202
Vinyl sulfones, 202
Vinyl sulfoxides, 202

W

Water, 41
Water as solvent in SE_1 reaction, 163
Wittig rearrangement, 165, 230, 232
Wolff-Kishner reaction, 33, 159,
 212, 213

X

Xanthane, 4, 43
m-Xylene, 27, 30, 31
o-Xylene, 27, 29, 30
p-Xylene, 27, 30, 31

Y

Ylides, 228

Z

Zwitterion in Favorskii rearrangement,
 248